Fuzzy Management Methods

Series Editors

Andreas Meier, Fribourg, Switzerland

Witold Pedrycz, Edmonton, Canada

Edy Portmann, Bern, Switzerland

With today's information overload, it has become increasingly difficult to analyze the huge amounts of data and to generate appropriate management decisions. Furthermore, the data are often imprecise and will include both quantitative and qualitative elements. For these reasons it is important to extend traditional decision making processes by adding intuitive reasoning, human subjectivity and imprecision. To deal with uncertainty, vagueness, and imprecision, Lotfi A. Zadeh introduced fuzzy sets and fuzzy logic. In this book series "Fuzzy Management Methods" fuzzy logic is applied to extend portfolio analysis, scoring methods, customer relationship management, performance measurement, web reputation, web analytics and controlling, community marketing and other business domains to improve managerial decisions. Thus, fuzzy logic can be seen as a management method where appropriate concepts, software tools and languages build a powerful instrument for analyzing and controlling the business.

More information about this series at http://www.springer.com/series/11223

Rédina Berkachy

The Signed Distance Measure in Fuzzy Statistical Analysis

Theoretical, Empirical and Programming Advances

 Springer

Rédina Berkachy 🆔
Applied Statistics And Modelling (ASAM)
Department of Informatics,
Faculty of Management, Economics
and Social Sciences
University of Fribourg
Fribourg, Switzerland

ISSN 2196-4130   ISSN 2196-4149   (electronic)
Fuzzy Management Methods
ISBN 978-3-030-76918-5   ISBN 978-3-030-76916-1   (eBook)
https://doi.org/10.1007/978-3-030-76916-1

This Springer imprint is published by the registered company Springer Nature Switzerland AG.
The registered company address is: Gewerbestrasse 11, 6330 Cham, Switzerland

*To my beloved mom Marcelle and dad
Georges, my precious sister Rana, my brother
Roy and my sister-in-law Mathilda, to
Georges and Maria, and to my wonderful
husband Salim,*

Foreword

At the beginning of her thesis work, Rédina Berkachy thoroughly examined the empirical distributions of defuzzified fuzzy quantities. In particular, she focused her attention on the signed distance measure as a defuzzification tool. She compared this measure with other methods. Although the signed distance measure was already well known, surprisingly, little was done to clearly understand the impact on the defuzzified distribution. For an applied statistician, it is essential to have in mind the potential repercussions of the tool used on data. The mean, variance, skewness and kurtosis were particularly observed. It was not very unexpected to find that this distance measure tends to normalise the data, in the sense that the crisp data obtained by defuzzification tend to have some characteristics of a normal distribution.

The signed distance measure was applied in global and individual evaluations of linguistic questionnaires. Based on the normality assumption, confirmed by the precedent results, testing hypotheses on individual evaluations could be made rightfully as in classical statistics. It was the start of promising research.

Several distance measures are commonly used in fuzzy statistics. The signed distance is one among them. It is not a new concept and has been widely applied, for example, in linguistic questionnaires analyses. Indeed, Rédina Berkachy proposed two new declinations of this measure, proved their properties and used them in different contexts or procedures.

Fuzzy inference methods are central in the work of Rédina Berkachy. It is worth mentioning the use of fuzzy logic principles in the formulation of fuzzy statistical hypotheses. Doing so, not only the data could be vague or imprecise but also the hypotheses. The development of fuzzy confidence intervals using the likelihood function is a remarkable achievement. The approach appears to be general because the technique is generalisable to many situations. By means of a bootstrap algorithm, one can simulate the distribution of the fuzzy likelihood ratio and construct a proper confidence interval. Then, this interval can be used, for example, to test hypotheses. Furthermore, Rédina Berkachy showed how to properly use the signed distance measure to defuzzify fuzzy decisions and fuzzy p-values and wrote the decision rules.

Another welcomed result is the proposed fuzzy analysis of variance, in which the method is generalised to the multi-ways analysis of variance. It is shown that the decomposition of fuzzy treatment effects is possible, and proper testing can be done after simulation of the distributions by a bootstrap algorithm. Particularly appealing, the application of the method is straightforward.

Rédina Berkachy's book proposes a lot of procedures and methods, which could discourage a reader. Fortunately, all of them are illustrated by numerous examples. An R package, named FuzzySTs, is now also available on CRAN. The reader can install it and run the functions to perform the analyses. On the other hand, the book is well structured, and since a pedagogical point of view was adopted in writing, its content is very accessible to everyone. My dearest hope is a warm welcome from the community, fast dissemination of the results and the acquisition of new fans of fuzzy statistics.

Prof. Dr. Laurent Donzé

Preface

This book intends to present some advances in fuzzy statistical analyses. A particular distance between fuzzy numbers, called the signed distance, seems to be appealing because of its directional property. It has the ability of describing the direction of travel between two fuzzy numbers. In addition, it has been often used as a fuzzy ranking tool or a defuzzification operator. Despite the fact that this distance appears to have interesting properties, it presents serious drawbacks. To overcome these problems, this book develops improved versions of it given by two L_2 metrics for which the first one is directional and preserves the properties of the signed distance, and the other one is non-directional. Both sophisticated distances have the advantage of taking into consideration the shapes of the fuzzy numbers and their possible irregularities. The core aim of this book is therefore to apply these novel distances in a series of statistical approaches defined on the set of fuzzy numbers.

Based on the proposed distances, this book provides a methodology of testing hypotheses in the fuzzy environment. This method of testing relies on the estimation of fuzzy confidence intervals, where we consider not only the data as fuzzy, but also the hypotheses. As such, the defended distances are used at different stages of the process of the inference test: in the conception of the models themselves or as defuzzification operators of the obtained fuzzy decisions. Moreover, since the traditional way of estimating fuzzy confidence intervals is in some sense limited in terms of the chosen parameters and the involved distributions, this book presents a practical procedure of estimation of such intervals based on the likelihood method. This new procedure is seen as general, since it can be used with any type of parameter and without the obligation of defining a particular distribution a priori. Analogously to this testing model, a definition of the fuzzy p-value described in the same setups with its corresponding decision rule is given. In this context, a defuzzification of this fuzzy p-value can be of good use to get a crisp interpretable decision.

Furthermore, this book presents two conceptually different applications in which the defended distances are involved. On a first stage, a novel methodology of assessment of linguistic questionnaires is developed on two distinct levels: the global and the individual ones. The proposed procedure has the great possibility of being able to treat the sampling weights and the eventual missingness occurring in the concerned data sets. The second application consists an extension of the multi-ways analysis of variance to the space of fuzzy sets. The decision related to the corresponding test statistic can be made according to two decision rules: a heuristic one preserving the fuzzy nature of the sums of squares, and another crisp one based on the defended distances.

To illustrate these approaches, multiple empirical and simulation studies are displayed using synthetic and real data sets. Note that a prominent objective of these studies is to investigate empirically some different statistical contexts, and explore theoretically the use of the newly introduced distances compared to known ones from one side and to the results obtained from the conventional statistical theory from another side. A recurrent general finding is that the classical approach seems to be a particular case of the fuzzy one. As such, the mentioned fuzzy approaches seem to be potentially promising.

A coherent R package covering all the previously mentioned concepts with complete documentation and some use cases are finally described. This package is developed from scratch in compliance with all the theoretical tools presented in a user-friendly comprehensive programming environment with a big panoply of calculation methods.

Fribourg, Switzerland   Rédina Berkachy

Acknowledgements

This book is a revised version of my Ph.D. thesis presented to the Faculty of Management, Economics and Social Sciences at the University of Fribourg. Enrolling in these Ph.D. studies has truly been a life-changing experience, and this would not have been possible without the support and guidance I received from many people.

Originating from a very beautiful village called Deir El Ahmar in the valley of Lebanon, I decided in 2013 to go beyond my frontiers and be adventurous as always. When I first embarked in Switzerland to start my Ph.D., I did not imagine my life will change drastically, leaving not only my books and my souvenirs in my beloved country, but also my heart and my thoughts.

First of all, let me express my gratitude to my mentor Prof. Dr. Laurent Donzé who gave me the opportunity to be part of the Applied Statistics and Modelling Group at the University of Fribourg. I would like to thank him for his constant support and patience, for all his motivation and encouragement. Thank you for cheering me up whenever I felt down, for all our friendly discussions around coffee breaks, and for believing in me.

I am also grateful to Prof. Dr. Marino Widmer who stood by me in every difficult step I faced. Thank you from the bottom of my heart for always motivating me, and for always telling me not to surrender. You gave me strength. Thank you.

I would like also to acknowledge the reviewers of this manuscript for accepting to do this task. Thank you for providing me with constructive and interesting enhancements of my book. In addition, I am particularly thankful to the dean's office in the Faculty of Management, Economics and Social Sciences at the University of Fribourg who encouraged me throughout the book process.

The publication of this manuscript is supported by the FMSquare association represented by Prof. Dr. Andreas Meier and Prof. Dr. Edy Portmann. I am very thankful for their support during the whole process.

My sincere thanks also go to Prof. Mathias Rossi and Prof. Pascale Voirin who encouraged me in the beginning of this thesis and helped to make it happen. I must also thank Dr. Jean-Vincent Le Bé from Nestlé beverages who trusted in me and collaborated with me on a very interesting project.

I would like to express my very profound gratitude to my friends Nayla Sokhn, Zeina Youssef-Aebli, Rana Bassil, Marielle Hayeck, Elie-Jacques Fares, Karl Daher, Rita Abboud, Clauda Mhanna and Joanne Mrad, and to my colleagues and friends, Christine Schaeren, Martina Raggi and Layal Lettry who all kept their sense of humour when I lost mine. Thank you for all the encouragement and support during this difficult period.

I would like finally to say a heartfelt thank you to my beloved family: my mom Marcelle and my dad Georges for their constant prayers and thoughts, for always believing in me and cheering me up. I am also indebted to my twin-sister Rana who did not miss any chance to come to Switzerland (and unfortunately travelled a lot because of me) whenever I felt down. I also thank my brother Roy and my sister-in-law Mathilda for always standing by me with their support in each step and their constant humour. Thank you to little Georges and Maria for giving me the inspiration and the patience. Seeing them growing up so fast is a blessing.

Last but not least, I would like to thank my wonderful husband Salim for his endless loyalty, for his unfailing love, support, and care. Thank you for cheering me up and standing by me especially when I completely lost hope. Because of you, I have gained self-confidence and found my life's passion. Thank you also for your invaluable advices throughout the writing process, they were crucial.

This accomplishment would not have been achieved without the encouragement and motivation extended by all these people. Thank you all from the bottom of my heart.

Contents

Conclusion Part I

List of Figures

List of Tables

Chapter 1
Introduction

In statistical analyses, we usually encode the human opinion by discrete or continuous variables. In both cases, these real-valued modelling choices of the opinion are assumed to be clear and precise. However, it is obvious that human perception is often involved by uncertainty and subjectivity. Conventional statistical methods appear to be limited in terms of capturing this uncertainty and describing it in a precise manner. Despite the fact that exact modelling methods are widely used, vagueness in human response cannot be avoided, especially when it comes to a verbal language. By verbal language, we mean the inherent semantics that a human would use to describe a given phenomenon, such as "approximately," "around," "maybe," "between good and bad," etc. In the same way, it is known that in traditional surveys, practitioners tend to use categorical variables for which the modalities are often given by Likert scales. For instance, the concept of linguistic questionnaires arises. This type of questionnaires is defined as a set of items, i.e. the questions, encoded by given categories. The modalities are then called "linguistics". Linguistic questionnaires are used in a wide number of fields, for example, in satisfaction surveys, in economic sciences, etc. It is clear to see that getting exact answers from such questionnaires is in some sense impossible. It could be because of the limited number of linguistics per question. Such questionnaires are strongly affected by imprecision.

The theory of fuzzy sets was firstly introduced by Zadeh (1965) to provide a framework to treat such imprecisions. In this context, Zadeh (1975) asserted that linguistic variables are *"variables whose values are not numbers but words or sentences in a natural or artificial language."*[1] Following Zadeh (1975), the advantage of using words is that linguistic variables are usually less specific than number-valued ones. Hence, a convenient "treating" method should be adopted in such cases. The idea behind fuzzy set theory is then to propose mathematical models which capture the ambiguity contained in a human language. The point is then

---

[1] Zadeh (1975, p. 201)

© The Author(s), under exclusive license to Springer Nature Switzerland AG 2021
R. Berkachy, *The Signed Distance Measure in Fuzzy Statistical Analysis*,
Fuzzy Management Methods, https://doi.org/10.1007/978-3-030-76916-1_1

to re-encode vague terms (eventually linguistics) in a more natural comprehensive way. By fuzzy logic, researchers tend to construct fuzzy rule-based systems to treat such linguistic data, and by that to aggregate the different vague answers related to a considered unit. Accordingly, by means of the so-called extension principle, traditional statistical methods were extended into the theory of fuzzy sets in order to be able to mine vague information by connecting them to a human approximate reasoning. This fact is seen as giving statistical analyses a new flexible facet using fuzzy sets.

In the context of modelling information by fuzzy set theory, the information modelled by fuzzy sets can be regarded by two main views as seen in Couso and Dubois (2014): the epistemic and the ontic ones. The epistemic view is a representation of a real-valued quantity assumed to be vague, as if the information is seen as incomplete, while the ontic view is a precise representation of a naturally vague entity. These two definitions induce two schools of approaches in fuzzy statistical analyses in terms of the definition of random variables in the fuzzy environment. Thus, two conceptually different approaches are defined in this context: the Kwakernaak (1979)–Kruse and Meyer (1987), and the Féron (1976a)–Puri and Ralescu (1986). The Kwakernaak–Kruse and Meyer approach corresponds to the epistemic view of objects. In other words, it is based on the fuzzy perception of an original real-valued entity. Contrariwise, the Féron–Puri and Ralescu corresponding to the ontic view assumes that the initial entity is fuzzy by nature. Following these definitions, various fuzzy statistical methods have been exposed.

Although fuzzy set theory is considered to be advantageous in terms of modelling uncertain information, it presents severe limitations, mainly reduced to the following two problems:

- The space of fuzzy sets presents a semi-linear structure, instead of a linear one. This semi-linearity in this space provokes difficulties in the computation of the difference arithmetic operation between fuzzy numbers;
- It presents a lack in total ordering between fuzzy numbers in the space of fuzzy sets.

These problems can be overcome by well adapted metrics defined on the space of fuzzy sets. In the Féron–Puri and Ralescu methodology such distances are particularly used due to their connection with the difference arithmetic. In this context, many distances have been introduced in the last decades. They are divided into directional and non-directional ones. Directional distances were less used than non-directional ones because of their lack in symmetry and separability. In other terms, such distances can have negative values, signalizing the direction of travel between two given fuzzy numbers. We are interested in a specific directional distance called the signed distance (SGD), given in Yao and Wu (2000). We have used this distance in many contexts, such as the evaluations of linguistic questionnaires shown in Berkachy and Donzé (2016b) and Berkachy and Donzé (2016a) as instance, in fuzzy hypotheses testing as proposed in Berkachy and Donzé (2017a), Berkachy and Donzé (2017c), etc. In addition to its simplicity and flexibility, the SGD appeared to have nice properties. It has not only been used as a ranking tool but also as a defuzzification operator. Yet, the SGD shows a serious drawback. In fact, this distance coincides with a central location measure. In other terms, the spreads and the shapes of the fuzzy numbers cannot be taken into consideration in this calculation. Only a middle tendency measure often contained in the core set will eventually be concerned, while the possible irregularities in the shape are not at all counted. An objective of this thesis is then to introduce an improved version of this distance denoted by the generalized signed distance (GSGD) to solve this problem. It will also be important to apply it in a range of statistical analyses in order to understand its advantages and disadvantages. As such, based on the SGD, we propose two L_2 metrics which take into consideration the shapes of the fuzzy numbers.

We intend to investigate the use of the aforementioned distances in hypotheses testing procedures. Grzegorzewski (2000) and Grzegorzewski (2001) provided a method of testing hypotheses by fuzzy confidence intervals, in which the data are supposed to be vague. By this method, one would obtain fuzzy decisions. The author proposed then to defuzzify them. As for us, we will extend this procedure in order to be able to take into consideration not only the vagueness of the data, but also of the hypotheses. This case is shown in Berkachy and Donzé (2019d). To complete our study, it appears that the use of traditional expressions of the fuzzy confidence intervals are somehow limited, since they are introduced for specific parameters with pre-defined distributions only. For this reason, we propose a practical methodology of estimation of a fuzzy confidence interval by the well-known likelihood ratio method. This latter can be used for any parameter, with any chosen distribution. The outputs of such hypotheses testing procedures are fuzzy decisions for which a defuzzification is often required. Thus, we would be interested to perform this step by the original SGD or the GSGD. Furthermore, from the conventional statistical theories, it is familiar to calculate the p-value in the context of testing hypotheses. It is the same in the fuzzy theory. Filzmoser and Viertl (2004) defined a fuzzy p-value in a way to suppose that the data is fuzzy, while Parchami et al. (2010) considered a definition where the fuzziness is rather a matter of hypotheses. In our method seen in Berkachy and Donzé (2019b), we suppose that the hypotheses and the data can simultaneously be vague. Since the fuzzy p-value is obviously of fuzzy type, then one could imagine to defuzzify it using a prudent metric in order to get an interpretable decision.

We would like after to explore the influence of the use of the defended distances in some other statistical contexts. For this purpose, we chose two conceptually different procedures: the evaluation of linguistic questionnaires as seen in Berkachy and Donzé (2016a), for example, and the Mult-FANOVA as presented in Berkachy and Donzé (2018b) for the univariate case. For the first case, we propose to aggregate the records of the units of a given linguistic questionnaire modelled by fuzzy numbers, on a global and an individual levels. These assessments can be done using the original and the improved signed distances, seen as ranking operators. By our method, one could take into account the sampling weights and the missing values in the data. For the missingness problem, indicators of information rate are also defined

on both levels. Their aim is to give indications of the quantity of missing values in
the answers of a given unit or in the complete data set. For the second case, in terms
of the defended distances, we recall the methods of the fuzzy multi-ways analysis of
variance. As such, we propose two exploratory decision rules: one based on using
a convenient distance (the original or the generalized signed distances as instance),
and another heuristic one. To illustrate these approaches, we provide applications
on real data sets from different contexts. A finding of this study is to confirm that in
our situations the involved classical statistical analyses seem to be a particular case
of the fuzzy analyses.

In terms of the usability of the aforementioned methodologies, we propose a
complete coherent R package composed by a plenty of functions covering the
contents of all the defined concepts. This user-friendly package is constructed on
the basis of the packages FuzzyNumbers shown in Gagolewski (2014), and
polynom given in Venables et al. (2019). The presented functions are validated
on several data sets, coming from multiple sources.

We can summarize the main contributions of this thesis in the following
manner:

- to introduce an improved version of the signed distance which preserve its
  properties, by taking into account the shape of the concerned fuzzy numbers;
- to propose the definitions of different statistical measures and point estimators
  with respect to the signed distance and its improved forms;
- to provide a practical procedure of computation of fuzzy confidence intervals by
  the likelihood ratio method, where any type of parameter can be used without the
  obligation of a priori defining the distribution of the data set. For this calculation,
  we propose to use the bootstrap technique to empirically estimate the required
  distribution and we introduce two different algorithms for the construction of the
  bootstrap samples;
- to expose a method of testing hypotheses by fuzzy confidence intervals where the
  data as well as the hypotheses are assumed to be fuzzy. The fuzzy decisions, i.e.
  outcome of such methods, can also be defuzzified by a convenient distance;
- to propose a procedure of calculation of the fuzzy p-value by considering
  simultaneously the fuzziness of the data and of the hypotheses, followed by the
  corresponding decision rule. In this case, it would be useful to defuzzify this
  fuzzy p-value in some situations;
- to introduce a general method of evaluations of linguistic questionnaires based
  on the signed distance and its generalized forms at the global and the individual
  levels, taking into account the sampling weights and the eventual missing values
  in the data set;
- to expose a fuzzy multi-ways analysis of variance in the fuzzy environment. For
  this purpose, two exploratory decision rules are given: one based on a chosen
  distance such as the signed distance and its improved version, and the other based
  on a heuristic procedure where the fuzzy nature is preserved;
- to apply the previous theoretical methodologies on synthetic and real data sets;

- to develop a coherent self-sustaining R package where all the prior theoretical methodologies are implemented in a user-friendly programming environment with very few dependencies.

This thesis, composed of 3 parts, is organized as follows. The first one divided into 5 chapters is devoted to the theoretical results where we expose in Chap. 2 the fundamental concepts of the theory of fuzzy sets. In Chap. 3, we briefly recall the idea behind the fuzzy rule-based systems composed by 3 steps: fuzzification, IF-THEN rules, and defuzzification. Chapter 4 covers several known distances with their properties, as well as the signed distance. Accordingly, the aim of this chapter is to develop two new L_2 metrics called the $d_{SGD}^{\theta^\star}$ and the generalized signed distance d_{GSGD}. Similarly to the signed distance, this latter is directional, but where the shapes of the considered fuzzy numbers are taken into account during the calculations. In Chap. 5, we recall the two approaches of random variables in the fuzzy environment: the Kwakernaak–Kruse and Meyer and the Féron–Puri and Ralescu. The concepts of distribution functions related to these approaches, the expectation, the variance, and the notion of estimators of the distribution parameters are also given. In this chapter we show how one can use the signed distance and its generalized form in the calculations of sample moments and some other statistical measures. Last but not least, Chap. 6 expands the ideas of the first four chapters onto the fuzzy hypotheses testing procedures. Thus, we first propose a practical methodology to compute fuzzy confidence intervals by the likelihood method, followed by a hypothesis testing procedure based on the estimated intervals. The idea is to consider that the data and the hypotheses can be assumed to be fuzzy at the same time, and consequently to construct fuzzy decisions related to not rejecting a given hypothesis of the model. In the same setups, i.e. both data and hypotheses are supposed to be fuzzy, we provide a method of calculating a fuzzy p-value with its decision rule. In common cases, the fuzzy decisions as well as the fuzzy p-value might need a defuzzification in order to get an interpretable decision, a particular operator can be used in this intention. The mentioned approaches are afterwards illustrated by detailed examples, in addition to various simulation studies. The objective of these latter is to investigate the use of the generalized signed distance compared to all the other evoked distances and aggregation operators. The classical and the fuzzy approaches are finally compared, and guidelines for their use are given.

The second part is devoted to some applications of the original and the generalized signed distance in statistical methods. These latter are chosen in the purpose of understanding the pertinence of such operators in various contexts. These approaches are the evaluations of linguistic questionnaires by these distances, and the fuzzy multi-ways analysis of variance. Therefore, we expose in Chap. 7 a procedure of evaluation of linguistic questionnaires made on two distinct levels: the global and the individual ones. In these calculations, a distance is needed. We note that the sampling weights can be considered. Our procedure takes into account also the missingness that could occur in data sets. The indicators of information rate are also shown. In Chap. 8, we show a Mult-FANOVA model in the fuzzy context, where a preliminary heuristic decision rule as well as one based on a considered metric are shown. Both chapters are illustrated by detailed applications with real data sets. For these approaches, the application of the signed distance and its improved form are our areas of interest.

The last part of this thesis consists on a detailed description of our R package called `FuzzySTs` shown in Berkachy and Donzé (2020), constructed from scratch in compliance with all the concepts presented in the first two parts. Numerous functions are exposed with a wide range of calculation methods.

References

Berkachy, R., & Donzé, L. (2016a). Individual and global assessments with signed distance defuzzification, and characteristics of the output distributions based on an empirical analysis. In *Proceedings of the 8th International Joint Conference on Computational Intelligence - Volume 1: FCTA* (pp. 75–82). ISBN: 978-989-758-201-1. https://doi.org/10.5220/0006036500750082

Berkachy, R., & Donzé, L. (2016b). Linguistic questionnaire evaluation: an application of the signed distance defuzzification method on different fuzzy numbers. The impact on the skewness of the output distributions. In *International Journal of Fuzzy Systems and Advanced Applications, 3*, 12–19.

Berkachy, R., & Donzé, L. (2017a). Defuzzification of a fuzzy hypothesis decision by the signed distance method. In *Proceedings of the 61st World Statistics Congress, Marrakech, Morocco.*

Berkachy, R., & Donzé, L. (2017c). Testing fuzzy hypotheses with fuzzy data and defuzzification of the fuzzy p-value by the signed distance method. In *Proceedings of the 9th International Joint Conference on Computational Intelligence (IJCCI 2017)* (pp. 255–264). ISBN: 978-989-758-274-5.

Berkachy, R., & Donzé, L. (2018b). Fuzzy one-way ANOVA using the signed distance method to approximate the fuzzy product. In Collectif LFA (Ed.), *LFA 2018 - Rencontres Francophones sur la Logique Floue et ses Applications,* Cépaduès (pp. 253–264).

Berkachy, R., & Donzé, L. (2019b). Defuzzification of the fuzzy p-value by the signed distance: Application on real data. In *Computational intelligence*. Studies in Computational Intelligence (vol. 829, pp. 77–97). Cham: Springer International Publishing. ISSN: 1860-949X. https://doi.org/10.1007/978-3-030-16469-0

Berkachy, R., & Donzé, L. (2019d). Testing hypotheses by Fuzzy methods: A comparison with the classical approach. In A. Meier, E. Portmann, & L. Terán (Eds.), *Applying Fuzzy logic for the digital economy and society* (pp. 1–22). Cham: Springer International Publishing. ISBN: 978-3-030-03368-2. https://doi.org/10.1007/978-3-030-03368-2_1

Berkachy, R., & Donzé, L. (2020). *FuzzySTs: Fuzzy statistical tools, R package.* https://CRAN.R-project.org/package=FuzzySTs

Couso, I., & Dubois, D. (2014). Statistical reasoning with set-valued information: Ontic vs. epistemic views. *International Journal of Approximate Reasoning, 55*(7), 1502–1518.

Féron, R. (1976a). Ensembles aléatoires flous. *Comptes Rendus de l'Académie des Sciences de Paris A, 282*, 903–906.

Filzmoser, P., & Viertl, R. (2004). Testing hypotheses with fuzzy data: The fuzzy p-value. In *Metrika* (vol. 59, pp. 21–29). Springer. ISSN: 0026-1335. https://doi.org/10.1007/s001840300269

Grzegorzewski, P. (2000). Testing statistical hypotheses with vague data. *Fuzzy Sets and Systems, 112*(3), 501–510. ISSN: 0165-0114. http://doi.org/10.1016/S0165-0114(98)00061-X. http://www.sciencedirect.com/science/article/pii/S016501149800061X

Grzegorzewski, P. (2001). Fuzzy tests - defuzzification and randomization. *Fuzzy Sets and Systems*, *118*(3), 437–446. ISSN: 0165-0114. http://doi.org/10.1016/S0165-0114(98)00462-X. http://www.sciencedirect.com/science/article/pii/S016501149800462X

Gagolewski, M. (2014). *FuzzyNumbers package: Tools to deal with fuzzy numbers in R*. http://FuzzyNumbers.rexamine.com/

Kwakernaak, H. (1979). Fuzzy random variables II. Algorithms and examples for the discrete case. *Information Sciences*, *17*(3), 253–278.ISSN: 0020-0255. https://doi.org/10.1016/0020-0255(79)90020-3. http://www.sciencedirect.com/science/article/pii/0020025579900203

Kruse, R., & Meyer, K. D. (1987). *Statistics with vague data* (Vol. 6). Netherlands: Springer.

Parchami, A., Taheri, S. M., & Mashinchi, M. (2010). Fuzzy p-value in testing fuzzy hypotheses with crisp data. *Stat Pap*, *51*(1), 209–226. ISSN: 0932-5026. https://doi.org/10.1007/s00362-008-0133-4

Puri, M. L., & Ralescu, D. A. (1986). Fuzzy random variables. In *Journal of Mathematical Analysis and Applications*, *114*(2), 409–422. ISSN: 0022-247X. https://doi.org/10.1016/0022-247X(86)90093-4. http://www.sciencedirect.com/science/article/pii/0022247X86900934

Venables, B., Hornik, K., & Maechler, M. (2019). *Polynom: A Collection of Functions to Implement a Class for Univariate Polynomial Manipulations, R Package*. https://CRAN.R-project.org/package=polynom

Yao, J.-S., & Wu, K. (2000). Ranking fuzzy numbers based on decomposition principle and signed distance. In *Fuzzy Sets and Systems*, *116*(2), 275–288.

Zadeh, L. A. (1965). Fuzzy sets. In *Information and Control*, *8*(3), 338–353. ISSN: 0019-9958. http://doi.org/10.1016/S0019-9958(65)90241-X. http://www.sciencedirect.com/science/article/pii/S001999586590241X

Zadeh, L. A. (1975). The concept of a linguistic variable and its application to approximate reasoning—I. In *Information Sciences*, *8*(3), 199–249. ISSN: 0020-0255. https://doi.org/10.1016/0020-0255(75)90036-5. http://www.sciencedirect.com/science/article/pii/0020025575900365

Part I
Theoretical Part

Once we assume that our data set is subject to uncertainty, and we decide to choose a convenient "treatment" tool such as fuzzy theory, corresponding statistical analyses should be well adapted to such situations. In the recent years, fuzzy methods seemed to be appealing in terms of the development of several statistical methods.

Based on known theories, we are interested in developing procedures of testing hypotheses in a fuzzy environment. These procedures require a clear definition of random variables. In the fuzzy context, two main approaches are often highlighted: the Kwakernaak–Kruse and Meyer and the Féron–Puri and Ralescu approaches. The difference between these approaches is mainly conceptual. In the methodology by Féron–Puri and Ralescu, a distance between fuzzy numbers is needed. This distance is often used to overcome the difficulty induced by the difference arithmetic seen as complicated in the fuzzy environment. In addition, a total ordering does not exist yet in the space of fuzzy numbers. Therefore, proposing well adapted metrics to these calculations in the space of fuzzy sets is welcome. Multiple distances have then been introduced. A directional distance called the signed distance seems to have interesting properties. This distance has also served as a defuzzification operator and as a ranking tool in different contexts. However, it appears that it presents a drawback since its outcome coincides with a measure of central location. Practically, the possible irregularities in the shape of the fuzzy numbers are not taken into account. Consequently, it could be very useful to propose an upgrade of this distance, in a way to preserve the same directional properties of the original signed distance.

The contributions of this part are various: In terms of measuring the distance between two fuzzy numbers, we propose two novel L_2 metrics called the $d_{SGD}^{\theta^\star}$ and the generalized signed distance d_{GSGD}. Namely, these metrics are based on the original signed distance, but where they could efficiently take into consideration the shape of the concerned fuzzy numbers. Based on the concept of Féron–Puri and Ralescu's random variables, these metrics will be afterward used in the calculation of the expectation, the variance, and some other statistical fuzzy measures and estimators. For the Kwakernaak–Kruse and Meyer approach, these distances can eventually be used as defuzzification operators since the outcome of such approaches is known to be fuzzy nature. All of the previously mentioned information will lead us to propose new methodologies of fuzzy hypotheses tests. It is known that the uncertainty can have two sources: the data and the hypotheses. We then propose a testing method where we assume that the fuzziness does not only come from the data but also from the hypotheses. This approach relies on fuzzy confidence intervals, for which we introduce a new practical procedure of calculation based on the known likelihood ratio method. The advantage of this latter is that we are now on able to calculate a sophisticated fuzzy confidence interval where the procedure is in some sense general and can be applied on any parameter. In addition, we do not have the obligation of defining a priori a particular distribution. This hypotheses test, mainly connected to an epistemic view, is in the same direction of the definition of fuzzy random variables of Kwakernaak–Kruse and Meyer. By this method, fuzzy decisions given by fuzzy numbers are obtained. These decisions are related to the null and the alternative hypotheses and can accordingly be defuzzified in the purpose of getting interpretable decisions. The defuzzified values are interpreted as the so-called degree of conviction, which means a degree of acceptability of the hypotheses. This measure indicates then the degree of "not rejecting" a given hypothesis. The decisions can accordingly be defuzzified in the purpose of getting interpretable decisions. From another side, for a given test statistic, the concept of the p-value can also be used. As such, we propose a general method to compute fuzzy p-values where we consider fuzzy data and fuzzy hypotheses. Thereafter, we would compare this fuzzy p-value to the pre-defined significance level. It is clear to see that these two entities might often overlap. Thus, we propose to defuzzify them by a suitable operator, in the purpose of obtaining a decision, easy to interpret.

To sum up, this theoretical part is composed of 5 chapters. We open this part in Chap. 2 by numerous fundamental concepts of fuzzy sets and fuzzy numbers and the related logical and arithmetic operations. The extension principle is also given. In Chap. 3, we expose a résumé of the fuzzy rule-based systems comprising the three following steps: the fuzzification, the IF-THEN rules, and the defuzzification. Chapter 4 is devoted to recall many well-known distances between fuzzy numbers such as the ρ_1, ρ_2, etc., as well as the signed distance. In this chapter, based on the signed distance measure, we propose two new metrics taking into account the shapes of the fuzzy numbers. These metrics are denoted by $d_{SGD}^{\theta^*}$ and d_{GSGD}. This latter is called the generalized signed distance and preserves the directionality of the signed distance. The aim of Chap. 5 is to describe two approaches related to defining a random variable in the fuzzy environment, i.e. the approaches by Kwakernaak–Kruse and Meyer and Féron–Puri and Ralescu. We also highlight the concepts of distribution functions associated with these random variables, the expectation and the variance, in addition to the empirical estimators of the distribution parameters. Finally, in Chap. 6 discussing the fuzzy statistical inference, we first introduce a practical procedure to estimate fuzzy confidence intervals based on the likelihood ratio method. We afterward expose a fuzzy hypotheses test where one could consider the data and/or the hypotheses as fuzzy. This test is based on the constructed fuzzy confidence intervals. In addition, a fuzzy p-value generalizing both cases with its decision rule is also introduced. All the proposed concepts are illustrated by detailed examples of real data sets. One of our aims is to investigate the possible advantages and drawbacks of the application of different metrics in such methodologies, in order to get a broad idea about their use. A synthesis in which we interpret the different outputs of the estimations is given at last. We close the part by a comparison between the classical and the fuzzy theory in the context of statistical inference tests, and we give the pros and cons of both theories.

Chapter 2
Fundamental Concepts on Fuzzy Sets

As a first step, it is important to postulate the fundamental concepts of fuzzy sets as proposed in Zadeh (1965). This chapter is then devoted to basic notions of fuzzy sets, the logical and arithmetic related-operations, and to the well-known extension principle.

We open this chapter by definitions of fuzzy sets and fuzzy numbers, followed by the definitions of some particular shapes of fuzzy numbers: the triangular, the trapezoidal, the gaussian, and the two-sided gaussian fuzzy numbers. We discuss also the notions of vectors of fuzzy numbers, and of fuzzy vector. The aim is to highlight the differences that exist between them. Then, we recall the operations on fuzzy sets notably the logical and arithmetic operations. These latter are the foundations of further fuzzy analyses. We will at last have the occasion to display the so-called extension principle. Detailed examples illustrating the different sections are given.

2.1 Fuzzy Sets and Fuzzy Numbers

When Zadeh (1965) firstly discussed the notion of fuzzy sets, his idea was a bit more philosophical than methodological. He intended to propose a new ontology to cases that often occur in the analysis of data obtained from real life problems, by modelling them using the so-called fuzzy sets.

In the initial phase, let us then expose the basic definitions and concepts of fuzziness.

Definition 2.1.1 (Fuzzy Set) If A is a collection of objects denoted generically by x, then a fuzzy set or class $\tilde{X}$ in A is a set of ordered pairs:

$$\tilde{X} = \left\{ (x, \mu_{\tilde{X}}(x)) \mid x \in A \right\}, \tag{2.1}$$

© The Author(s), under exclusive license to Springer Nature Switzerland AG 2021
R. Berkachy, *The Signed Distance Measure in Fuzzy Statistical Analysis*,
Fuzzy Management Methods, https://doi.org/10.1007/978-3-030-76916-1_2

where the mapping $\mu_{\tilde{X}}$ representing the "grade of membership" is a crisp real valued function such that

$$\mu_{\tilde{X}} : \mathbb{R} \to [0; 1]$$

$$x \mapsto \mu_{\tilde{X}}(x)$$

is called the membership function (or generalized characteristic function).

For conventional researchers, the concept of membership characterizing fuzzy sets might often be confusing, since it coincides by far to some features of known notions of the classical theory. The subsequent remarks intend to clarify this idea.

Remarks 2.1.1

1. For sake of clarification, we mainly use the expression "membership function (MF)" instead of "characteristic function."
2. Note that the MF $\mu_{\tilde{X}}$ of a fuzzy set $\tilde{X}$ describing the uncertainty of a given observation resembles to the probability function of a stochastic entity X. However, some main conceptual differences exist between these two notions. As such, many references discussed of these differences. We note, for example, Laviolette et al. (1995) and Zadeh (1995), and others.

To illustrate these concepts, we propose to consider the subsequent scenario:

We are interested in the variable age in a given population. If we ask a person about his age, many types of possible answers can be given. He could, for example, answer by a number (20, 21, 22 . . .), or a linguistic term (such as very young, young, old, very old), etc. In the case of a number-valued answer, if we would like to assign for each number a class of aging, the task is difficult since the interpretation of each number is subjective and can be different between humans. As instance, for a 80 years old person, an individual having 20 years old is very young. However, for a person of 10 years old age, the same person having 20 years is seen to be old. From another side, in the case of a linguistic term, it is obvious that choosing between different categories is vague. Thus, defining precisely the age regarding the category is complicated.

Example 2.1.1 (The age Example) Consider the statement "Paul is very young." The term "very young" is uncertain since Paul can be 5 or 20 years old and still be very young. In fact, depending on our perception, one could be "very young" between 0 and 30, young between 25 and 50, "old" between 45 and 70 and "very old" between 65 and upper. We model the term very young by a fuzzy set denoted by $\tilde{X}$ with its corresponding MF $\mu_{\tilde{X}}$ shown in Fig. 2.1. This latter shows that the 5 year-old Paul is at 100% very young ($\mu_{\tilde{X}}(5) = 1$), and if he has 20 years old, then the grade of membership is roughly around 0.67 i.e. $\mu_{\tilde{X}}(20) = 0.67$. He is then at 67% very young.

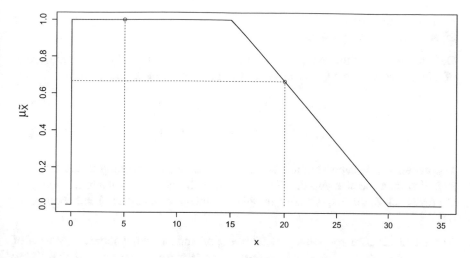

Fig. 2.1 Membership function of $\tilde{X}$ describing the term very young—Example 2.1.1

If the MF of a given fuzzy set attains the highest grade, i.e. 1, we define the so-called normal fuzzy set as follows:

Definition 2.1.2 (Normal Fuzzy Set) A fuzzy set $\tilde{X}$ is said to be normal iff

$$\sup \mu_{\tilde{X}}(x) = 1. \tag{2.2}$$

In the same way, it is important to define the so-called convex and strongly convex fuzzy sets as proposed by Zadeh (1965) and recalled by Drewniak (1987) and others. This concept has been introduced in a way preserving the properties of convexity of the classical sets. The definition can be written by:

Definitions 2.1.1 (Convex and Strongly Convex Fuzzy Sets)

- A fuzzy set $\tilde{X}$ is said to be convex iff its membership function $\mu_{\tilde{X}}$ is given such that

$$\mu_{\tilde{X}}(tx_1 + (1-t)x_2) \geq \min\left(\mu_{\tilde{X}}(x_1), \mu_{\tilde{X}}(x_2)\right), \quad \forall x_1, x_2 \in \mathbb{R} \text{ and } t \in [0; 1]. \tag{2.3}$$

- A fuzzy set $\tilde{X}$ is said to be strongly convex iff its membership function $\mu_{\tilde{X}}$ is given such that

$$\mu_{\tilde{X}}(tx_1 + (1-t)x_2) > \min\left(\mu_{\tilde{X}}(x_1), \mu_{\tilde{X}}(x_2)\right), \quad \forall x_1, x_2 \in \mathbb{R} \text{ and } t \in (0; 1). \tag{2.4}$$

It is useful to expose the support and the kernel of a given fuzzy set. They are given in the following manner:

Definitions 2.1.2 (Support and Kernel of a Fuzzy Set) The support and the kernel of a fuzzy set $\tilde{X}$ denoted, respectively, by supp $\tilde{X}$ and core $\tilde{X}$, are given by

$$\text{supp } \tilde{X} = \{x \in \mathbb{R} \mid \mu_{\tilde{X}}(x) > 0\}, \tag{2.5}$$

$$\text{core } \tilde{X} = \{x \in \mathbb{R} \mid \mu_{\tilde{X}}(x) = 1\}. \tag{2.6}$$

In other terms, the support of a fuzzy set $\tilde{X}$ is a crisp set containing all the elements such that their membership function is not zero. In the same manner, the kernel of the fuzzy set $\tilde{X}$ is a crisp set containing all elements with degree of membership of one.

Example 2.1.2 (The age Example) The support and the kernel of the fuzzy set $\tilde{X}$ of Example 2.1.1 describing the uncertainty in the term very young are as follows:

$$\text{supp } \tilde{X} = \{x \in \mathbb{R} \mid \mu_{\tilde{X}}(x) > 0\} = \;]0; 30[,$$

$$\text{core } \tilde{X} = \{x \in \mathbb{R} \mid \mu_{\tilde{X}}(x) = 1\} = [0; 15].$$

We denote now by $\mathbb{F}(\mathbb{R})$ the class of all non-empty normal fuzzy sets on the real line $\mathbb{R}$, and by $\mathbb{F}_c^\star(\mathbb{R})$ the class of the non-empty compact, convex and normal fuzzy sets on $\mathbb{R}$. Thereafter, a fuzzy number can be defined as follows:

Definition 2.1.3 (Fuzzy Number) A fuzzy number $\tilde{X}$ is a normal and convex fuzzy subset of A on $\mathbb{F}_c^\star(\mathbb{R})$ which obeys the following conditions:

1. $\exists x_0$ in $\mathbb{R}$ for which $\mu_{\tilde{X}}(x_0) = 1$;
2. $\mu_{\tilde{X}}(x)$ is upper semi-continuous;
3. its support is bounded.

A given fuzzy set is afterwards characterized by a collection of crisp sets called the α-level sets, as seen in the following definition:

Definition 2.1.4 (α-Level Set or α-Cut) An α-level set $\tilde{X}_\alpha$ of the fuzzy set $\tilde{X}$ is the (crisp) set of elements such that:

$$\tilde{X}_\alpha = \{x \in A \mid \mu_{\tilde{X}}(x) \geq \alpha\}. \tag{2.7}$$

The α-level set is a closed bounded and non-empty interval denoted generally by $\left[\tilde{X}_\alpha^L; \tilde{X}_\alpha^R\right]$ where for $\forall \alpha \in [0; 1]$, $\tilde{X}_\alpha^L$ and $\tilde{X}_\alpha^R$ are the left and right hand sides of $\tilde{X}_\alpha$ called, respectively, the left and right α-cuts such that:

$$\tilde{X}_\alpha^L = \inf\{x \in \mathbb{R} \mid \mu_{\tilde{X}}(x) \geq \alpha\} \text{ and } \tilde{X}_\alpha^R = \sup\{x \in \mathbb{R} \mid \mu_{\tilde{X}}(x) \geq \alpha\}. \tag{2.8}$$

We add that since the considered α-level sets are assumed to be bounded and not empty, for the values $\alpha = 0$ and $\alpha = 1$, the required α-cuts exist. We highlight that for $\alpha = 0$, the 0-level set $\tilde{X}_{\alpha=0}$ is called the closure of the set $\left\{ x \in \mathbb{R} \mid \mu_{\tilde{X}}(x) > 0 \right\}$.

One could express the α-level sets of a fuzzy number by means of an arbitrary function, the so-called the "representative of a class." The functions of α, denoted by $\tilde{X}_{\alpha}^{L}(\alpha)$ and $\tilde{X}_{\alpha}^{R}(\alpha)$, are said to be the representatives of the left and right α-cuts of the class containing $\tilde{X}$.

Furthermore, a fuzzy number $\tilde{X}$, also called Left-Right (L-R) fuzzy number, can be represented by the family set of his α-cuts $\left\{ \tilde{X}_{\alpha} \mid \alpha \in [0; 1] \right\}$. This set is a union of finite compact and bounded intervals $\left[\tilde{X}_{\alpha}^{L}(\alpha); \tilde{X}_{\alpha}^{R}(\alpha) \right]$ such that, $\forall \alpha \in [0; 1]$,

$$\tilde{X} = \bigcup_{0 \leq \alpha \leq 1} \left[\tilde{X}_{\alpha}^{L}(\alpha); \tilde{X}_{\alpha}^{R}(\alpha) \right], \tag{2.9}$$

where $\tilde{X}_{\alpha}^{L}(\alpha)$ and $\tilde{X}_{\alpha}^{R}(\alpha)$ are the functions of the left and right hand sides of $\tilde{X}$.

The topic of α-level sets has gained lots of attention in the definition of fuzzy numbers. It would then be also interesting to define the concepts of the strong α-level set, and the one of its indicator functions.

Definition 2.1.5 (Strong α-Level Set or Strong α-Cut) An α-cut $\tilde{X}'_{\alpha}$ of the fuzzy number $\tilde{X}$ is said to be strong if it verifies the following property:

$$\tilde{X}'_{\alpha} = \left\{ x \in A \mid \mu_{\tilde{X}}(x) > \alpha \right\}. \tag{2.10}$$

Definition 2.1.6 (Indicator Function of an α-Cut) The Indicator function $I_{\tilde{X}_{\alpha}}(x)$ of the α-cuts of the fuzzy number $\tilde{X}$ is defined in the following manner:

$$I_{\tilde{X}_{\alpha}}(x) = \begin{cases} 1 & \text{if} \quad \mu_{\tilde{X}}(x) \geq \alpha, \\ 0 & \text{otherwise.} \end{cases} \tag{2.11}$$

Therefore, from the least-upper bound property generalized to ordered sets, the MF of $\tilde{X}$ can be written as

$$\mu_{\tilde{X}}(x) = \max \left\{ \alpha I_{\tilde{X}_{\alpha}}(x) : \alpha \in [0, 1] \right\}. \tag{2.12}$$

Finally, if two fuzzy numbers are equal, the following condition is applied:

Definition 2.1.7 Two fuzzy numbers $\tilde{X}$ and $\tilde{Y}$ are equal if and only if their membership functions are equal, i.e. $\mu_{\tilde{X}}(x) = \mu_{\tilde{Y}}(x)$, for $\forall x \in \mathbb{R}$.

2.2   Common Fuzzy Numbers

When uncertainty occurs in data set, useful modelling tools should be available. Historically, different ways were proposed to overcome this problem. In the fuzzy environment, when Zadeh (1965) introduced the fuzzy logic, his contribution was more into philosophy than into detailed suggestions on how to model fuzziness contained into the information. In later years, specific shapes of fuzzy numbers were suggested. However, some practitioners persisted to model the vagueness by their own shapes. In other terms, they preferred to draw them intentionally at their own sense. Nevertheless, the necessary and sufficient conditions for the construction of a fuzzy number such as the continuity, the monotony—the convexity—should be insured. We note as instance a study from a questionnaire called TIMSS-PIRLS presented in Gil et al. (2015) where 9 years old children were asked to draw their own fuzzy numbers preserving the previously mentioned conditions. The influence of the shapes of fuzzy numbers was discussed. Thus, we can assert that the type of membership functions is chosen according to the user experience. Conversely, the analyst should be aware that the choice of a given MF could eventually influence the modelling process. Moreover, in literature, particular shapes of fuzzy numbers were frequently used to model the uncertainty.

As an illustration, we define the trapezoidal, triangular, gaussian and two-sided gaussian fuzzy numbers. The preference of researchers for each of these shapes is due to different reasons: for example, the trapezoidal, and evidently the triangular fuzzy numbers are easy to handle, to draw and to interpret, while the gaussian, and the two-sided gaussian ones are more elaborated, and thus, more demanding from a computational point of view. These fuzzy numbers with their respective membership functions and α-cuts are shown in the subsequent definitions.

Definition 2.2.1 (Triangular Fuzzy Number) A triangular fuzzy number $\tilde{X}$ is a fuzzy number often represented by the tuple of three values p, q, and r, i.e. $\tilde{X} = (p, q, r)$, where $p \leq q \leq r \in \mathbb{R}$. Its MF has a form of a triangle and is given by

$$\mu_{\tilde{X}}(x) = \begin{cases} \frac{x-p}{q-p} & \text{if} \quad p < x \leq q, \\ \frac{x-r}{q-r} & \text{if} \quad q < x \leq r, \\ 0 & \text{elsewhere.} \end{cases} \tag{2.13}$$

From a horizontal view, the left and right α-cuts $\tilde{X}_\alpha^L$ and $\tilde{X}_\alpha^R$ are, respectively, expressed by

$$\begin{cases} \tilde{X}_\alpha^L(\alpha) = p + (q - p)\alpha, \\ \tilde{X}_\alpha^R(\alpha) = r - (r - q)\alpha, \end{cases} \quad , \quad \forall \alpha \in [0; 1]. \tag{2.14}$$

Definition 2.2.2 (Trapezoidal Fuzzy Number) A trapezoidal fuzzy number $\tilde{X}$ is written as the quadruple p, q, r, and s, i.e. $\tilde{X} = (p, q, r, s)$, where $p \leq q \leq r \leq$

$s \in \mathbb{R}$. Its MF $\mu_{\tilde{X}}$ is given by

$$
\mu_{\tilde{X}}(x) = \begin{cases} \frac{x-p}{q-p} & \text{if} \quad p < x \leq q, \\ 1 & \text{if} \quad q < x \leq r, \\ \frac{x-s}{r-s} & \text{if} \quad r < x \leq s, \\ 0 & \text{elsewhere.} \end{cases} \tag{2.15}
$$

For all α in the interval $[0; 1]$, the left and right α-cuts $\tilde{X}_{\alpha}^{L}$ and $\tilde{X}_{\alpha}^{R}$ are the following:

$$
\begin{cases} \tilde{X}_{\alpha}^{L}(\alpha) = p + (q - p)\alpha, \\ \tilde{X}_{\alpha}^{R}(\alpha) = s - (s - r)\alpha. \end{cases} \tag{2.16}
$$

For the subsequent gaussian and two-sided gaussian cases, we have to mention that the support sets of the corresponding shapes are not bounded. Therefore, they do not obey the bounded property of fuzzy numbers. However, these shapes are described in this context since they are often used in fuzzy programming techniques.

Definition 2.2.3 (Gaussian Fuzzy Number) We denote by $\tilde{X}$ a gaussian shaped fuzzy number represented by the couple of parameters (μ, σ). The MF of a gaussian fuzzy number is nothing but the function of the gaussian distribution given as:

$$
\mu_{\tilde{X}}(x) = \exp\left[-\frac{(x - \mu)^2}{2\sigma^2} \right], \forall x \in \mathbb{R}, \ \mu \in \mathbb{R} \text{ and } \sigma > 0. \tag{2.17}
$$

This formula leads to the following left and right α-cuts $\tilde{X}_{\alpha}^{L}$ and $\tilde{X}_{\alpha}^{R}$ of $\tilde{X}$:

$$
\begin{cases} \tilde{X}_{\alpha}^{L}(\alpha) = \mu - \sqrt{-2\sigma^2 \ln \alpha}, \\ \tilde{X}_{\alpha}^{R}(\alpha) = \mu + \sqrt{-2\sigma^2 \ln \alpha}, \end{cases} \tag{2.18}
$$

with $-2\sigma^2 \ln(\alpha) \geq 0$ under the condition $0 < \alpha \leq 1$.

Definition 2.2.4 (Two-Sided Gaussian Fuzzy Number) For a two-sided gaussian fuzzy number $\tilde{X}$ represented by the four parameters $(\mu_1, \sigma_1, \mu_2, \sigma_2)$, the couples of parameters (μ_1, σ_1) and (μ_2, σ_2) correspond to the gaussian curves defining, respectively, the left and right sides of the fuzzy number $\tilde{X}$. For $\mu_1 \leq \mu_2 \in \mathbb{R}$, and σ_1 and σ_2 both strictly positive, the MF of such fuzzy numbers is written as:

$$
\mu_{\tilde{X}}(x) = \begin{cases} \exp\left[-\frac{(x-\mu_1)^2}{2\sigma_1^2} \right] & \text{if } x \leq \mu_1, \\ 1 & \text{if } \mu_1 < x \leq \mu_2, \\ \exp\left[-\frac{(x-\mu_2)^2}{2\sigma_2^2} \right] & \text{if } x > \mu_2. \end{cases} \tag{2.19}
$$

The left and right α-cuts can be expressed by

$$\begin{cases} \tilde{X}_\alpha^L(\alpha) = \mu_1 - \sqrt{-2\sigma_1^2 \ln \alpha}, \\ \tilde{X}_\alpha^R(\alpha) = \mu_2 + \sqrt{-2\sigma_2^2 \ln \alpha}, \end{cases} \tag{2.20}$$

for which, for $\forall \alpha \in [0; 1]$, the values $-2\sigma_1^2 \ln \alpha$ and $-2\sigma_2^2 \ln \alpha$ are always positive.

Direct relations exist between the four types of the previously defined fuzzy numbers, as expressed in the following remark:

Remark 2.2.1 A trapezoidal fuzzy number can be treated as a triangular one when $q = r$. Analogously, a two-sided gaussian fuzzy number which has the particularity $\mu_1 = \mu_2$ and $\sigma_1 = \sigma_2$, is nothing but a gaussian one.

Furthermore, although we know that the choice of a specific MF is according to the user experience, the utility of different types should be known. In particular, the triangular and trapezoidal fuzzy numbers have simple shapes. Thus, from an operational point of view, they are convenient. Yet, the gaussian and two-sided gaussian ones, seen as more elaborated shapes than the trapezoidal and the triangular ones as instance, are computationally heavier, but more reliable. An example of these shapes is provided as follows:

Example 2.2.1 A person is asked to give an information about the approximate distance between his work and his home. His answer was "almost 2 km." This assertion is uncertain. One could use different shapes to model the uncertainty contained in this information. Some possible fuzzy numbers are shown in Fig. 2.2.

2.3 Vector of Fuzzy Numbers and Fuzzy Vector

For multivariate cases, one has to differentiate between a fuzzy vector and a vector of fuzzy numbers. For instance, the terminology of a vector of fuzzy entities relates to the case where a vector of multiple fuzzy elements is considered, while a fuzzy vector is seen to be a fuzzy version of a crisp vector of elements. Both concepts are defined as follows:

Definition 2.3.1 (Vector of Fuzzy Numbers) A n-dimensional vector of fuzzy numbers $(\tilde{X}_1, \ldots, \tilde{X}_n)$ is a vector of n uncertain elements. Thus, each element alone is fuzzy. The MF of the vector is consequently determined by each of the n membership functions of $\tilde{X}_i$, $i = 1, \ldots, n$, as follows:

$$\mu_{(\tilde{X}_1, \ldots, \tilde{X}_n)} : \qquad \mathbb{R}^n \to [0; 1]^n$$

$$(x_1, \ldots, x_n) \mapsto \left(\mu_{\tilde{X}_1}(x_1), \ldots, \mu_{\tilde{X}_n}(x_n)\right). \tag{2.21}$$

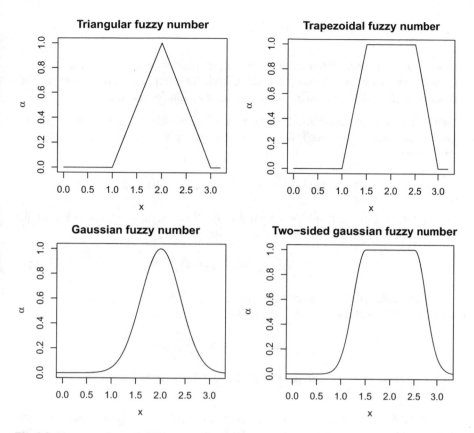

Fig. 2.2 Fuzzy numbers of different symmetrical shapes—Example 2.2.1

Definition 2.3.2 (Fuzzy Vector) A fuzzy vector $\underline{\tilde{X}}$ is the fuzzy version of a given vector $\underline{X} = (X_1, \ldots, X_2)$, such that

$$\underline{\tilde{X}} = (\widetilde{X_1, \ldots, X_2}). \tag{2.22}$$

Its MF $\mu_{\underline{\tilde{X}}}$ is given by a function of n variables $(x_1, \ldots, x_n)$ on $\mathbb{R}$. This function is often called the vector-characterizing function and is written as:

$$\mu_{\underline{\tilde{X}}}: \qquad \mathbb{R}^n \to [0; 1]$$
$$\underline{x} = (x_1, \ldots, x_n) \mapsto \mu_{\underline{\tilde{X}}}(x_1, \ldots, x_n). \tag{2.23}$$

Similarly to Eq. 2.7, its α-cut set is expressed by:

$$\underline{\tilde{X}}_\alpha = \left\{ \underline{x} \in \mathbb{R}^n \mid \mu_{\underline{\tilde{X}}}(\underline{x}) \geq \alpha \right\}. \tag{2.24}$$

We note that the Eq. 2.12 can be consequently deduced.

2.4 Logical Operations on Fuzzy Sets

Logical operations are from the most used operations in fuzzy processes. We show in this section the intersection, the union and the complement of two fuzzy sets denoted by $\tilde{X}$ and $\tilde{Y}$ with their respective membership functions $\mu_{\tilde{X}}$ and $\mu_{\tilde{Y}}$.

Definition 2.4.1 (Intersection Between Two Fuzzy Sets) The membership function of the intersection (logical and, i.e. ∩) between two fuzzy sets $\tilde{X}$ and $\tilde{Y}$ is defined as:

$$\mu_{\tilde{X} \cap \tilde{Y}}(x) = \min\left(\mu_{\tilde{X}}(x), \mu_{\tilde{Y}}(x)\right), \quad \forall x \in \mathbb{R}. \tag{2.25}$$

Definition 2.4.2 (Union of Two Fuzzy Sets) The membership function of the union (logical or, i.e. ∪) of $\tilde{X}$ and $\tilde{Y}$ is written by

$$\mu_{\tilde{X} \cup \tilde{Y}}(x) = \max\left(\mu_{\tilde{X}}(x), \mu_{\tilde{Y}}(x)\right), \quad \forall x \in \mathbb{R}. \tag{2.26}$$

Definition 2.4.3 (Complement of a Fuzzy Set) The membership function of the complement (negation ¬) of the fuzzy set $\tilde{X}$ is expressed by

$$\mu_{\tilde{X}^c}(x) = 1 - \mu_{\tilde{X}}(x), \quad \forall x \in \mathbb{R}. \tag{2.27}$$

We propose now an example illustrating the three described logical operations. It is given by the following:

Example 2.4.1 (The age Example) Relating to Example 2.1.1, consider the trapezoidal and triangular fuzzy numbers denoted, respectively, by $\tilde{X} = (25, 30, 40, 50)$ and $\tilde{Y} = (45, 60, 70)$, modelling the categories young and old, with their respective membership functions $\mu_{\tilde{X}}$ and $\mu_{\tilde{Y}}$ as seen in Fig. 2.3.

We would like to calculate the intersection and union of the two fuzzy sets when $x = 47$. They are given as follows:

- Intersection between $\tilde{X}$ and $\tilde{Y}$: $\mu_{\tilde{X} \cap \tilde{Y}}(47) = \min\left(\mu_{\tilde{X}}(47), \mu_{\tilde{Y}}(47)\right) = \min(0.3, 0.134) = 0.134$.
- Union of $\tilde{X}$ and $\tilde{Y}$: $\mu_{\tilde{X} \cup \tilde{Y}}(47) = \max\left(\mu_{\tilde{X}}(47), \mu_{\tilde{Y}}(47)\right) = \max(0.3, 0.134) = 0.3$.

Note that the intersection and union shown in Fig. 2.3 are not surfaces, but the segments of the MF resulting from both operations.

For the complement of the fuzzy set $\tilde{X}$ at the value $x = 47$, we have that:

$$\mu_{\tilde{X}^c}(47) = 1 - \mu_{\tilde{X}}(47) = 1 - 0.3 = 0.7.$$

For the interpretation of this complement, we can say that a given person having 47 years is at 30% "young" and at 70% "not young."

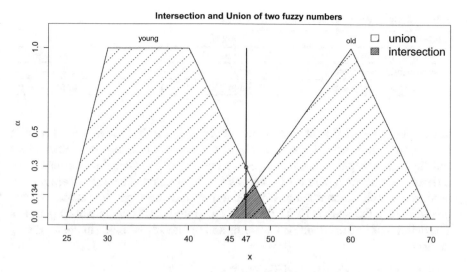

Fig. 2.3 Example with logical operations—Example 2.4.1

For the properties of the logical operations of fuzzy sets, one could easily remark that these properties are very similar to the ones of crisp sets. Thus, since the membership grades of a fuzzy or a crisp set are both contained in the interval [0; 1], the classical sets might be regarded as special cases of fuzzy sets. We denote by ∅ an empty fuzzy set such that its membership function is always null. Some important properties of these operations are given as follows:

Properties 2.4.1 (Properties of the Intersection and the Union of Fuzzy Sets)
Let $\tilde{X}$, $\tilde{Y}$, and $\tilde{Z}$ be fuzzy sets on $\mathbb{F}(\mathbb{R})$. The operations of intersection and union between these fuzzy sets have the following respective properties:

- Commutativity, such that $\mu_{\tilde{X}\cup\tilde{Y}} = \mu_{\tilde{Y}\cup\tilde{X}}$ and $\mu_{\tilde{X}\cap\tilde{Y}} = \mu_{\tilde{Y}\cap\tilde{X}}$;
- Associativity, such that $\mu_{\tilde{X}\cup(\tilde{Y}\cup\tilde{Z})} = \mu_{(\tilde{X}\cup\tilde{Y})\cup\tilde{Z}}$ and $\mu_{\tilde{X}\cap(\tilde{Y}\cap\tilde{Z})} = \mu_{(\tilde{X}\cap\tilde{Y})\cap\tilde{Z}}$;
- Distributivity, such that $\mu_{\tilde{X}\cup(\tilde{Y}\cap\tilde{Z})} = \mu_{(\tilde{X}\cup\tilde{Y})\cap(\tilde{X}\cup\tilde{Z})}$ and $\mu_{\tilde{X}\cap(\tilde{Y}\cup\tilde{Z})} = \mu_{(\tilde{X}\cap\tilde{Y})\cup(\tilde{X}\cap\tilde{Z})}$;
- Idempotency, such that $\mu_{\tilde{X}\cup\tilde{X}} = \mu_{\tilde{X}}$ and $\mu_{\tilde{X}\cap\tilde{X}} = \mu_{\tilde{X}}$;
- Identity, such that $\mu_{\tilde{X}\cup\emptyset} = \mu_{\tilde{X}}$ and $\mu_{\tilde{X}\cap\emptyset} = \mu_{\emptyset}$.

As aforementioned, many of the properties of the theory of classical sets are still valid in the case of fuzzy sets. However, in this latter, the completion of sets cannot be assured as seen as follows:

$$\mu_{\tilde{X}\cup\neg\tilde{X}} \neq \mu_{\tilde{X}} \text{ and } \mu_{\tilde{X}\cap\neg\tilde{X}} \neq \mu_{\emptyset}.$$

Nevertheless, in the theory of fuzzy sets, if multiple elements belong to a given set with different degrees, then it would not be possible to clearly specify whether these elements belong or not to a specific subset of this set, as in the classical

Table 2.1 Some examples of known t-norms and t-conorms

Chosen approach	T-norm	T-conorm
The Zadeh's approach	$\min(x, y)$	$\max(x, y)$
The probabilistic approach	xy	$x + y - xy$
The Lukasiewicz approach	$\max(0, x + y - 1)$	$\min(1, x + y)$

case. Thus, an extension of these logical operations has been reviewed as seen in Bouchon-Meunier and Marsala (2003). It is based on a family of mathematical functions mainly related to the intersection and union, so-called the t-norms for the intersection operation, and the t-conorms for the reunion. These latter are mainly used in approximate reasoning, with the aim of transforming two sets to a single one defined in the interval [0; 1] also. They are then seen as aggregation operators. The definitions and the traditional t-norms and t-conorms are given in the following manner:

Definition 2.4.4 (T-Norms and T-Conorms)

- A t-norm is a function $T : [0; 1] \times [0; 1] \to [0; 1]$ such that T is commutative, associative, monotone, and 1 is its neutral element;
- A t-conorm is a function $S : [0; 1] \times [0; 1] \to [0; 1]$ such that S is commutative, associative, monotone, and 0 is its neutral element.

For all values x and y belonging to the interval [0; 1], three main used forms of t-norms and t-conorms are summarized in Table 2.1. Additional forms and properties of t-norms can be found in Klement et al. (2000).

2.5 Extension Principle

The so-called extension principle was introduced by Zadeh (1965). It is the basis of all the studies treating of the transition between the "crisp" classical concepts to the fuzzy ones. Its aim is to extend the traditional mathematical operations of the classical theory, i.e. the real space theory—as instance a given arbitrary function $f : A \to B$ defined on the space of real values—to the fuzzy set one. The idea is to calculate the membership grade of different elements of fuzzy sets resulting from fuzzy operations (or functions on fuzzy sets). It is done in the following manner:

Extension Principle
Consider a function $f : A \mapsto B$, where A and B are two collections of objects. Let $\tilde{X}$ be a fuzzy element of A with its MF $\mu_{\tilde{X}} : A \mapsto [0; 1]$. The image $\tilde{Y} = f(\tilde{X})$ of the fuzzy element $\tilde{X}$ by f is a fuzzy subset of B. Its MF $\mu_{\tilde{Y}}$ is given by

$$\mu_{\tilde{Y}}(y) = \begin{cases} \sup\left\{\mu_{\tilde{X}}(x) \mid x \in A, f(x) = y\right\} & \text{if } \exists x : f(x) = y, \\ 0 & \text{if } \nexists x : f(x) = y. \end{cases} \tag{2.28}$$

The MF by the extension principle can analogously be written as:

$$\mu_{\tilde{Y}}(y) = \begin{cases} \sup \mu_{\tilde{X}}(x) & \text{if } f^{-1}(y) \neq \emptyset, \\ 0 & \text{if } f^{-1}(y) = \emptyset, \end{cases} \tag{2.29}$$

where $f^{-1}(y) = \{x \mid f(x) = y\}$ is the pre-image of y. Note that if f is bijective, f^{-1} will be the inverse function of f.

We are now able to give the extension principle of functions of two variables. The purpose is to propose the arithmetic operations between fuzzy numbers in the next section.

Definition 2.5.1 Consider a function $f : A \times B \mapsto C$. Let $\tilde{X}_1$ and $\tilde{X}_2$ be fuzzy subsets of A and B with their respective membership functions $\mu_{\tilde{X}_1}$ and $\mu_{\tilde{X}_2}$. The membership function of $\tilde{Y} = f(\tilde{X}_1, \tilde{X}_2)$ can be written by

$$\mu_{\tilde{Y}}(y) = \begin{cases} \sup \min(\mu_{\tilde{X}_1}(x_1), \mu_{\tilde{X}_2}(x_2)) & \text{if } f^{-1}(y) \neq \emptyset, \\ 0 & \text{if } f^{-1}(y) = \emptyset, \end{cases} \tag{2.30}$$

where $f^{-1}(y) = \{(x_1, x_2) \mid f(x_1, x_2) = y\}$.

The application of this previous principle can be seen in the subsequent example:

Example 2.5.1 Consider $f : \mathbb{R} \times \mathbb{R} \mapsto \mathbb{R}$ such that $f(x_1, x_2) = x_1 + x_2$. We suppose that the membership functions of the fuzzy subsets $\tilde{X}_1$ and $\tilde{X}_2$ are given in Fig. 2.4. We would like to find the MF $\mu_{\tilde{Y}}$ of the image $\tilde{Y}$ by f of $(\tilde{X}_1, \tilde{X}_2)$. This has to be done by considering every value of the definition domain $\mathbb{R}$, i.e. $\forall y \in \mathbb{R}$. We will consider only the finite number of values marked by circles in Fig. 2.4, i.e. the values $x_1 = 1$ or $x_1 = 2$, and $x_2 = 5$ or $x_2 = 6$.
For this example, let us compute the membership grade of $y = 7$ only. It is given by the following:

$$\mu_{\tilde{Y}}(7) = \sup \min \left(\mu_{\tilde{X}_1}(x_1), \mu_{\tilde{X}_2}(x_2) \right)$$

$$= \max \left(\min \left(\mu_{\tilde{X}_1}(1), \mu_{\tilde{X}_2}(6) \right), \min \left(\mu_{\tilde{X}_1}(2), \mu_{\tilde{X}_2}(5) \right) \right)$$

$$= \max \left(\min(0, 1), \min(1, 0) \right)$$

$$= \max \left(0, 0 \right) = 0.$$

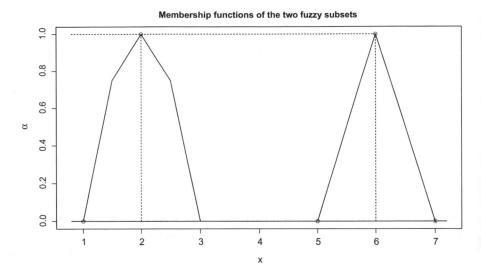

Fig. 2.4 Example of the extension principle—Example 2.5.1

2.6 Arithmetic Operations on Fuzzy Sets

Now we introduce the arithmetic operations defined on the space of fuzzy sets $\mathbb{F}(\mathbb{R})$, inspired by the Minkowski addition, i.e. the natural sum between intervals. The extension principle of Sect. 2.5 is used in the definition of each of them. We have to mention that the properties of these operations (commutativity, associativity, etc.) can be easily deduced by the ones related to the logical operations given in Properties 2.4.1.

Let $\tilde{X}$ and $\tilde{Y}$ be two fuzzy numbers with their respective membership functions $\mu_{\tilde{X}}$ and $\mu_{\tilde{Y}}$. The different arithmetic operations are expressed by:

Definition 2.6.1 (Sum of Fuzzy Numbers) The sum of two fuzzy numbers $\tilde{X}$ and $\tilde{Y}$, denoted by $\tilde{X} \oplus \tilde{Y}$ is the fuzzy number $\tilde{Z}$ such that its membership function $\mu_{\tilde{Z}}$ is given by

$$\mu_{(\tilde{X}\oplus\tilde{Y})}(z) = \sup_{M=\{(x,y)|x+y=z\}} \min\left[\mu_{\tilde{X}}(x), \mu_{\tilde{Y}}(y)\right]. \tag{2.31}$$

In addition, the sum of two L-R fuzzy numbers can be expressed by the α-level sets as follows:

$$(\tilde{X} \oplus \tilde{Y})_\alpha = \left[\tilde{X}_\alpha^L + \tilde{Y}_\alpha^L; \tilde{X}_\alpha^R + \tilde{Y}_\alpha^R\right], \quad \forall \alpha. \tag{2.32}$$

The sum of two trapezoidal fuzzy numbers can be expressed in terms of the quadruples of both fuzzy numbers as given in the following proposition:

Proposition 2.6.1 *If $\tilde{X}$ and $\tilde{Y}$ are trapezoidal fuzzy numbers and are represented, respectively, by (p_1, q_1, r_1, s_1) and (p_2, q_2, r_2, s_2), then the fuzzy addition $\tilde{X} \oplus \tilde{Y}$ is a trapezoidal fuzzy number represented by $(p_1 + p_2, q_1 + q_2, r_1 + r_2, s_1 + s_2)$.*

The extension principle of Zadeh (see Zadeh, 1965 and Zimmermann, 2001) gave a definition of the product of two fuzzy numbers.

Definition 2.6.2 (Product of Fuzzy Numbers) The product of two fuzzy numbers $\tilde{X}$ and $\tilde{Y}$ with their respective membership functions $\mu_{\tilde{X}}$ and $\mu_{\tilde{Y}}$ denoted by $\tilde{X} \otimes \tilde{Y}$, is the fuzzy number $\tilde{Z}$, such that its membership function is written by the extension principle as:

$$\mu_{(\tilde{X} \otimes \tilde{Y})}(z) = \sup_{M = \{(x,y) \mid x \cdot y = z\}} \min \left[\mu_{\tilde{X}}(x), \mu_{\tilde{Y}}(y) \right]. \tag{2.33}$$

The product can also be expressed by its α-cuts in the following manner:

$$\left(\tilde{X} \otimes \tilde{Y} \right)_\alpha = \Big[\min \{ \tilde{X}_\alpha^L \cdot \tilde{Y}_\alpha^L, \ \tilde{X}_\alpha^L \cdot \tilde{Y}_\alpha^R, \ \tilde{X}_\alpha^R \cdot \tilde{Y}_\alpha^L, \ \tilde{X}_\alpha^R \cdot \tilde{Y}_\alpha^R \};$$

$$\max \{ \tilde{X}_\alpha^L \cdot \tilde{Y}_\alpha^L, \ \tilde{X}_\alpha^L \cdot \tilde{Y}_\alpha^R, \ \tilde{X}_\alpha^R \cdot \tilde{Y}_\alpha^L, \ \tilde{X}_\alpha^R \cdot \tilde{Y}_\alpha^R \} \Big], \quad \forall \alpha. \tag{2.34}$$

This calculation is seen to be a complicated task from a computational point of view. An efficient way to reduce this complexity is to use convenient estimations, as discussed in further chapters.

Definition 2.6.3 (Product of a Fuzzy Number by a Scalar) The product of a fuzzy number $\tilde{X}$ by a scalar $c \in \mathbb{R}$ noted as $c \otimes \tilde{X}$ is the fuzzy number $\tilde{Z}$ with its membership function of $\tilde{Z}$ given by

$$\mu_{(c \otimes \tilde{X})}(z) = \sup_{M = \{x \mid c \cdot x = z\}} \mu_{\tilde{X}}(x) = \begin{cases} \mu_{\tilde{X}}(\frac{z}{x}) & \text{if} \quad c \neq 0 \\ \mathbb{1}_{\{0\}}(z) & \text{if} \quad c = 0 \end{cases}, \quad \forall x \in \mathbb{R}. \tag{2.35}$$

$\forall x \in \mathbb{R}$. The product $c \otimes \tilde{X}$ is equivalent to considering the product of an interval by a scalar, level wise. In addition, the fuzzy number $\tilde{Z}$ can also be written in terms of its α-cuts as follows:

$$\left(c \otimes \tilde{X} \right)_\alpha = \begin{cases} [c \cdot \tilde{X}_\alpha^L; \ c \cdot \tilde{X}_\alpha^R] & \text{if} \quad c \geq 0 \\ [c \cdot \tilde{X}_\alpha^R; \ c \cdot \tilde{X}_\alpha^L] & \text{if} \quad c < 0 \end{cases}, \quad \forall \alpha. \tag{2.36}$$

The quotient of two fuzzy numbers is also proposed in terms of the extension principle. This operation is given by:

Definition 2.6.4 (Quotient of Fuzzy Numbers) The quotient of two fuzzy numbers $\tilde{X}$ and $\tilde{Y}$, is the fuzzy number $\tilde{Z} = \tilde{X} \oslash \tilde{Y}$ such that its MF is given by

$$\mu_{(\tilde{X} \oslash \tilde{Y})}(z) = \sup_{M=\{(x,y)|x/y=z\}} \min\left[\mu_{\tilde{X}}(x), \mu_{\tilde{Y}}(y)\right], \qquad (2.37)$$

with the necessary condition that $0 \notin \text{supp}\,(\tilde{Y})$. The quotient of the fuzzy numbers $\tilde{X}$ and $\tilde{Y}$ can then be given by its α-cuts as follows:

$$\left(\tilde{X} \oslash \tilde{Y}\right)_\alpha = \Big[\min\left\{\tilde{X}_\alpha^L/\tilde{Y}_\alpha^L,\ \tilde{X}_\alpha^L/\tilde{Y}_\alpha^R,\ \tilde{X}_\alpha^R/\tilde{Y}_\alpha^L,\ \tilde{X}_\alpha^R/\tilde{Y}_\alpha^R\right\};$$

$$\max\left\{\tilde{X}_\alpha^L/\tilde{Y}_\alpha^L,\ \tilde{X}_\alpha^L/\tilde{Y}_\alpha^R,\ \tilde{X}_\alpha^R/\tilde{Y}_\alpha^L,\ \tilde{X}_\alpha^R/\tilde{Y}_\alpha^R\right\} \Big], \quad \forall \alpha. \qquad (2.38)$$

Several researches have supposed that the product between two fuzzy numbers is nothing but the respective product between the left and right hand sides of both fuzzy numbers. However, this operation is not always relevant. In many cases, the obtained fuzzy number could then be considered as an approximation of the true product of the fuzzy numbers. This idea is shown in the following remark:

Remark 2.6.1 The product of two fuzzy numbers $\tilde{X}$ and $\tilde{Y}$ cannot be always written as follows:

$$(\tilde{X} \otimes \tilde{Y})_\alpha = \left[\tilde{X}_\alpha^L \cdot \tilde{Y}_\alpha^L ;\ \tilde{X}_\alpha^R \cdot \tilde{Y}_\alpha^R\right].$$

It is the same for the quotient.

The sum, the product and the quotient operations of fuzzy numbers are calculated in the following simple example:

Example 2.6.1 (Sum, Product and Quotient of Fuzzy Numbers) Let $\tilde{X} = (1, 2, 3)$ be a triangular fuzzy number, and $\tilde{Y} = (4, 5, 6, 8)$ be a trapezoidal one. Using Definitions 2.2.1 and 2.2.2, their α-cuts are given, respectively, by the following:

$$\tilde{X}_\alpha = \begin{cases} \tilde{X}_\alpha^L(\alpha) = 1 + \alpha, \\ \tilde{X}_\alpha^R(\alpha) = 3 - \alpha, \end{cases} \quad \text{and} \quad \tilde{Y}_\alpha = \begin{cases} \tilde{Y}_\alpha^L(\alpha) = 4 + \alpha, \\ \tilde{Y}_\alpha^R(\alpha) = 8 - 2\alpha. \end{cases} \qquad (2.39)$$

The sum and product of $\tilde{X}$ and $\tilde{Y}$, seen in Figs. 2.5 and 2.6, can be expressed by

$$(\tilde{X} \oplus \tilde{Y})_\alpha = \left[(1 + \alpha) + (4 + \alpha);\ (3 - \alpha) + (8 - 2\alpha)\right] = \left[5 + 2\alpha;\ 11 - 3\alpha\right].$$

$$(\tilde{X} \otimes \tilde{Y})_\alpha = \Big[\min\left\{((1 + \alpha) \cdot (4 + \alpha)),\ ((1 + \alpha) \cdot (8 - 2\alpha)),\ ((3 - \alpha) \cdot (4 + \alpha)),\right.$$

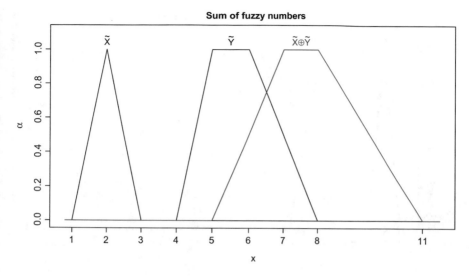

Fig. 2.5 Sum of the fuzzy numbers $\tilde{X}$ and $\tilde{Y}$—Example 2.6.1

$$\left((3-\alpha)\cdot(8-2\alpha)\right)\Big\};\ \max\Big\{\left((1+\alpha)\cdot(4+\alpha)\right),\left((1+\alpha)\cdot(8-2\alpha)\right),$$

$$\left((3-\alpha)\cdot(4+\alpha)\right),\left((3-\alpha)\cdot(8-2\alpha)\right)\Big\}\Big],$$

$$=\Big[(1+\alpha)\cdot(4+\alpha)\ ;\ (3-\alpha)\cdot(8-2\alpha)\Big].$$

We note that in this case $(\tilde{X}\otimes\tilde{Y})_\alpha=\big[(\tilde{X})_\alpha^L\cdot(\tilde{Y})_\alpha^L\ ;\ (\tilde{X})_\alpha^R\cdot(\tilde{Y})_\alpha^R\big]$, but this expression cannot be generalized to all possible cases.

   We could also calculate the quotient between both fuzzy numbers. It can then be written as follows:

$$(\tilde{X}\oslash\tilde{Y})_\alpha=\Big[\ \min\Big\{\left((1+\alpha)/(4+\alpha)\right),\left((1+\alpha)/(8-2\alpha)\right),\left((3-\alpha)/(4+\alpha)\right),$$

$$\left((3-\alpha)/(8-2\alpha)\right)\Big\};\ \max\Big\{\left((1+\alpha)/(4+\alpha)\right),\left((1+\alpha)/(8-2\alpha)\right),$$

$$\left((3-\alpha)/(4+\alpha)\right),\left((3-\alpha)/(8-2\alpha)\right)\Big\}\Big],$$

$$=\Big[(1+\alpha)/(8-2\alpha);\ (3-\alpha)/(4+\alpha)\Big].$$

The expressions of the α-cuts of the product and the quotient of the fuzzy numbers $\tilde{X}$ and $\tilde{Y}$ are not trapezoidal, though we cannot clearly recognize them from Figs. 2.6 and 2.7.

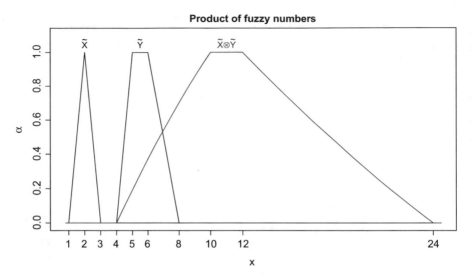

Fig. 2.6 Product of the fuzzy numbers $\tilde{X}$ and $\tilde{Y}$—Example 2.6.1

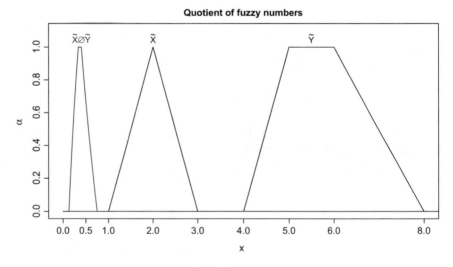

Fig. 2.7 Quotient of the fuzzy numbers $\tilde{X}$ and $\tilde{Y}$—Example 2.6.1

A counter-example of the product between fuzzy numbers can be given by calculating the product between the following two fuzzy numbers

$$\tilde{X} = (3, 4, 5, 6) \quad \text{and} \quad \tilde{Y} = (-2, -1, 6, 7),$$

with their respective α-cuts expressed by

$$\tilde{X}_\alpha = \begin{cases} \tilde{X}_\alpha^L(\alpha) = 3 + \alpha, \\ \tilde{X}_\alpha^R(\alpha) = 6 - \alpha, \end{cases} \quad \text{and} \quad \tilde{Y}_\alpha = \begin{cases} \tilde{Y}_\alpha^L(\alpha) = -2 + \alpha, \\ \tilde{Y}_\alpha^R(\alpha) = 7 - \alpha. \end{cases} \tag{2.40}$$

The product of these fuzzy numbers expressed by their α-cuts is then given by

$$(\tilde{X} \otimes \tilde{Y})_\alpha = \Big[\min \Big\{ \big((3 + \alpha) \cdot (-2 + \alpha)\big), \big((3 + \alpha) \cdot (7 - \alpha)\big), \big((6 - \alpha) \cdot (7 - \alpha)\big),$$

$$\big((6 - \alpha) \cdot (-2 + \alpha)\big) \Big\}; \ \max \Big\{ \big((3 + \alpha) \cdot (-2 + \alpha)\big), \big((3 + \alpha) \cdot (7 - \alpha)\big),$$

$$\big((6 - \alpha) \cdot (7 - \alpha)\big), \big((6 - \alpha) \cdot (-2 + \alpha)\big) \Big\} \Big],$$

$$= \Big[(6 - \alpha) \cdot (-2 + \alpha); \ (6 - \alpha) \cdot (7 - \alpha) \Big].$$

However, the product between the left and right hand sides of $\tilde{X}$ and $\tilde{Y}$ is written by $\Big[(3 + \alpha) \cdot (-2 + \alpha); \ (6 - \alpha) \cdot (7 - \alpha) \Big]$. Consequently, we can clearly see that

$$(\tilde{X} \otimes \tilde{Y})_\alpha \neq \Big[(3 + \alpha) \cdot (-2 + \alpha); \ (6 - \alpha) \cdot (7 - \alpha) \Big].$$

Definition 2.6.5 (Difference Between Fuzzy Numbers) The difference between two fuzzy numbers $\tilde{X}$ and $\tilde{Y}$, denoted by $\tilde{Z}$, is the fuzzy number $\tilde{Z} = \tilde{X} \ominus \tilde{Y}$ with its following membership function:

$$\mu_{(\tilde{X} \ominus \tilde{Y})}(z) = \sup_{M = \{(x,y) \mid x - y = z\}} \min \Big[\mu_{\tilde{X}}(x), \mu_{\tilde{Y}}(y) \Big]. \tag{2.41}$$

The space of fuzzy sets has not a linear but semi-linear structure. This is due to the lack of symmetrical elements. In other terms, the equation

$$\tilde{X} \ominus \tilde{X} = \tilde{0}$$

does not hold. Thus, a well defined difference between fuzzy numbers in $\mathbb{F}(\mathbb{R})$ such that

$$\tilde{X} \oplus (\tilde{Y} \ominus \tilde{X}) = \tilde{Y}, \quad \forall \tilde{X}, \tilde{Y} \in \mathbb{F}(\mathbb{R}),$$

cannot be stated levelwise. One way to overcome this disadvantage is by using suitable metrics on $\mathbb{F}(\mathbb{R})$. These metrics will be discussed in details in the subsequent chapters.

One could briefly define the minimum and maximum operations of two fuzzy numbers using the extension principle seen in Definition 2.5.1. The definitions are written as follows:

Definition 2.6.6 (Minimum and Maximum Between Fuzzy Numbers) The minimum and maximum between two fuzzy numbers $\tilde{X}$ and $\tilde{Y}$, denoted, respectively, by $\tilde{Z}_1$ and $\tilde{Z}_2$, are the fuzzy numbers $\tilde{Z}_1 = \min(\tilde{X}, \tilde{Y})$ and $\tilde{Z}_2 = \max(\tilde{X}, \tilde{Y})$ with their respective membership functions:

$$\mu_{(\min(\tilde{X},\tilde{Y}))}(z_1) = \sup_{M=\{(x,y)|\min(x,y)=z_1\}} \min\left[\mu_{\tilde{X}}(x), \mu_{\tilde{Y}}(y)\right], \tag{2.42}$$

$$\mu_{(\max(\tilde{X},\tilde{Y}))}(z_2) = \sup_{M=\{(x,y)|\max(x,y)=z_2\}} \min\left[\mu_{\tilde{X}}(x), \mu_{\tilde{Y}}(y)\right], \tag{2.43}$$

and their corresponding α-cuts $(\tilde{Z}_1)_\alpha$ and $(\tilde{Z}_2)_\alpha$ given, respectively, by

$$(\tilde{Z}_1)_\alpha = \left[\min(\tilde{X}_\alpha^L, \tilde{Y}_\alpha^L); \min(\tilde{X}_\alpha^R, \tilde{Y}_\alpha^R) \right], \tag{2.44}$$

$$(\tilde{Z}_2)_\alpha = \left[\max(\tilde{X}_\alpha^L, \tilde{Y}_\alpha^L); \max(\tilde{X}_\alpha^R, \tilde{Y}_\alpha^R) \right]. \tag{2.45}$$

To sum up, the fuzzy numbers are not only the basis of the theory of fuzzy sets but also essential for modelling uncertainty contained in the human perception. However, identifying a fuzzy number is often considered as a complex task. For this reason, researchers defined two ways for it: by the vertical (the x-axis) and the horizontal view (the y-axis). Characterizing the fuzzy number by its MF is said to be the vertical view. Analogously, by the horizontal view, we mean the association of the fuzzy number to its α-level sets.

References

Bouchon-Meunier, B., & Marsala, C. (2003). *Logique floue, principes, aide à la la décision.* Hermés Science Publications.

Drewniak, J. (1987). Convex and strongly convex fuzzy sets. *Journal of Mathematical Analysis and Applications, 126*(1), 292–300. ISSN: 0022-247X. https://doi.org/10.1016/0022-247X(87)90093-X. http://www.sciencedirect.com/science/article/pii/0022247X8790093X

Gil, M. A., Lubiano, M. A., De Sáa, S. D. L. R., & Sinova, B. (2015). Analyzing data from a fuzzy rating scale-based questionnaire. A case study. *Psicotherma, 27*(2), 182–191.

Klement, E. P., Mesiar, R., & Pap, E. (2000). *Triangular norms.* Dordrecht: Springer. ISBN: 978-90-481-5507-1. https://doi.org/10.1007/978-94-015-9540-7

Laviolette, M., Seaman, J. W., Barrett, J. D., & Woodall, W. H. (1995). A probabilistic and statistical view of fuzzy methods. *Technometrics, 37*(3), 249–261. ISSN: 00401706. http://www.jstor.org/stable/1269905

Zadeh, L. A. (1965). Fuzzy sets. *Information and Control, 8*(3), 338–353. ISSN: 0019-9958. http://doi.org/10.1016/S0019-9958(65)90241-X. http://www.sciencedirect.com/science/article/pii/S001999586590241X

Zadeh, L. A. (1995). Discussion: Probability theory and fuzzy logic are complementary rather than competitive. *Technometrics, 37*(3), 271–276. http://doi.org/10.1080/00401706.1995.10484330

Zimmermann, H. J. (2001). The extension principle and applications. In *Fuzzy set theory—and its applications* (pp. 55–69). Netherlands: Springer. ISBN: 978-94-010-0646-0. https://doi.org/10.1007/978-94-010-0646-0_5

Chapter 3
Fuzzy Rule-Based Systems

Different methods of processing fuzzy information exist. We note, as instance, the Sugeno method (Takagi & Sugeno, 1985) which allows to simplify the calculations based on an aggregation of the information, in the purpose of obtaining an interpretable solution. This method is often used in real time applications where the problem of computational time arises. Another regularly used one is the Mamdani approach shown in Mamdani and Assilian (1975). This one will be developed in this chapter. The Mamdani fuzzy process, also called the Mamdani fuzzy inference system or the fuzzy logic system (FLS), consists of three main steps: fuzzification, IF-THEN rules, and defuzzification. The architecture of a given FLS composed by these three components is given in Fig. 3.1, as well as, by the following algorithm provided for such system:

Algorithm:

1. Convert uncertain data to fuzzy values by defining the membership functions of each linguistic. FUZZIFICATION
2. Evaluate the fuzzy rules for the proposed inputs, and combine them into a fuzzy set. INFERENCE RULES
3. Convert the resulting fuzzy set to a crisp value, by a convenient operator. DEFUZZIFICATION

The Traffic Scenario To illustrate this chapter, we will be interested in the following simple scenario: understanding the situation of the traffic on a chosen highway at a given time of the day. For this purpose, the idea is to ask a number of drivers from different spots of the highway about several aspects of the driving journey. Our aim is to have information about the expected delay on a meeting that can be caused by the traffic jam. We suppose that this information can be explained by the two variables "approximate

© The Author(s), under exclusive license to Springer Nature Switzerland AG 2021
R. Berkachy, *The Signed Distance Measure in Fuzzy Statistical Analysis*,
Fuzzy Management Methods, https://doi.org/10.1007/978-3-030-76916-1_3

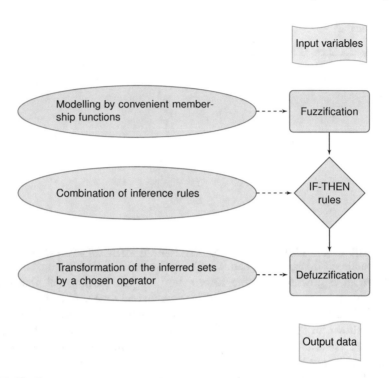

Fig. 3.1 The fuzzy process

Table 3.1 Possible linguistics of the input and output variables for the traffic scenario

Input variables		Output variable
Waiting time on a traffic light	Estimated number of vehicles	Expected delay for a meeting
Very low		
Low	Low traffic	Low delay
Moderate	Moderate traffic	Moderate delay
High	High traffic	High delay
Very high		

waiting time on a traffic light" and "estimated number of surrounding vehicles at the traffic light stop." For these variables, the drivers could answer by an approximate value, which can also be given by a so-called linguistic. For example, the possible linguistics for the approximate waiting time on a traffic light are: very low, low, moderate, high, and very high. The different linguistics of the input and output variables are given in Table 3.1. Regarding the answers of the group of chosen drivers, the perception of each one of them about the traffic conditions could be in many cases severely different. This could be due to many variants, such as personal, cultural, etc. Thus, waiting for approximately 5 min at a traffic light could eventually be long for a driver, and moderate for another one. Therefore, these variables are seen to be uncertain, and will be processed by fuzzy rule-based systems through all the chapter.

3.1 Fuzzification

The fuzzification is the operation of transforming the input data into a fuzzy one using membership functions, often called the fuzzifiers: uncertain information are gathered and converted to fuzzy sets by choosing a particular type of membership functions quantifying the linguistic terms. This choice of functions is often context-dependent and can differ from user to another. We remind that the strength of this procedure is to get for every numerical real value, multiple membership grades related to the linguistic, as such a given value belongs to multiple sets at the same time.

Example 3.1.1 (Traffic Situation—the Fuzzification) We will first fuzzify the variable approximate waiting time on a traffic light. For the information about the time spent on a traffic light, we suppose that the possible linguistics are: very low, low, moderate, high, and very high. They are shown in Table 3.2. In this case, we chose trapezoidal fuzzy numbers to model the uncertainty contained in the linguistics.

If the variable the waiting time on a traffic light has a range of 0–4.5 min, the fuzzification step consists of associating a membership grade to each proposed linguistic and consequently converts the uncertain initial data set to a fuzzy one with its characterizing membership levels.

For instance, consider a person asserting that he waited approximately 2 min at each traffic light. Then, according to Table 3.2 and Fig. 3.2, we have that:

$$\mu_{\text{Very low}}(2) = 0, \quad \mu_{\text{Low}}(2) = 0, \quad \mu_{\text{Moderate}}(2) = 1, \quad \mu_{\text{High}}(2) = 0, \quad \mu_{\text{Very high}}(2) = 0.$$

In this case, one could say that by fuzzification of the uncertain data, i.e. getting the corresponding membership grades related to each linguistic term, the answer "approximately 2 min" is considered to be at 100% moderate since the membership

Table 3.2 Linguistics of the variable the waiting time on each traffic light and their corresponding membership functions—Example 3.1.1

Linguistic	Corresponding fuzzy number
Very low	(0,0,0.5,1)
Low	(0.5,1,1.5,2)
Moderate	(1.5,2,2.5,3)
High	(2.5,3,3.5,4)
Very high	(3.5,4,4,4)

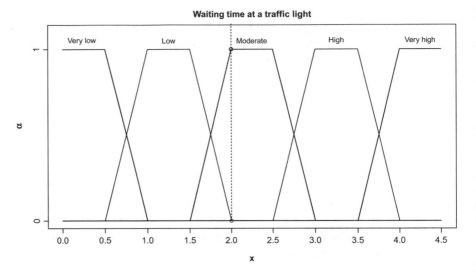

Fig. 3.2 Membership functions of the variable `waiting time at a traffic light`

grade at this linguistic is the maximum, i.e. 1, and 0% very low, low, high, and very high.

3.2   Inference Rules

This step of the process is also called "IF-THEN rules." It is named as such since it is based on the so-called fuzzy rule. In order to make the inference reliable, this step aims to define potential realistic rules according to the expert knowledge. The inference rules step is seen to be the most difficult one, because it depends on a subjective set of rules based on fuzzy logical operations. Thus, the choice of rules is not innocent, and can vary between contexts and users. From another side, once the membership functions and the rules are exposed, the next move is to combine and evaluate these rules, specifically in the case of multiple input variables.

Definition 3.2.1 (Fuzzy Rule) A fuzzy rule is a statement of the form:

$$\text{IF } X \text{ is } a \text{ THEN } Y \text{ is } b,$$

where X and Y are variables, and a and b are given linguistic terms determined by the variables X and Y.

This definition of the fuzzy rule can be illustrated by the following two simple examples about the traffic situation:

Table 3.3 Linguistics of the variable `estimated number of surrounding vehicles at the traffic light stop` and their corresponding membership functions—Example 3.2.2

Linguistic	Corresponding fuzzy number
Low	(0,0,3,6)
Moderate	(3,5,8,10)
High	(8,10,15,15)

Example 3.2.1 (Traffic Situation) We are interested in the expected delay for a meeting that can be caused by the traffic jam at a given schedule. Consider the statement `the waiting time on a traffic light` is "very high." The linguistic variable `the expected delay for a meeting` has the linguistic "high." An example of a fuzzy rule is:

IF `the waiting time on a traffic light` is "very high," THEN `the expected delay for a meeting` is "high."

Example 3.2.2 (Traffic Situation—IF-THEN Rules) We consider now the variables `waiting time on a traffic light` and `estimated number of surrounding vehicles at the traffic light stop`, previously introduced. Drivers are then asked to give the time spent at a traffic light and the approximate number of vehicles surrounding in the traffic jam. This latter is considered to be uncertain as well as the first one and will be modelled by the membership functions shown in Table 3.3.

Many inference rules exist for every process. The construction of these rules is based on two components: the logical operation chosen for the aggregation—OR or AND—, and on considering every combination of possibilities between the studied variables and the related output variable. For this example, the union denoted by the operator OR is used. For sake of simplification of the example, the different possibilities of rules are reduced to the following three inference rules:

* **Rule 1**: IF `the waiting time on a traffic light` is "very low" OR `the estimated number of surrounding vehicles at the traffic light stop` is "low," THEN `the expected delay on a meeting` is "low."
* **Rule 2**: IF `the waiting time on a traffic light` is "moderate" OR `the estimated number of surrounding vehicles at the traffic light stop` is "moderate," THEN `the expected delay on a meeting` is "moderate."
* **Rule 3**: IF `the waiting time on a traffic light` is "very high" OR `the estimated number of surrounding vehicles at the traffic light stop` is "high," THEN `the expected delay on a meeting` is "high."

For every observation, the idea is to evaluate the aforementioned rules and find the corresponding membership grades for the associated linguistics. As in-

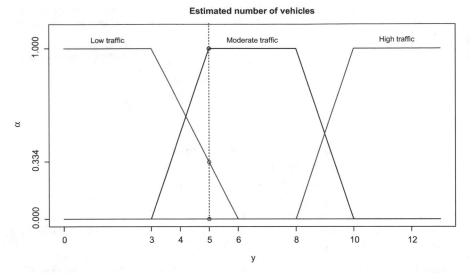

Fig. 3.3 Membership functions of the variable estimated number of surrounding vehicles at the traffic light stop

stance, a given driver answered that he waited approximately 2 min at a traffic light, and he saw 5 vehicles surrounding his one at a traffic light stop. For the variable waiting time at a traffic light, the grades of membership at the value approximately $x = 2$ are given in Example 3.1.1. In addition, for the variable estimated number of surrounding vehicles at the traffic light stop, the membership grades for approximately $y = 5$ are seen in Fig. 3.3. They are given as follows:

$$\mu_{\text{Low}}(5) = 0.334, \quad \mu_{\text{Moderate}}(5) = 1, \quad \mu_{\text{High}}(5) = 0.$$

The next step is to evaluate the proposed inference rules using these membership grades. The logical operator used is the inclusive OR, defined as the maximum. Thus, we get from Fig. 3.4 that:

- For **Rule 1**, $\max(\mu_{X=\text{very low}}(2), \mu_{Y=\text{low}}(5)) = \max(0, 0.334) = 0.334$.
- For **Rule 2**, $\max(\mu_{X=\text{moderate}}(2), \mu_{Y=\text{moderate}}(5)) = \max(1, 1) = 1$.
- For **Rule 3**, $\max(\mu_{X=\text{very high}}(2), \mu_{Y=\text{high}}(5)) = \max(0, 0) = 0$.

The resulting combined fuzzy set is shown in Fig. 3.5. It is represented by the curve composed by the three levels of membership due to the combination of the three pre-defined inference rules.

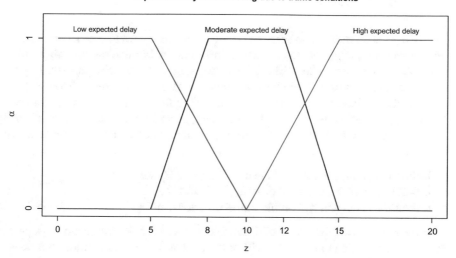

Fig. 3.4 Membership functions of the output variable `expected delay for a meeting` `caused by traffic conditions`

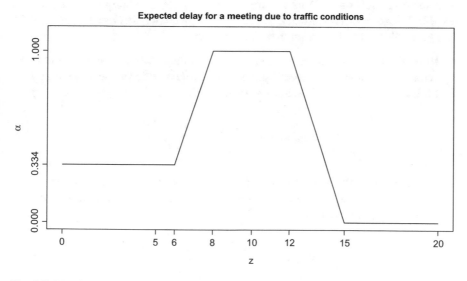

Fig. 3.5 Membership functions of the resulting combined fuzzy set

3.3 Defuzzification

The defuzzification is the process of converting a fuzzy set to a crisp value. It is widely seen as a way provoking a loss of information. For this reason, this method has attracted less researchers than the first two stages of the fuzzy process. Indeed, fuzzy data contains much more information than crisp data. However, fuzzy entities are not always interpretable. Therefore, a defuzzification is suggested. This can be done by a convenient operator. Furthermore, according to Leekwijck and Kerre (1999), different criteria should be respected, as such a defuzzification should verify the following:

- it should contain the kernel of the considered fuzzy set;
- it should preserve the natural ordering in $\mathbb{F}(\mathbb{R})$;
- it should be coherent with affine transformations, etc.

From another side, two types of defuzzification exist: a defuzzification to a point and a defuzzification to a set. We will only consider the defuzzification to a point. A corresponding operator D is of the form:

$$D: \quad \mathbb{F}(\mathbb{R}) \quad \mapsto \quad \mathbb{R}.$$

It maps fuzzy sets on $\mathbb{F}(\mathbb{R})$ into elements of the space $\mathbb{R}$. Multiple operators were used in literature (see Leekwijck & Kerre, 1999, Runkler, 1997 and Yager, 1996). We give below a brief description of few of them:

- **the centroid method**: this method is based on the center of gravity of the fuzzy set. For a continuous membership function, the defuzzified value $X_{centroid}$ by the centroid method applied on a fuzzy set $\tilde{X}$ is given by

$$X_{centroid} = \frac{\int_{\text{supp}(\tilde{X})} x \cdot \mu_{\tilde{X}}(x)dx}{\int_{\text{supp}(\tilde{X})} \mu_{\tilde{X}}(x)dx}, \tag{3.1}$$

where $\mu_{\tilde{X}}(x)$ is the membership function of $\tilde{X}$, and supp $(\tilde{X})$ is its support set.

Note that for a discrete membership function, the integral is replaced by a sum. Thus, Eq. 3.1 can be expressed as:

$$X_{centroid} = \frac{\sum_{i=1}^{j} X_{centroid}^{i} \cdot A_{S_i}}{\sum_{i=1}^{j} A_{S_i}}, \tag{3.2}$$

where the total surface under the membership function of $\tilde{X}$ is decomposed into j sub-surfaces S_i, $i = 1, \ldots, j$, with A_{S_i} the area of each sub-surface S_i and $X_{centroid}^{i}$ the centroid point of A_{S_i}.

- **the bisector method**: the bisector operator consists of finding the position under the membership function of $\tilde{X}$, such that the areas on the left and right hand sides at this position are equal. In other terms, the defuzzified value $X_{bisector}$ should verify the following equation:

$$\int_{\beta_1}^{X_{bisector}} \mu_{\tilde{X}}(x)dx = \int_{X_{bisector}}^{\beta_2} \mu_{\tilde{X}}(x)dx, \qquad (3.3)$$

with $\beta_1 = \min \operatorname{supp}(\tilde{X})$ and $\beta_2 = \max \operatorname{supp}(\tilde{X})$. Similarly to Eq. 3.2, one can re-write the previous equation for the case of a discrete membership function.

- **the maxima methods**: they are composed by **the smallest of maximum (SOM)**, **the largest of maximum (LOM)**, and **the mean of maximum (MOM)** methods. The operations by the maxima methods are based on the maximum membership grades. For instance, the smallest and largest of maximum are determined, respectively, by the smallest and largest values of the definition domain associated with the maximum membership grade.

When more than one element attain the maximum membership grade, we define the so-called mean of maximum. It is expressed by the mean of the values associated with the maximum membership, as follows:

$$X_{MOM} = \frac{\sum_{x \in \text{core}(\tilde{X})} x}{|\text{core}(\tilde{X})|}, \qquad (3.4)$$

where core $(\tilde{X})$ is the kernel set of $\tilde{X}$, and $|\text{core}(\tilde{X})|$ is its cardinality.

Defuzzifying a vector of fuzzified observations leads to a real-valued vector of crisp data, ready to be dealing with similarly to any other vector of observations in the classical real space theory. Therefore, the application of one or another of these operators can probably influence the obtained crisp distributions. Choosing carefully between the operators is recommended. However, and in this same perspective, some distances present promising properties specifically when used in the process of defuzzification. Such metrics will be discussed in further chapters.

Example 3.3.1 (Traffic Situation—the Defuzzification) We would like now to defuzzify the fuzzy set obtained by the combination of the fuzzy rules for the observation described in Example 3.2.2. For this purpose, we apply the operators discussed above. For our case, the defuzzified values are as follows:

- By the centroid operator: we decompose the total surface into 5 sub-surfaces S_1, S_2, S_3, S_4 and S_5, as seen in Fig. 3.6. According to Eq. 3.2 and to the calculations exposed in Table 3.4, we have that

$$X_{centroid} = \frac{\sum_{i=1}^{5} X_{centroid}^i \cdot A_{S_i}}{\sum_{i=1}^{5} A_{S_i}} = \frac{75.0557}{8.8334} = 8.4968.$$

- By the bisector operator: solving Eq. 3.3 for our case gives

$$X_{bisector} = 9.5.$$

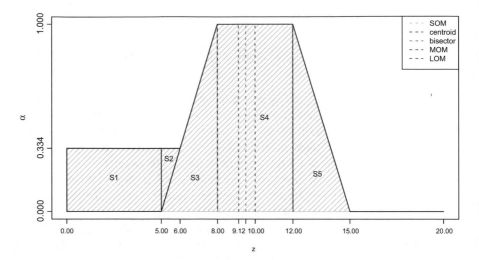

Fig. 3.6 Decomposition of the membership surface in the process of defuzzification by the centroid

- By the SOM operator: the defuzzified value by the SOM can be defined as the minimum of the core set of the combined fuzzy set, i.e.

$$X_{SOM} = 8.$$

- By the LOM operator: the one by the LOM is seen as the maximum value of the core set, i.e.

$$X_{LOM} = 12.$$

- By the MOM operator: according to Eq. 3.4, the defuzzified value by the MOM for this case can be written as:

$$X_{MOM} = \frac{X_{SOM} + X_{LOM}}{2} = \frac{8 + 12}{2} = 10.$$

The defuzzified values found are seen in Fig. 3.6. A remark is that by the proposed defuzzification operators, the obtained results are different. Therefore, a general conclusion can be made: the choice of an operator is prominent, investigating the properties of each one and/or proposing another efficient tool for defuzzifying could be of good use.

Table 3.4 Calculations used in the defuzzification by the centroid—Example 3.2.2

Surface (S_i)	Area (A_{S_i})	Centroid ($X^i_{centroid}$)	Area × Centroid
S_1	$5 \times \frac{1}{3} = 1.6667$	$\frac{0+5}{2} = 2.5$	$1.6667 \times 2.5 = 4.1668$
S_2	$\frac{1}{2}(1 \times \frac{1}{3}) = 0.1667$	$\frac{5+5+6}{3} = 5.334$	$0.1667 \times 5.334 = 0.8889$
S_3	$\frac{1}{2}(3 \times 1) = 1.5$	$\frac{5+8+8}{3} = 7$	$1.5 \times 7 = 10.5$
S_4	$4 \times 1 = 4$	$\frac{8+12}{2} = 10$	$4 \times 10 = 40$
S_5	$\frac{1}{2}(3 \times 1) = 1.5$	$\frac{12+12+15}{3} = 13$	$1.5 \times 13 = 19.5$
Sum	8.8334		75.0557

References

Leekwijck, W., & Kerre, E. E. (1999). Defuzzification: criteria and classification. *Fuzzy Sets and Systems, 108*(2), 159–178. ISSN: 0165-0114. https://doi.org/10.1016/S0165-0114(97)00337-0

Mamdani, E. H., & Assilian, S. (1975). An experiment in linguistic synthesis with a fuzzy logic controller. *International Journal of Man-Machine Studies, 7*(1), 1–13. ISSN: 0020-7373. https://doi.org/10.1016/S0020-7373(75)80002-2. http://www.sciencedirect.com/science/article/pii/S0020737375800022

Runkler, T. A. (1997). Selection of appropriate defuzzification methods using application specific properties. *IEEE Transactions on Fuzzy Systems, 5*, 72–79. https://doi.org/10.1109/91.554449

Takagi, T., & Sugeno, M. (1985). Sugeno, M.: Fuzzy identification of systems and its applications to modeling and control. IEEE Transactions on Systems, Man, and Cybernetics SMC-15(1), 116-132. *IEEE Transactions Systems, Man and Cybernetics, 15*, 116–132. https://doi.org/10.1109/TSMC.1985.6313399

Yager, R. R. (1996). Knowledge-based defuzzification. *Fuzzy Sets and Systems, 80*(2), 177–185

Chapter 4
Distances Between Fuzzy Sets

Calculating the difference between fuzzy numbers is often seen as a complicated operation, mainly because of the semi-linear nature of $\mathbb{F}(\mathbb{R})$. In addition, in the space of fuzzy numbers a universally acceptable method of total ordering does not yet exist. An efficient solution for these problems is to consider a suitable metric. The proposed metrics are distances defined on the space of fuzzy sets. On the other hand, some of these distances have served as defuzzification operators and/or ranking operators. Properties of such metrics are appealing to investigate. In this chapter, we give a chronological review of some known distances (see Fig. 4.1) and some of their properties. We finally expose a new L_2 metric based on the so-called signed distance, paving the way to the construction of an updated version of it. This new version, called the "generalized signed distance," preserves the directionality of the original signed distance. We finally highlight that we will consider the case of a one-dimensional setting only, in other words, a given fuzzy component $\tilde{X}$ is associated with a single dimension x with the corresponding α-level set.

From decades, the Hausdorff distance has been one of the most used distances as a basis for the development of metrics in the fuzzy environment. We express this distance in an informative perspective only.

Consider the space $\mathcal{K}_c(\mathbb{R}^n)$ of non-empty compact and convex subsets of $\mathbb{R}^n$. In the interval-valued context, this distance defined over $\mathbb{R}^n$ was briefly described in Puri and Ralescu (1985) as follows:

Definition 4.0.1 (Hausdorff Distance) The Hausdorff distance between X and Y $\in \mathcal{K}_c(\mathbb{R}^n)$ denoted by $\delta_H(X, Y)$ is given by

$$\delta_H(X, Y) = \max\left[\sup_{x \in X}\left\{ \inf_{y \in Y} \|x - y\|\right\}, \sup_{y \in Y}\left\{ \inf_{x \in X} \|x - y\|\right\}\right], \qquad (4.1)$$

where $\|\cdot\|$ is the Euclidean norm.

© The Author(s), under exclusive license to Springer Nature Switzerland AG 2021
R. Berkachy, *The Signed Distance Measure in Fuzzy Statistical Analysis*,
Fuzzy Management Methods, https://doi.org/10.1007/978-3-030-76916-1_4

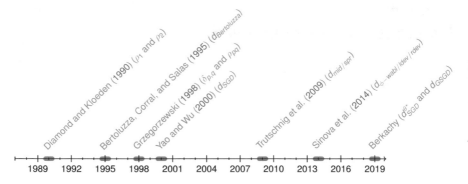

Fig. 4.1 Chronological review of distances of fuzzy quantities

Accordingly, the metric δ_H induces the metric space $(\mathcal{K}_c(\mathbb{R}^n), \delta_H)$. One can say that by the Hausdorff distance, two sets are close to each other if every element of either set is close to an element of the other set.

In a fuzzy context, the Hausdorff metric has been extended by defining a distance between two fuzzy sets $\tilde{X}$ and $\tilde{Y} \in \mathbb{F}_c^\star(\mathbb{R}^n)$. This distance denoted by $\delta_\infty(\tilde{X}, \tilde{Y})$ is the supremum of all the Hausdorff distances applied between corresponding level sets of both fuzzy numbers as follows:

$$\delta_\infty(\tilde{X}, \tilde{Y}) = \sup_{\alpha > 0} \delta_H(\tilde{X}_\alpha, \tilde{Y}_\alpha). \tag{4.2}$$

The metric space $(\mathbb{F}_c^\star(\mathbb{R}^n), \delta_\infty)$ of normal convex fuzzy sets over $\mathbb{R}^n$ is often involved in applications of fuzzy theory. We note, for example, Kaleva (1987) who used it in fuzzy differential equations, and Puri and Ralescu (1985) in fuzzy random variables, etc. However, the Hausdorff distance has said to be having handicaps when used in the fuzzy environment. Indeed, it is seen as not very realistic because it disadvantages fuzzy sets compared to interval ones. Another drawback of this distance is that the space $(\mathbb{F}_c^\star(\mathbb{R}^n), \delta_\infty)$ fails to be separable (Klement et al., 1986). Consequently, the topology related to the Hausdorff metric failed to be extended in the fuzzy context. Introducing other suitable metrics defined on $\mathbb{F}_c^\star(\mathbb{R}^n)$ has been a necessity. Researchers tended to develop this distance or to propose new ones in order to better fit into the fuzzy context.

Many references have discussed of fuzzy metric spaces. For instance, Ivan Kramosil (1975) introduced a fuzzy metric space. Years later, George and Veeramani (1994) updated this concept and gave a Hausdorff topology of a fuzzy metric space. Remark that we will consider the case of a metric of fuzzy quantities rather than of a fuzzy metric which is directly related to the definition of a metric in the classical theory, but with a particular extension to the fuzzy environment.

In this thesis, the proposed distances are mainly defined on the real space, in a one-dimensional setting. For this purpose, we give a definition of a metric of fuzzy sets with its image as a real crisp value.

Definition 4.0.2 (Metric of Fuzzy Quantities) Let $\tilde{X}$, $\tilde{Y}$, and $\tilde{Z}$ be fuzzy sets of the class $\mathbb{F}_c^{\star}(\mathbb{R})$ of normal bounded fuzzy sets. Consider a mapping d on $\mathbb{F}_c^{\star}(\mathbb{R})^2$ such that $d : \mathbb{F}_c^{\star}(\mathbb{R}) \times \mathbb{F}_c^{\star}(\mathbb{R}) \mapsto \mathbb{R}^+$. The function d is said to be a metric of fuzzy quantities, and the space $(\mathbb{F}_c^{\star}(\mathbb{R}), d)$ a metric space, if the following holds:

- $d(\tilde{X}, \tilde{X}) = 0$ (reflexive),
- $d(\tilde{X}, \tilde{Y}) \geq 0$ (non-negative),
- $d(\tilde{X}, \tilde{Y}) = 0 \Leftrightarrow \tilde{X} = \tilde{Y}$ (non-degenerate),
- $d(\tilde{X}, \tilde{Y}) = d(\tilde{Y}, \tilde{X})$ (symmetric),
- $d(\tilde{X}, \tilde{Z}) \leq d(\tilde{X}, \tilde{Y}) + d(\tilde{Y}, \tilde{Z})$ (triangular inequality).

4.1 The ρ_1 and ρ_2 Distances

Diamond and Kloeden (1990) introduced two families of metrics defined on $\mathbb{F}_c^{\star}(\mathbb{R})^2$, the first one denoted by d_p and the second one by ρ_p. The purpose of the metric d_p is to propose a metric based on the extension of the Hausdorff distance given in Eqs. 4.1 and 4.2. This class includes the metrics corresponding to $p = 1$ and $p = \infty$, denoted by d_1 and d_∞. In the same perspective, the second class of metrics designated by ρ_p is based on the family of L_p metrics defined for the so-called support functions characterizing a convex compact set. Accordingly, the spaces $(\mathbb{F}_c^{\star}(\mathbb{R}), d_p)$ and $(\mathbb{F}_c^{\star}(\mathbb{R}), \rho_p)$ are metric spaces $\forall p \leq \infty$. One of the main findings of this article is that $\forall p \leq \infty$, the metric spaces $(\mathbb{F}_c^{\star}(\mathbb{R}), d_p)$ are separable, complete, and locally compact. In addition, for a given value of p, these spaces are topologically equivalent to $(\mathbb{F}_c^{\star}(\mathbb{R}), \rho_p)$.

Moreover, Diamond and Kloeden (1990) emphasized the importance of the ρ_p class of metrics for the reason that it includes the case for which $p = 2$, i.e. the distance ρ_2, inspired by the Euclidean metric. They asserted that this "L$_2$ case induces additive variance." As a consequence, in this section, we are mainly interested in ρ_p class of metrics, and specifically in the cases $p = 1$ and $p = 2$, i.e. ρ_1 and ρ_2. The definition is given by:

Definition 4.1.1 (The ρ_1 and ρ_2 Distances) Consider two fuzzy numbers $\tilde{X}$ and $\tilde{Y}$ of the class of non-empty compact bounded fuzzy sets, with their respective α-cuts $\tilde{X}_\alpha = [\tilde{X}_\alpha^L, \tilde{X}_\alpha^R]$ and $\tilde{Y}_\alpha = [\tilde{Y}_\alpha^L, \tilde{Y}_\alpha^R]$. The distances ρ_1 and ρ_2 between $\tilde{X}$ and $\tilde{Y}$ can be written in terms of their left and right α-cuts by

$$\rho_1(\tilde{X}, \tilde{Y}) = \frac{1}{2} \int_0^1 |\tilde{X}_\alpha^L - \tilde{Y}_\alpha^L| d\alpha + \frac{1}{2} \int_0^1 |\tilde{X}_\alpha^R - \tilde{Y}_\alpha^R| d\alpha, \qquad (4.3)$$

$$\rho_2(\tilde{X}, \tilde{Y}) = \sqrt{\frac{1}{2} \int_0^1 (\tilde{X}_\alpha^L - \tilde{Y}_\alpha^L)^2 d\alpha + \frac{1}{2} \int_0^1 (\tilde{X}_\alpha^R - \tilde{Y}_\alpha^R)^2 d\alpha}. \qquad (4.4)$$

4.2   The Bertoluzza Distance

Bertoluzza et al. (1995) asserted that the Hausdorff-based metric and the ones by Diamond and Kloeden (1990) present serious drawbacks. The authors saw that the distance between fuzzy numbers is "disadvantaged" compared to the distance between crisp numbers. In fact, these metrics take into consideration the distance between the extreme points of the support sets of the studied fuzzy numbers only. This calculation is not recommendable in fuzzy set theory, distances between different intermediate levels are more convenient. Moreover, the authors assigned the problem of the Diamond and Kloeden (1990) metrics to the fact that these latter are implicitly conceived in a way to have intervals centered on the same value. We could eventually face the case where the intervals would be left or right shifted. A more realistic distance should then be considered.

The aim of Bertoluzza et al. (1995) is to introduce a metric based on the distance between every convex combination of the infinum and the supremum values corresponding to every α-level. They proposed a versatile class of distances between fuzzy numbers regarded as a weighted mean of the distances between their α-cuts. According to the authors, following this procedure, they managed to remove the aforementioned disadvantages. Therefore, the strength of this L_2 metric is that it considers every inner point of the interval defining the α-cuts of a given fuzzy number, and obviously not only the extreme points. In addition, it is nowadays well adapted in the development of statistical methods with fuzzy numbers due to its operational flexibility.

In order to construct their generalized metric, the authors suggested a parametrization of the studied intervals. This transformation was described later on by Dubois and Prade (2008) as the so-called gradual numbers, defined by:

Definition 4.2.1 (Gradual Number) Consider a fuzzy number $\tilde{X}$ with its respective left and right α-cuts $\tilde{X}_\alpha^L$ and $\tilde{X}_\alpha^R$. A gradual number is of the form $(\alpha, \tilde{X}_\alpha^{[\lambda]})$, where $\tilde{X}_\alpha^{[\lambda]}$ is a linear mapping, function of $\lambda \in [0; 1]$ such that

$$\tilde{X}_\alpha^{[\lambda]} = \lambda \tilde{X}_\alpha^R + (1 - \lambda) \tilde{X}_\alpha^L. \tag{4.5}$$

Furthermore, the authors proposed to consider two normal weighting measures γ and ϕ on the measurable space $([0; 1], \mathcal{B}_{[0;1]})$. These weights are measuring entities. Thus, they will be used for the integration process only. We note that they do not necessarily have distribution laws in the pure probability sense. They are introduced for a weighting purpose only. These weights are defined as follows:

- The weight $\gamma(\lambda)$ is supposed to be a probability function related to a non-degenerate distribution of λ. This weight intends to express a preference for a given hand side (left or right) of a fuzzy number. For example, if our preference for both sides is the same, the distribution given by γ is symmetric such that $\gamma(\lambda) = \gamma(1 - \lambda), \forall \lambda \in [0; 1]$.
- The weight $\phi(\alpha)$ is a probability density function which expresses the influence of each level of a given fuzzy entity. This weight takes into account the degree

of fuzziness. It is associated with a strictly increasing and absolutely continuous distribution function. The Lebesgue measurable function is often used for $\phi(\alpha)$.

The Bertoluzza distance can now on be expressed. It is given by the following definition:

Definition 4.2.2 (The Bertoluzza Distance) Let $\tilde{X}$ and $\tilde{Y}$ be two fuzzy sets of $\mathbb{F}^\star_c(\mathbb{R})$ with their corresponding left and right α-cuts. The Bertoluzza L_2-type distance between $\tilde{X}$ and $\tilde{Y}$ can be expressed by

$$d_{Bertoluzza}(\tilde{X}, \tilde{Y}) = \sqrt{\int_0^1 \int_0^1 \left(\tilde{X}^{[\lambda]}_\alpha - \tilde{Y}^{[\lambda]}_\alpha\right)^2 \mathrm{d}\gamma(\lambda)\mathrm{d}\phi(\alpha)}, \qquad (4.6)$$

where $(\alpha, \tilde{X}^{[\lambda]}_\alpha)$ and $(\alpha, \tilde{Y}^{[\lambda]}_\alpha)$ are gradual numbers.

4.3 The $\delta^\star_{p,q}$ and $\rho^\star_p$ Distances

In the same way, Grzegorzewski (1998) generalized the cases of L_2 families of distances and proposed two classes $\delta^\star_{p,q}$ and $\rho^\star_p$ of distances of type L_p with the corresponding metric spaces $(\mathbb{F}^\star_c(\mathbb{R}), \delta^\star_{p,q})$ and $(\mathbb{F}^\star_c(\mathbb{R}), \rho^\star_p)$. The author proposed a mean of ordering fuzzy numbers in decision making procedures based on these two metrics and consequently used those metrics in hypothesis testing.

These distances are parametrized basically by a parameter p. The distance $\delta^\star_{p,q}$ has two parameters p and q for which $1 \leq p \leq \infty$ and $0 \leq q \leq 1$. Philosophically, the interpretation of the parameter q of $\delta^\star_{p,q}$ corresponds to the sense of λ of the previous section. For instance, if one gives higher weight for the left hand side of a given fuzzy number, the parameter q would be between 0 and $\frac{1}{2}$. For the case of the preference of the right hand side, q would then be between $\frac{1}{2}$ and 1. Consequently, if no preferences are depicted, the parameter q is of value $\frac{1}{2}$, i.e. $\delta^\star_{p,\frac{1}{2}}$. We add that whenever $q = \frac{1}{2}$, both classes $\delta^\star_{p,\frac{1}{2}}$ and $\rho^\star_p$ are equally recommended.

Definition 4.3.1 (The $\delta^\star_{p,q}$ Distance) Let $\tilde{X}$ and $\tilde{Y}$ be two fuzzy numbers of $\mathbb{F}^\star_c(\mathbb{R})$. Consider the two parameters p and q such that $1 \leq p \leq \infty$ and $0 \leq q \leq 1$. The $\delta^\star_{p,q}$ distance between $\tilde{X}$ and $\tilde{Y}$ is written by

$$\delta^\star_{p,q}(\tilde{X}, \tilde{Y}) = \begin{cases} \sqrt[p]{(1-q)\int_0^1 |\tilde{Y}^L_\alpha - \tilde{X}^L_\alpha|^p \mathrm{d}\alpha + q \int_0^1 |\tilde{Y}^R_\alpha - \tilde{X}^R_\alpha|^p \mathrm{d}\alpha}, & \text{if } 1 \leq p < \infty, \\ (1-q) \sup_{0<\alpha\leq 1} \left(|\tilde{Y}^L_\alpha - \tilde{X}^L_\alpha|\right) + q \sup_{0<\alpha\leq 1} \left(|\tilde{Y}^R_\alpha - \tilde{X}^R_\alpha|\right), & \text{for } p = \infty, \end{cases}$$

$$(4.7)$$

where $\tilde{X}_\alpha = [\tilde{X}^L_\alpha, \tilde{X}^R_\alpha]$ and $\tilde{Y}_\alpha = [\tilde{Y}^L_\alpha, \tilde{Y}^R_\alpha]$ are the α-cuts of $\tilde{X}$ and $\tilde{Y}$.

On the relation between $\delta^{\star}_{p,q}$ from one side, and ρ_1 and ρ_2 from another one, the following remark is interesting to mention.

Remark 4.3.1 If $q = \frac{1}{2}$, when $p = 1$ and $p = 2$, the distance $\delta^{\star}_{p,q}$ is, respectively, equivalent to the ρ_1 and ρ_2 metrics described in Sect. 4.1.

Definition 4.3.2 (The $\rho^{\star}_p$ Distance) Consider again the two fuzzy numbers $\tilde{X}$ and $\tilde{Y}$ of the class of all fuzzy numbers, with their respective α-cuts and $1 \leq p \leq \infty$. The distance ρ_p can be expressed by

$$\rho^{\star}_p(\tilde{X}, \tilde{Y}) = \begin{cases} \max\left\{ \sqrt[p]{\int_0^1 |\tilde{Y}^L_\alpha - \tilde{X}^L_\alpha|^P d\alpha}, \ \sqrt[p]{\int_0^1 |\tilde{Y}^R_\alpha - \tilde{X}^R_\alpha|^P d\alpha} \right\}, & \text{if } 1 \leq p < \infty, \\ \max\left\{ \sup_{0 < \alpha \leq 1} (|\tilde{Y}^L_\alpha - \tilde{X}^L_\alpha|), \ \sup_{0 < \alpha \leq 1} (|\tilde{Y}^R_\alpha - \tilde{X}^R_\alpha|) \right\}, & \text{for } p = \infty. \end{cases}$$

$$(4.8)$$

4.4   The "mid/spr" Distance

Trutschnig et al. (2009) recalled an extension of the Bertoluzza et al. (1995) metric. Their distance is based on the center location or the middle position of a fuzzy number denoted by "mid," and on its spread, also called the radius and denoted by "spr". The authors asserted that their metric is a generalization of the previously mentioned ones (δ_∞, ρ_2, $d_{Bertoluzza}$), but in terms of generalized mids and spreads. Thus, their metric is defined in a multidimensional space. In addition, it possesses good mathematical and intuitive properties and can be easily calculated and interpreted. Their metric is of class L_2 and could be very promising in the statistical context. For our situation, we will be mainly interested in the one-dimensional setting, with a Lebesgue measured weight. We define first the mid and spread of a given fuzzy number in order to be able to propose the "mid/spr" distance later on.

Definition 4.4.1 (Center and Radius of a Fuzzy Number) The center "mid" and radius "spr" of a fuzzy number $\tilde{X} \in \mathbb{F}^{\star}_c(\mathbb{R})$ are, respectively, given by

$$\text{mid}\tilde{X} = \frac{1}{2}(\tilde{X}^L_\alpha + \tilde{X}^R_\alpha), \tag{4.9}$$

$$\text{spr}\tilde{X} = \frac{1}{2}(\tilde{X}^R_\alpha - \tilde{X}^L_\alpha). \tag{4.10}$$

Definition 4.4.2 (The "mid/spr" Distance) The "mid/spr" distance between two fuzzy numbers $\tilde{X}$ and $\tilde{Y}$ of $\mathbb{F}^{\star}_c(\mathbb{R})$ is written by

$$d_{mid/spr}(\tilde{X}, \tilde{Y}) = \sqrt{\int_0^1 \left(\text{mid}\tilde{X} - \text{mid}\tilde{Y}\right)^2 d\phi(\alpha) + \theta_t \int_0^1 \left(\text{spr}\tilde{X} - \text{spr}\tilde{Y}\right)^2 d\phi(\alpha)},$$

$$(4.11)$$

with the two weighting parameters ϕ and θ_t, where ϕ is a function of α, and θ_t is a function of a weighting measure $W(t)$, such that

$$\theta_t = \int_0^1 (2t - 1)^2 \mathrm{d}W(t). \tag{4.12}$$

Interesting remarks related to the connection of the mid/spr metric to the Bertoluzza one are given in the following manner:

Remarks 4.4.1

1. If θ_t is a Lebesgue measure, then we have that $\mathrm{d}W(t) = \mathrm{d}t$ and θ_t will be of value $\frac{1}{3}$.
2. Note that the weights θ_t and $\phi(\alpha)$ coincide, respectively, with the meanings of the Bertoluzza weights $\gamma(\lambda)$ and $\phi(\alpha)$ of Sect. 4.2.
3. In a one-dimensional setting, the mid/spr metric is equivalent to the Bertoluzza distance, see Trutschnig et al. (2009).

Consequently, since we are considering a one-dimensional setting only, the calculation of the Bertoluzza distance will be used for an informative perspective only. We will rather use the mid/spr distance in further calculations.

4.5   The "ϕ-wabl/ldev/rdev" Distance

Based on the Bertoluzza et al. (1995) and Trutschnig et al. (2009) metrics, Sinova et al. (2014) introduced a new family of L_2 metrics denoted by "ϕ-wabl/ldev/rdev". This name is due to the fact that it features the "deviations in the shape" of the fuzzy numbers levelwise given by "ldev" and "rdev", and the measure of the central location given by "wabl". By their family of metrics, Sinova et al. (2014) insured the necessary and sufficient characterizing properties of the corresponding fuzzy numbers.

First of all, based on the Definition 4.4.1 of the mid, we define the different entities "wabl", "ldev", and "rdev" related to a given fuzzy number. They are written as follows:

Definition 4.5.1 Let $\tilde{X}$ be a fuzzy number of the class of bounded fuzzy numbers $\mathbb{F}_c^\star(\mathbb{R})$ with its α-level set $\tilde{X}_\alpha$. Consider ϕ an absolutely continuous probability function on $([0; 1], \mathcal{B}_{[0; 1]})$. The measure of central location "wabl" of $\tilde{X}$ can be written as:

$$\mathrm{wabl}^\phi(\tilde{X}) = \int_0^1 \mathrm{mid}\, \tilde{X}_\alpha \mathrm{d}\phi(\alpha). \tag{4.13}$$

In the same way, the left and right deviations in the shape associated with the weight ϕ and functions of the level α are denoted by "ldev$^\phi$" and "rdev$^\phi$". They are given by

$$\mathrm{ldev}^{\phi}\tilde{X}(\alpha) = \mathrm{wabl}^{\phi}(\tilde{X}) - \tilde{X}_{\alpha}^{L}(\alpha), \qquad (4.14)$$

$$\mathrm{rdev}^{\phi}\tilde{X}(\alpha) = \tilde{X}_{\alpha}^{R}(\alpha) - \mathrm{wabl}^{\phi}(\tilde{X}). \qquad (4.15)$$

One can remark that the α-cuts of a given fuzzy number $\tilde{X}$ can be written in terms of the previous measures by

$$\tilde{X}_{\alpha} = \left[\mathrm{wabl}^{\phi}(\tilde{X}) - \mathrm{ldev}^{\phi}\tilde{X}(\alpha),\, \mathrm{wabl}^{\phi}(\tilde{X}) + \mathrm{rdev}^{\phi}\tilde{X}(\alpha)\right].$$

We are now able to give the ϕ-wabl/ldev/rdev metric expressed in the following manner:

Definition 4.5.2 (The "ϕ-wabl/ldev/rdev" Distance) Consider two fuzzy numbers $\tilde{X}$ and $\tilde{Y}$ of $\mathbb{F}_{c}^{\star}(\mathbb{R})$, a parameter θ_{t} such that $0 \leq \theta_{t} \leq 1$ as given in Eq. 4.12 and the weighting factor $\phi(\alpha)$ on the measurable space $([0; 1], \mathcal{B}_{[0;1]})$. The ϕ-wabl/ldev/rdev distance is the following application

$$d_{\phi - wabl/ldev/rdev} : \mathbb{F}_{c}^{\star}(\mathbb{R}) \times \mathbb{F}_{c}^{\star}(\mathbb{R}) \rightarrow \mathbb{R}^{+}$$

$$\tilde{X} \times \tilde{Y} \qquad \mapsto d_{\phi - wabl/ldev/rdev}(\tilde{X}, \tilde{Y}),$$

such that

$$d_{\phi - wabl/ldev/rdev}(\tilde{X}, \tilde{Y}) = \left[\left(\mathrm{wabl}^{\phi}(\tilde{X}) - \mathrm{wabl}^{\phi}(\tilde{Y})\right)^{2} \right.$$

$$+ \left. \theta_{t}\int_{0}^{1}\left(\frac{1}{2}\left(\mathrm{ldev}^{\phi}\tilde{X} - \mathrm{ldev}^{\phi}\tilde{Y}\right)^{2} + \frac{1}{2}\left(\mathrm{rdev}^{\phi}\tilde{X} - \mathrm{rdev}^{\phi}\tilde{Y}\right)^{2}\right)d\phi(\alpha)\right]^{\frac{1}{2}}. \quad (4.16)$$

Note that the space $(\mathbb{F}_{c}^{\star}(\mathbb{R}), d_{\phi - wabl/ldev/rdev})$ is metric, and the L_{2} distance $d_{\phi - wabl/ldev/rdev}$ is translation and scale invariant. The proof of these properties can be found in Sinova et al. (2014).

4.6 The Signed Distance

The signed distance in its current form was first described by Yao and Wu (2000) in the context of ranking fuzzy numbers. It has been introduced few years before by Dubois and Prade (1987) as an expected value of a particular fuzzy number. This distance has been used extensively in different contexts, such as evaluations of linguistic questionnaires or hypotheses testing and others. It has been known for its simplicity and most of all for its directional property. Thus, one could expect to get a negative distance between two fuzzy numbers. This fact made the SGD an appealing one to investigate. As instance, it is seen as an efficient tool in fuzzy ranking approaches (see Abbasbandy and Asady 2006). In addition, this distance is affordable in terms of computations.

The SGD can be defined in the following way:

Definition 4.6.1 (Signed Distance of a Real Value)   The signed distance measured from the origin $d_0(a, 0)$ for $a \in \mathbb{R}$ is a itself, i.e. $d_0(a, 0) = a$.

Definition 4.6.2 (Signed Distance Between Two Real Values)   The signed distance between a and $b \in \mathbb{R}$ is $d(a, b) = a - b$.

Consider now $\tilde{X}$ and $\tilde{Y}$ two fuzzy sets of the class of fuzzy sets $\mathbb{F}(\mathbb{R})$, with their respective α-cuts $\tilde{X}_\alpha$ and $\tilde{Y}_\alpha$. We remind that the left and right α-cuts of $\tilde{X}$ and $\tilde{Y}$, denoted, respectively, by $\tilde{X}_\alpha^L$, $\tilde{X}_\alpha^R$, $\tilde{Y}_\alpha^L$, and $\tilde{Y}_\alpha^R$ are integrable for all $\alpha \in [0; 1]$. The signed distance between the fuzzy numbers $\tilde{X}$ and $\tilde{Y}$ is defined as:

Definition 4.6.3 (Signed Distance Between Two Fuzzy Sets)   The signed distance between $\tilde{X}$ and $\tilde{Y}$ denoted by d_{SGD} is the mapping

$$d_{SGD} : \mathbb{F}_c^\star(\mathbb{R}) \times \mathbb{F}_c^\star(\mathbb{R}) \to \mathbb{R}$$

$$\tilde{X} \times \tilde{Y} \quad \mapsto d_{SGD}(\tilde{X}, \tilde{Y}),$$

such that

$$d_{SGD}(\tilde{X}, \tilde{Y}) = \frac{1}{2} \int_0^1 \left[\tilde{X}_\alpha^L(\alpha) + \tilde{X}_\alpha^R(\alpha) - \tilde{Y}_\alpha^L(\alpha) - \tilde{Y}_\alpha^R(\alpha) \right] d\alpha. \qquad (4.17)$$

It is useful to mention some important properties of the SGD in order to be used in further sections. They are given by:

Properties 4.6.1

- The signed distance d_{SGD}: $\mathbb{F}_c^\star(\mathbb{R}) \times \mathbb{F}_c^\star(\mathbb{R}) \to \mathbb{R}$ is continuous and differentiable.
- The signed distance is reflexive, non-degenerate and the triangular inequality holds.
- The signed distance is commutative, associative, distributive, and transitive.
- The signed distance is translation invariant and scale invariant.

The proofs of these properties are mostly direct. Further information can be found in Yao and Wu (2000).

We are now able to propose a simple example illustrating the calculation of the SGD of two fuzzy numbers. It is given by:

Example 4.6.1   Let $\tilde{X}$ and $\tilde{Y}$ be two fuzzy numbers given by their respective left and right α-cuts as follows:

$$\tilde{X}_\alpha = [2 + \alpha; \ 5 - \alpha],$$

$$\tilde{Y}_\alpha = [1 + \alpha; \ 4 - \alpha].$$

We would like to calculate the SGD between $\tilde{X}$ and $\tilde{Y}$ as described in Definition 4.6.3. It is expressed by

$$d_{SGD}(\tilde{X}, \tilde{Y}) = \frac{1}{2} \int_0^1 \left[\tilde{X}_\alpha^L(\alpha) + \tilde{X}_\alpha^R(\alpha) - \tilde{Y}_\alpha^L(\alpha) - \tilde{Y}_\alpha^R(\alpha) \right] d\alpha$$

$$= \frac{1}{2} \int_0^1 \left[(2 + \alpha) + (5 - \alpha) - (1 + \alpha) - (4 - \alpha) \right] d\alpha = \frac{1}{2} \int_0^1 2 = 1.$$

The so-called fuzzy origin $\tilde{0}$ is defined as the fuzzy number for which the support and the core sets are reduced to the singleton $\{0\}$. The SGD of a fuzzy set measured from the fuzzy origin $\tilde{0}$ seems to be an interesting tool in terms of ranking fuzzy numbers. In our context, this distance will be useful in further calculations. We highlight that this measure has been introduced as the expected value of a fuzzy number as proposed in Dubois and Prade (1987) and Heilpern (1992). The signed distance measured from the fuzzy origin is then written as:

Definition 4.6.4 (Signed Distance of a Fuzzy Set) The signed distance of the fuzzy set $\tilde{X}$ measured from the fuzzy origin $\tilde{0}$ is

$$d_{SGD}(\tilde{X}, \tilde{0}) = \frac{1}{2} \int_0^1 \left[\tilde{X}_\alpha^L(\alpha) + \tilde{X}_\alpha^R(\alpha) \right] d\alpha. \qquad (4.18)$$

The SGD between two fuzzy numbers can then be expressed in terms of the distance of each one of them to the fuzzy origin $\tilde{0}$. This relation is given as follows:

Proposition 4.6.1 *For $\tilde{X}$ and $\tilde{Y} \in \mathbb{F}_c^\star(\mathbb{R})$, we have that*

$$d_{SGD}(\tilde{X}, \tilde{Y}) = d_{SGD}(\tilde{X}, \tilde{0}) - d_{SGD}(\tilde{Y}, \tilde{0}). \qquad (4.19)$$

Proof The proof is direct from Definitions 4.6.3 and 4.6.4.   □

The SGD could be related to other presented distances, notably the "ϕ-wabl/ldev/rdev" one as seen in the following remark:

Remark 4.6.1 For a fuzzy number $\tilde{X} \in \mathbb{F}_c^\star(\mathbb{R})$, the $\text{wabl}^\phi(\tilde{X})$ metric described in Definition 4.5.1 coincides with the signed distance of $\tilde{X}$ measured from the fuzzy origin $\tilde{0}$, i.e.

$$\text{wabl}^\phi(\tilde{X}) = d_{SGD}(\tilde{X}, \tilde{0}),$$

where the weight ϕ is Lebesgue measured.

While most known distances are defined on $\mathbb{R}^+$ (said to be non-directional), a particularity of the signed distance is that it is actually defined on $\mathbb{R}$. Some references call it as a directional distance since its signed characteristic indicates the direction. As a consequence, the separability and the symmetry properties given by $\forall \tilde{X}, \tilde{Y} \in \mathbb{F}_c^\star(\mathbb{R})$, $d_{SGD}(\tilde{Y}, \tilde{X}) \geq 0$ and $d_{SGD}(\tilde{X}, \tilde{Y}) = d_{SGD}(\tilde{Y}, \tilde{X})$, respectively, are not present. However, the symmetry property is not fully lost. A "partial" symmetry property can be proposed.

Proposition 4.6.2 (Partial Symmetry of the Signed Distance) *The signed distance d_{SGD} is partially symmetric such that*

$$d_{SGD}(\tilde{X}, \tilde{Y}) = -d_{SGD}(\tilde{Y}, \tilde{X}). \tag{4.20}$$

In the same perspective, one can define a directional separability property of the signed distance relative to the sign of the distance, this latter according to the position of $\tilde{X}$ and $\tilde{Y}$ on the x-axis in a one-dimensional setting. This particularity makes this distance recommendable for ranking methods. Practically, by adopting a signed procedure in the process of calculating a distance, one can quickly understand the direction between $\tilde{X}$ and $\tilde{Y}$. For instance, if we consider the example of rating the imprecise performance of two machines A and B, the information obtained from calculating a non-directional distance between both performances could not be enough since we cannot know which one of machines A and B is better performing. One has to look closely to the modelling fuzzy sets, which becomes a tedious task. Introducing a directional distance on such cases in order to easily know which machine performs better is desirable. As such, using distances having the directional particularity could be eventually influencing from a statistical point of view. In the subsequent chapters, we will then be interested in investigating the effect of using such directional distances in statistical methods.

The defended distance can also be expressed in terms of the ρ_1 metric. The relation is given by:

Corollary 4.6.1 *We have that*

$$- \rho_1(\tilde{X}, \tilde{Y}) \leq d_{SGD}(\tilde{X}, \tilde{Y}) \leq \rho_1(\tilde{X}, \tilde{Y}). \tag{4.21}$$

The proof can be found in Appendix A.

Following up the Sect. 2.2 treating of particular types of fuzzy numbers, it is often practical to expose their signed distances.

Proposition 4.6.3 (Signed Distance of Common Fuzzy Sets) *The signed distance $d_{SGD}(\tilde{X}, \tilde{0})$ of a particular fuzzy set $\tilde{X}$ measured from the fuzzy origin $\tilde{0}$ is:*

- *for a triangular fuzzy number $\tilde{X} = (p, q, r)$ given in Definition 2.2.1,*

$$d_{SGD}(\tilde{X}, \tilde{0}) = \frac{1}{2} \int_0^1 \left[p + (q-p)\alpha + r - (r-q)\alpha \right] d\alpha = \frac{1}{4}(p + 2q + r); \tag{4.22}$$

- *for a trapezoidal fuzzy number $\tilde{X} = (p, q, r, s)$ seen in Definition 2.2.2,*

$$d_{SGD}(\tilde{X}, \tilde{0}) = \frac{1}{2} \int_0^1 \left[p + (q - p)\alpha + s - (s - r)\alpha \right] d\alpha = \frac{1}{4}(p + q + r + s); \tag{4.23}$$

- *for a gaussian fuzzy number $\tilde{X}$ given by the parameters μ and σ as described in Definition 2.2.3,*

$$d_{SGD}(\tilde{X}, \tilde{0}) = \frac{1}{2} \int_0^1 \left[\tilde{X}_\alpha^L(\alpha) + \tilde{X}_\alpha^R(\alpha) \right] d\alpha = \frac{1}{2} \int_0^1 2\mu d\alpha = \mu; \qquad (4.24)$$

- *for a two-sided gaussian fuzzy number $\tilde{X}$ given by the parameters μ_1, σ_1, μ_2 and σ_2 as shown in Definition 2.2.4,*

$$
\begin{aligned}
d_{SGD}(\tilde{X}, \tilde{0}) &= \frac{1}{2} \int_0^1 \left[\tilde{X}_\alpha^L(\alpha) + \tilde{X}_\alpha^R(\alpha) \right] d\alpha \\
&= \frac{1}{2} \int_0^1 \left[\mu_1 - \sqrt{-2\sigma_1^2 \ln \alpha} + \mu_2 + \sqrt{-2\sigma_2^2 \ln \alpha} \right] d\alpha \\
&= \frac{1}{2} (\mu_1 + \mu_2) - \frac{1}{2} \sigma_1 \int_0^1 \sqrt{-2 \ln \alpha} d\alpha + \frac{1}{2} \sigma_2 \int_0^1 \sqrt{-2 \ln \alpha} d\alpha
\end{aligned}
$$

and since $\int_0^1 \sqrt{-2 \ln \alpha} d\alpha = \sqrt{\frac{\pi}{2}}$, we have that

$$d_{SGD}(\tilde{X}, \tilde{0}) = \frac{1}{2}(\mu_1 + \mu_2) + \frac{1}{2}\sqrt{\frac{\pi}{2}}(\sigma_2 - \sigma_1). \qquad (4.25)$$

Besides the directional property of the signed distance, this latter presents a drawback since it coincides with a central location measure. When we adopt this approach to measure the distance between two fuzzy numbers, only the extreme values are mainly considered. Thus, we are conducted to a middle tendency between both numbers, without taking into consideration the weight that could be given to inner points between the extremes, i.e. the variability between the minima and the maxima in a given fuzzy set, or to the weight given to the shape of the studied fuzzy numbers.

4.7 The Generalized Signed Distance

The signed distance given in Sect. 4.6 seems to have nice properties and is eventually interesting to investigate in further statistical procedures. However, this distance presents several drawbacks. First, since it is seen as a measure of central tendency, by calculating the SGD we tend to neglect the influence of the shape of the fuzzy numbers and their possible irregularities. Second, this distance lacks in total separability and symmetry. Thus, it cannot be defined as a full metric, and by that induce a metric space. To solve these problems (separately or not), multiple steps further are required. We propose in this section a generalization of the SGD inspired by the well-known distances previously described. We will first introduce a L$_2$ metric denoted by $d_{SGD}^{\theta^\star}$, depending on a weight parameter $\theta^\star$. The strength of this new metric is the ability to take into account not only a central location measure, but the deviation in the shapes of the studied numbers. In addition, it has all the necessary and sufficient conditions to define a metric of fuzzy quantities. However, since the original signed distance has an advantage of its efficiency in ranking methods, we would be interested in revisiting it. We propose a ranking distance based on the signed distance and the so-called generalized signed distance denoted by d_{GSGD}. The GSGD is a version of the L$_2$ metric $d_{SGD}^{\theta^\star}$ which preserves the directionality of the signed distance.

Let us first give the functions corresponding to the deviations of the shape. They can be written in terms of the signed distance as follows:

Definition 4.7.1 Consider a fuzzy number $\tilde{X} \in \mathbb{F}_c^\star(\mathbb{R})$ with its α-level set $\tilde{X}_\alpha = [\tilde{X}_\alpha^L, \tilde{X}_\alpha^R]$. The left and right deviations of the shape of $\tilde{X}$ denoted by dev$^L \tilde{X}$ and dev$^R \tilde{X}$ are such that:

$$\mathrm{dev}^L \tilde{X}(\alpha) = d_{SGD}(\tilde{X}, \tilde{0}) - \tilde{X}_\alpha^L, \tag{4.26}$$

$$\mathrm{dev}^R \tilde{X}(\alpha) = \tilde{X}_\alpha^R - d_{SGD}(\tilde{X}, \tilde{0}). \tag{4.27}$$

These deviations in the shape could be connected to the deviations "ldev" and "rdev" introduced in the "ϕ-wabl/ldev/rdev" distance. A relation is given in the following sense:

Remark 4.7.1 When the weighting parameter ϕ is Lebesgue measured, since wabl$^\phi(\tilde{X}) = d_{SGD}(\tilde{X}, \tilde{0})$ as seen in Remark 4.6.1, one can easily see that the previous deviations are equivalent to the ones given in Definition 4.5.1, such that

$$\mathrm{ldev}^\phi \tilde{X} = \mathrm{dev}^L \tilde{X},$$

$$\mathrm{rdev}^\phi \tilde{X} = \mathrm{dev}^R \tilde{X}.$$

We are now able to define the new class of distances as follows:

Definition 4.7.2 (The $d_{SGD}^{\theta^\star}$ Distance) Consider two fuzzy numbers $\tilde{X}$ and $\tilde{Y}$ of the class of non-empty compact and bounded fuzzy numbers, and $\theta^\star$ the weight on the shape of the fuzzy numbers such that $0 \le \theta^\star \le 1$. Based on the signed distance between $\tilde{X}$ and $\tilde{Y}$, the L$_2$ metric $d_{SGD}^{\theta^\star}$ is the mapping

$$d_{SGD}^{\theta^\star} : \mathbb{F}_c^\star(\mathbb{R}) \times \mathbb{F}_c^\star(\mathbb{R}) \to \mathbb{R}^+$$

$$\tilde{X} \times \tilde{Y} \quad \mapsto d_{SGD}^{\theta^\star}(\tilde{X}, \tilde{Y}),$$

such that

$$d_{SGD}^{\theta^\star}(\tilde{X}, \tilde{Y}) = \Big((d_{SGD}(\tilde{X}, \tilde{Y}))^2$$

$$+ \; \theta^\star \Big(\int_0^1 \max \big(\mathrm{dev}^R \tilde{Y}(\alpha) - \mathrm{dev}^L \tilde{X}(\alpha), \mathrm{dev}^R \tilde{X}(\alpha) - \mathrm{dev}^L \tilde{Y}(\alpha) \big) d\alpha \Big)^2 \Big)^{\frac{1}{2}}.$$

$$(4.28)$$

We highlight that for the construction of this distance, our idea is to fully cover the surface between $\tilde{X}$ and $\tilde{Y}$. Furthermore, we intentionally proposed to use the maxima between the differences in the deviations of the fuzzy numbers $\tilde{X}$ and $\tilde{Y}$. The main reason is that we do not have any exact information about the ordering between these fuzzy numbers. Therefore, the maximum between the two differences in deviations is crucial. In addition, we exposed a parameter $\theta^\star$ interpreted as the weight given to the shape of the fuzzy numbers, compared to their signed distance. If $\theta^\star = 1$, it means that we would like to equally weight the shape and the central location measure. As such, if $\theta^\star = 0$, we get back to the absolute value of the signed distance.

It is important to show that the proposed distance is a metric of fuzzy quantities as defined in Definition 4.0.2. Note that for the subsequent Propositions 4.7.1–4.7.7, the proofs can be found in Appendix A.

Proposition 4.7.1 $(\mathbb{F}_c^\star(\mathbb{R}), d_{SGD}^{\theta^\star})$ *is a metric space, with* $d_{SGD}^{\theta^\star}$ *of type* L^2, *such that:*

1. $d_{SGD}^{\theta^\star}$ *is reflexive:* $d_{SGD}^{\theta^\star}(\tilde{X}, \tilde{X}) = 0$.
2. $d_{SGD}^{\theta^\star}$ *is non-negative (separability):* $d_{SGD}^{\theta^\star}(\tilde{X}, \tilde{Y}) \geq 0$, $\forall \tilde{X}, \tilde{Y} \in \mathbb{F}_c^\star(\mathbb{R})$.
3. $d_{SGD}^{\theta^\star}$ *is non-degenerate:* $d_{SGD}^{\theta^\star}(\tilde{X}, \tilde{Y}) = 0 \Leftrightarrow \tilde{X} = \tilde{Y}$.
4. $d_{SGD}^{\theta^\star}$ *is symmetric:* $d_{SGD}^{\theta^\star}(\tilde{X}, \tilde{Y}) = d_{SGD}^{\theta^\star}(\tilde{Y}, \tilde{X})$.
5. *The triangular inequality holds:* $d_{SGD}^{\theta^\star}(\tilde{X}, \tilde{Y}) \leq d_{SGD}^{\theta^\star}(\tilde{X}, \tilde{Z}) + d_{SGD}^{\theta^\star}(\tilde{Z}, \tilde{Y})$.

Proposition 4.7.2 $d_{SGD}^{\theta^\star}$ *is translation invariant and scale invariant, i.e.* $d_{SGD}^{\theta^\star}((\tilde{X} + k), (\tilde{Y} + k)) = d_{SGD}^{\theta^\star}(\tilde{X}, \tilde{Y})$, *and* $d_{SGD}^{\theta^\star}((k \cdot \tilde{X}), (k \cdot \tilde{Y})) = |k| \cdot d_{SGD}^{\theta^\star}(\tilde{X}, \tilde{Y})$, $\forall k \in \mathbb{R}$.

As discussed before and despite the case where $\theta^\star = 0$, one could clearly see that the $d_{SGD}^{\theta^\star}$ metric can give back the original signed distance in specific conditions, as seen in the following proposition:

Proposition 4.7.3 *For* $\theta^\star \neq 0$,

$$d_{SGD}^{\theta^\star}(\tilde{X}, \tilde{Y}) = |d_{SGD}(\tilde{X}, \tilde{Y})| \quad iff \quad spr(\tilde{X}) = spr(\tilde{Y}), \quad\quad (4.29)$$

where $|d_{SGD}(\tilde{X}, \tilde{0})| \leq |d_{SGD}(\tilde{Y}, \tilde{0})|$.

Furthermore, we would like now to position the L_2 metric $d_{SGD}^{\theta^\star}$ metric between the other known metrics. First, let us compare it to the ρ_1 and ρ_2 ones:

Proposition 4.7.4

$$\rho_1(\tilde{X}, \tilde{Y}) \leq d_{SGD}^{\theta^\star}(\tilde{X}, \tilde{Y}) \leq \rho_2(\tilde{X}, \tilde{Y}). \tag{4.30}$$

Similarly, when the weight measure ϕ is a measure of Lebesgue, $d_{SGD}^{\theta^\star}$ is connected as well to the metrics $d_{mid/spr}$, $d_{bertoluzza}$ and $d_{\phi-wabl/ldev/rdev}$ by the following inequalities:

Proposition 4.7.5

1. *For the mid/spr metric, we have that:*

$$d_{SGD}^{\theta^\star}(\tilde{X}, \tilde{Y}) \leq d_{mid/spr}(\tilde{X}, \tilde{Y}), \tag{4.31}$$

and since the Bertoluzza metric is equivalent to the mid/spr one in the considered conditions, then it is evident to write that:

$$d_{SGD}^{\theta^\star}(\tilde{X}, \tilde{Y}) \leq d_{Bertoluzza}(\tilde{X}, \tilde{Y}). \tag{4.32}$$

2. *For the $d_{\phi-wabl/ldev/rdev}$ metric, the following inequality holds:*

$$d_{SGD}^{\theta^\star}(\tilde{X}, \tilde{Y}) \leq d_{\phi-wabl/ldev/rdev}(\tilde{X}, \tilde{Y}). \tag{4.33}$$

Finally, we find interesting to mention the subsequent propositions discussing of the nearest trapezoidal symmetrical fuzzy numbers. The aim of presenting them is to show that with respect to the $d_{SGD}^{\theta^\star}$, one would revert to the original signed distance seen as an optimum of $d_{SGD}^{\theta^\star}$ in particular circumstances. By using the concept of the nearest fuzzy number, we have to note that these propositions could be eventually useful in fuzzy approximations. We add that these latter are needed in the implementation process of multiple statistical measures presented in Chap. 9.

Proposition 4.7.6 (Nearest Trapezoidal Fuzzy Number) *The nearest symmetrical trapezoidal fuzzy number $\tilde{S}$ given by the quadruple $\tilde{S} = [s_0 - 2\epsilon, s_0 - \epsilon, s_0 + \epsilon, s_0 + 2\epsilon]$ to a fuzzy number $\tilde{X}$ with respect to the metric $d_{SGD}^{\theta^\star}$ is given such that*

$$s_0 = d_{SGD}(\tilde{X}, \tilde{0}), \tag{4.34}$$

$$\epsilon = \frac{9}{14} d_{SGD}(\tilde{X}, \tilde{0}) - \frac{3}{7} \int_0^1 \tilde{X}_\alpha^L (2 - \alpha) d\alpha. \tag{4.35}$$

Proposition 4.7.7 (Nearest Mid-Way Trapezoidal Fuzzy Number) *The nearest symmetrical mid-way trapezoidal fuzzy number $\tilde{S}$ given by the quadruple $\tilde{S} = [s_0 - 2\epsilon, s_0 - \epsilon, s_0 + \epsilon, s_0 + 2\epsilon]$ between two fuzzy numbers $\tilde{X}$ and $\tilde{Y}$ with the same weight*

$\theta^\star$ *with respect to the metric $d^{\theta^\star}_{SGD}$ can be written as such*

$$s_0 = \frac{1}{2} d_{SGD}(\tilde{X}, \tilde{0}) + \frac{1}{2} d_{SGD}(\tilde{Y}, \tilde{0}) \tag{4.36}$$

$$\epsilon = \frac{9}{28} d_{SGD}(\tilde{X}, \tilde{0}) - \frac{9}{28} d_{SGD}(\tilde{Y}, \tilde{0})$$

$$-\frac{3}{14} \left[\int_0^1 \tilde{X}^L_\alpha (2 - \alpha) d\alpha - \int_0^1 \tilde{Y}^R_\alpha (2 - \alpha) d\alpha \right]. \tag{4.37}$$

The signed distance is seen to be a performant tool in ranking fuzzy numbers. In the process of improving this operator to better fit the ranking procedures, we would like to define a generalized version of it. The point is then to be able to preserve its directional property. Accordingly and based on the metric $d^{\theta^\star}_{SGD}$, we introduce the direction of travel between $\tilde{X}$ and $\tilde{Y}$ (or the sign) given by the following function:

Definition 4.7.3 (The Sign Function) Let $\tilde{X}$ and $\tilde{Y}$ be two fuzzy numbers in $\mathbb{F}^\star_c(\mathbb{R})$. The sign function denoted by δ_{SGD} is given by

$$\delta_{SGD}(\tilde{X}, \tilde{Y}) = \begin{cases} 1 & \text{if} \quad d_{SGD}(\tilde{X}, \tilde{Y}) \geq 0, \\ -1 & \text{if} \quad d_{SGD}(\tilde{X}, \tilde{Y}) < 0. \end{cases} \tag{4.38}$$

We update now the metric $d^{\theta^\star}_{SGD}$ using the sign function and we get the GSGD written as:

Definition 4.7.4 (The Generalized Signed Distance) Consider two fuzzy numbers $\tilde{X}$ and $\tilde{Y}$ in $\mathbb{F}^\star_c(\mathbb{R})$. The generalized signed distance is the mapping

$$d_{GSGD} : \mathbb{F}^\star_c(\mathbb{R}) \times \mathbb{F}^\star_c(\mathbb{R}) \to \mathbb{R}$$
$$\tilde{X} \times \tilde{Y} \quad \mapsto d_{GSGD}(\tilde{X}, \tilde{Y}),$$

such that

$$d_{GSGD}(\tilde{X}, \tilde{Y}) = \delta_{SGD}(\tilde{X}, \tilde{Y}) \cdot d^{\theta^\star}_{SGD}(\tilde{X}, \tilde{Y}). \tag{4.39}$$

By effect of the metric $d^{\theta^\star}_{SGD}$, the GSGD is reflexive, non-degenerate, for which the triangular inequality holds. The proofs are direct. This construction of the distance is intentional in order to keep the link between the signed distance and its generalized form. As a consequence, the same properties of the original distance such as the direction are preserved, but with a better consideration regarding the shape of the involved fuzzy numbers. We add that the partial symmetry and partial separability properties hold in this case. We remind that the definitions of such properties can be found in (McCulloch, 2016, p. 63).

Proposition 4.7.8 *The generalized signed distance d_{GSGD} is partially separable.*

Proof The proof is directly deduced from the following:

$$\begin{cases} d_{GSGD}(\tilde{X}, \tilde{Y}) \geq 0 & \text{if} & d_{SGD}(\tilde{X}, \tilde{Y}) \geq 0, \\ d_{GSGD}(\tilde{X}, \tilde{Y}) \leq 0 & \text{if} & d_{SGD}(\tilde{X}, \tilde{Y}) < 0. \end{cases} \tag{4.40}$$

$\square$

Proposition 4.7.9 *The generalized signed distance d_{GSGD} is partially symmetric.*

Proof It is obvious that $\delta_{SGD}(\tilde{X}, \tilde{Y}) = -\delta_{SGD}(\tilde{Y}, \tilde{X})$, and since $d_{SGD}^{\theta^\star}(\tilde{X}, \tilde{Y}) = d_{SGD}^{\theta^\star}(\tilde{Y}, \tilde{X})$, we can easily deduce that

$$d_{GSGD}(\tilde{X}, \tilde{Y}) = -d_{GSGD}(\tilde{Y}, \tilde{X}).$$

$\square$

In addition, some other properties of the GSGD are required. As such, one could see that this distance is directly related to the $d_{SGD}^{\theta^\star}$ metric given in the subsequent remark:

Remark 4.7.2 The Propositions 4.7.6 and 4.7.7 related to the $d_{SGD}^{\theta^\star}$ metric can be extended to the generalized signed distance. The proof is direct since d_{GSGD} could practically take either the value $-d_{SGD}^{\theta^\star}$, or the value $d_{SGD}^{\theta^\star}$.

In order to illustrate the calculations of the previously mentioned distances, let us consider the following detailed example showing the use of each of these distances.

Example 4.7.1 Consider a trapezoidal fuzzy number $\tilde{X}$ given by the quadruple $(1, 2, 2.3, 8)$ and a triangular one $\tilde{Y}$ given by the tuple $(9, 9.8, 16)$. We would like to calculate the distance between the fuzzy numbers $\tilde{X}$ and $\tilde{Y}$. We will use the previously described distances.

First, we give the α-cuts of the involved fuzzy numbers. They are given as follows:

$$\begin{cases} \tilde{X}_\alpha^L(\alpha) = 1 + \alpha, \\ \tilde{X}_\alpha^R(\alpha) = 8 - 5.7\alpha, \end{cases} \quad \text{and} \quad \begin{cases} \tilde{Y}_\alpha^L(\alpha) = 9 + 0.8\alpha, \\ \tilde{Y}_\alpha^R(\alpha) = 16 - 6.2\alpha. \end{cases}$$

The distance between $\tilde{X}$ and $\tilde{Y}$ using the different approaches is calculated as follows:

- The signed distance of the fuzzy numbers $\tilde{X}$ and $\tilde{Y}$ can be computed as seen in Definition 4.6.3 as follows:

$$\begin{aligned} d_{SGD}(\tilde{X}, \tilde{Y}) &= \frac{1}{2} \int_0^1 \left[\tilde{X}_\alpha^L(\alpha) + \tilde{X}_\alpha^R(\alpha) - \tilde{Y}_\alpha^L(\alpha) - \tilde{Y}_\alpha^R(\alpha) \right] d\alpha \\ &= \frac{1}{2} \int_0^1 \left[(1 + \alpha) + (8 - 5.7\alpha) - (9 + 0.8\alpha) - (16 - 6.2\alpha) \right] d\alpha \\ &= -7.825. \end{aligned}$$

We add that the signed distances of $\tilde{X}$ and $\tilde{Y}$ measured from the fuzzy origin $\tilde{0}$ are given, respectively, by

$$d_{SGD}(\tilde{X}, \tilde{0}) = \frac{1}{2} \int_0^1 \left[(1+\alpha) + (8 - 5.7\alpha) \right] d\alpha = 3.325,$$

$$d_{SGD}(\tilde{Y}, \tilde{0}) = \frac{1}{2} \int_0^1 \left[(9 + 0.8\alpha) + (16 - 6.2\alpha) \right] d\alpha = 11.15.$$

- For the metric $d^{\theta^\star}_{SGD}(\tilde{X}, \tilde{Y})$ shown in Definition 4.7.2, the deviations of their shapes are given by

$$\begin{cases} \text{dev}^L \tilde{X}(\alpha) = d_{SGD}(\tilde{X}, \tilde{0}) - \tilde{X}^L_\alpha = 3.325 - 1 - \alpha = 2.325 - \alpha, \\ \text{dev}^R \tilde{X}(\alpha) = \tilde{X}^R_\alpha - d_{SGD}(\tilde{X}, \tilde{0}) = 8 - 5.7\alpha - 3.325 = 4.675 - 5.7\alpha, \end{cases}$$

and $\begin{cases} \text{dev}^L \tilde{Y}(\alpha) = d_{SGD}(\tilde{Y}, \tilde{0}) - \tilde{Y}^L_\alpha = 11.15 - 9 - 0.8\alpha = 2.15 - 0.8\alpha, \\ \text{dev}^R \tilde{Y}(\alpha) = \tilde{Y}^R_\alpha - d_{SGD}(\tilde{Y}, \tilde{0}) = 16 - 6.2\alpha - 11.15 = 4.85 - 6.2\alpha. \end{cases}$

Then, the differences of the deviations are as follows:

$$\text{dev}^R \tilde{Y}(\alpha) - \text{dev}^L \tilde{X}(\alpha) = 4.85 - 6.2\alpha - 2.325 + \alpha = 2.525 - 5.2\alpha,$$

$$\text{dev}^R \tilde{X}(\alpha) - \text{dev}^L \tilde{Y}(\alpha) = 4.675 - 5.7\alpha - 2.15 + 0.8\alpha = 2.525 - 4.9\alpha.$$

For $0 \leq \alpha \leq 1$, we can see that $2.525 - 4.9\alpha > 2.525 - 5.2\alpha$, then we get that

$$\max\left(\text{dev}^R \tilde{X}(\alpha) - \text{dev}^L \tilde{Y}(\alpha), \text{dev}^R \tilde{Y}(\alpha) - \text{dev}^L \tilde{X}(\alpha) \right) = \text{dev}^R \tilde{X}(\alpha) - \text{dev}^L \tilde{Y}(\alpha).$$

We sum up these information in order to get the distance $d^{\theta^\star}_{SGD}(\tilde{X}, \tilde{Y})$. Suppose $\theta^\star = 1$, i.e. the central tendency and the shapes of the fuzzy numbers are equally weighted. This distance can be expressed by

$$d^{\theta^\star}_{SGD}(\tilde{X}, \tilde{Y}) = \sqrt{ \left(d_{SGD}(\tilde{X}, \tilde{Y}) \right)^2 + \theta^\star \left(\int_0^1 \max\left(\text{dev}^R \tilde{Y} - \text{dev}^L \tilde{X}, \text{dev}^R \tilde{X} - \text{dev}^L \tilde{Y} \right) d\alpha \right)^2 }$$

$$= \sqrt{ \left(d_{SGD}(\tilde{X}, \tilde{Y}) \right)^2 + \theta^\star \left(\int_0^1 \left(\text{dev}^R \tilde{X} - \text{dev}^L \tilde{Y} \right) d\alpha \right)^2 }$$

$$= \sqrt{ \left(-7.825 \right)^2 + \left(\int_0^1 \left(2.525 - 4.9\alpha \right) d\alpha \right)^2 }$$

$$= 7.825359.$$

- For the GSGD given in Definition 4.7.4, we have to use the previously calculated distance $d_{SGD}^{\theta^{\star}}(\tilde{X}, \tilde{Y})$ and the sign function. Therefore, since $d_{SGD}(\tilde{X}, \tilde{Y}) \leq 0$, then $\delta_{SGD}(\tilde{X}, \tilde{Y}) = -1$. We are now able to calculate the GSGD between both fuzzy numbers. It is written by

$$d_{GSGD}(\tilde{X}, \tilde{Y}) = \delta_{SGD}(\tilde{X}, \tilde{Y}) \cdot d_{SGD}^{\theta^{\star}}(\tilde{X}, \tilde{Y}) = -1 \cdot 7.825359 = -7.825359.$$

$$(4.41)$$

- For the ϕ-wabl/ldev/rdev distance seen in Definition 4.5.2, we suppose that the parameter ϕ is a measure of Lebesgue. Thus, by Remarks 4.6.1 and 4.7.1, we know that the wabl measure is equivalent to the signed distance measured from the fuzzy origin such that

$$\left(\text{wabl}^{\phi}(\tilde{X}) - \text{wabl}^{\phi}(\tilde{Y})\right)^2 = \left(d_{SGD}(\tilde{X}, \tilde{0}) - d_{SGD}(\tilde{Y}, \tilde{0})\right)^2.$$

In addition, the following expressions can be written

$$\int_0^1 \frac{1}{2}\left(\text{ldev}^{\phi}\tilde{X} - \text{ldev}^{\phi}\tilde{Y}\right)^2 = \int_0^1 \frac{1}{2}\left(\text{dev}^L\tilde{X} - \text{dev}^L\tilde{Y}\right)^2$$

$$= \int_0^1 \frac{1}{2}\left(2.325 - \alpha - 2.15 + 0.8\alpha\right)^2$$

$$= \int_0^1 \frac{1}{2}\left(0.175 - 0.2\alpha\right)^2 = 0.004479167,$$

$$\int_0^1 \frac{1}{2}\left(\text{rdev}^{\phi}\tilde{X} - \text{rdev}^{\phi}\tilde{Y}\right)^2 = \int_0^1 \frac{1}{2}\left(\text{dev}^R\tilde{X} - \text{dev}^R\tilde{Y}\right)^2$$

$$= \int_0^1 \frac{1}{2}\left(4.675 - 5.7\alpha - 4.85 + 6.2\alpha\right)^2$$

$$= \int_0^1 \frac{1}{2}\left(-0.175 + 0.5\alpha\right)^2 = 0.01322917.$$

Accordingly, the ϕ-wabl/ldev/rdev distance between $\tilde{X}$ and $\tilde{Y}$ is

$$d_{\phi-wabl/ldev/rdev}(\tilde{X}, \tilde{Y}) = \left[\left(\text{wabl}^{\phi}(\tilde{X}) - \text{wabl}^{\phi}(\tilde{Y})\right)^2\right.$$

$$+ \left. \theta_t \int_0^1 \left(\frac{1}{2}\left(\text{ldev}^{\phi}\tilde{X} - \text{ldev}^{\phi}\tilde{Y}\right)^2 + \frac{1}{2}\left(\text{rdev}^{\phi}\tilde{X} - \text{rdev}^{\phi}\tilde{Y}\right)^2\right)d\phi(\alpha)\right]^{\frac{1}{2}}$$

$$= \left[\left(-7.825\right)^2 + \theta_t \times (0.004479167 + 0.01322917)\right]^{\frac{1}{2}}$$

$$= 7.825377,$$

with $\theta_t = \frac{1}{3}$ as seen in Remark 4.4.1.1.

- The mid/spr distance shown in Definition 4.4.2 can be computed in the following manner. First, we give the mid measure of the fuzzy numbers $\tilde{X}$ and $\tilde{Y}$, and the difference between them. They are given by

$$\mathrm{mid}\tilde{X} = \frac{1}{2}(\tilde{X}_\alpha^L + \tilde{X}_\alpha^R)$$

$$= \frac{1}{2}\big((1+\alpha) + (8 - 5.7\alpha)\big) = 4.5 - 2.35\alpha, \tag{4.42}$$

$$\mathrm{mid}\tilde{Y} = \frac{1}{2}(\tilde{Y}_\alpha^L + \tilde{Y}_\alpha^R)$$

$$= \frac{1}{2}\big((9 + 0.8\alpha) + (16 - 6.2\alpha)\big) = 12.5 - 2.7\alpha, \tag{4.43}$$

while

$$\mathrm{mid}\tilde{X} - \mathrm{mid}\tilde{Y} = 4.5 - 2.35\alpha - 12.5 + 2.7\alpha = -8 + 0.35\alpha.$$

Similarly, the spread of both fuzzy numbers and their difference can be calculated as follows:

$$\mathrm{spr}\tilde{X} = \frac{1}{2}(\tilde{X}_\alpha^R - \tilde{X}_\alpha^L)$$

$$= \frac{1}{2}\big((8 - 5.7\alpha) - (1 + \alpha)\big) = 3.5 - 3.35\alpha, \tag{4.44}$$

$$\mathrm{spr}\tilde{Y} = \frac{1}{2}(\tilde{Y}_\alpha^R - \tilde{Y}_\alpha^L)$$

$$= \frac{1}{2}\big((16 - 6.2\alpha) - (9 + 0.8\alpha)\big) = 3.5 - 3.5\alpha. \tag{4.45}$$

with

$$\mathrm{spr}\tilde{X} - \mathrm{spr}\tilde{Y} = 3.5 - 3.35\alpha - 3.5 + 3.5\alpha = 0.15\alpha.$$

By considering a Lebesgue measure, we have that $\theta_t = \frac{1}{3}$, then the mid/spr distance between $\tilde{X}$ and $\tilde{Y}$ is given by

$$d_{mid/spr}(\tilde{X}, \tilde{Y}) = \sqrt{\int_0^1 \big(\mathrm{mid}\tilde{X} - \mathrm{mid}\tilde{Y}\big)^2 d\phi(\alpha) + \theta_t \int_0^1 \big(\mathrm{spr}\tilde{X} - \mathrm{spr}\tilde{Y}\big)^2 d\phi(\alpha)}$$

$$= \sqrt{\int_0^1 \big(-8 + 0.35\alpha\big)^2 d\alpha + \frac{1}{3}\int_0^1 (0.15\alpha)^2 d\alpha}$$

$$= 7.825812.$$

- For the ρ_1 and ρ_2 distances defined in Definition 4.1.1, the application is direct. It is given by

$$
\begin{aligned}
\rho_1(\tilde{X}, \tilde{Y}) &= \frac{1}{2} \int_0^1 |\tilde{X}_\alpha^L - \tilde{Y}_\alpha^L| d\alpha + \frac{1}{2} \int_0^1 |\tilde{X}_\alpha^R - \tilde{Y}_\alpha^R| d\alpha \\
&= \frac{1}{2} \int_0^1 |1 + \alpha - 9 - 0.8\alpha| d\alpha + \frac{1}{2} \int_0^1 |8 - 5.7\alpha - 16 + 6.2\alpha| d\alpha \\
&= \frac{1}{2} \int_0^1 |-8 + 0.2\alpha| d\alpha + \frac{1}{2} \int_0^1 |-8 + 0.5\alpha| d\alpha \\
&= 7.825,
\end{aligned}
$$

$$
\begin{aligned}
\rho_2(\tilde{X}, \tilde{Y}) &= \sqrt{\frac{1}{2} \int_0^1 \left(\tilde{X}_\alpha^L - \tilde{Y}_\alpha^L\right)^2 d\alpha + \frac{1}{2} \int_0^1 \left(\tilde{X}_\alpha^R - \tilde{Y}_\alpha^R\right)^2 d\alpha}, \\
&= \sqrt{\frac{1}{2} \int_0^1 \left(1 + \alpha - 9 - 0.8\alpha\right)^2 d\alpha + \frac{1}{2} \int_0^1 \left(8 - 5.7\alpha - 16 + 6.2\alpha\right)^2 d\alpha}, \\
&= \sqrt{\frac{1}{2} \int_0^1 \left(-8 + 0.2\alpha\right)^2 d\alpha + \frac{1}{2} \int_0^1 \left(-8 + 0.5\alpha\right)^2 d\alpha}, \\
&= 7.826131.
\end{aligned}
$$

- For the $\delta_{p,q}^\star$ and $\rho_p^\star$ distances shown in Sect. 4.3, we consider the cases of $p = 1$ and $p = 2$ only. For instance, from Remark 4.3.1, we know that $\delta_{1,\frac{1}{2}}^\star(\tilde{X}, \tilde{Y}) = \rho_1(\tilde{X}, \tilde{Y})$ and $\delta_{2,\frac{1}{2}}^\star(\tilde{X}, \tilde{Y}) = \rho_2(\tilde{X}, \tilde{Y})$, for $q = \frac{1}{2}$ and $p = 1$ and $p = 2$, respectively. We obtain that:

$$
\delta_{1,\frac{1}{2}}^\star(\tilde{X}, \tilde{Y}) = 7.825 \quad \text{and} \quad \delta_{2,\frac{1}{2}}^\star(\tilde{X}, \tilde{Y}) = 7.826131.
$$

On the other hand, by replacing conveniently the α-cuts of the fuzzy numbers, the distances $\rho_1^\star(\tilde{X}, \tilde{Y})$ and $\rho_2^\star(\tilde{X}, \tilde{Y})$ can be given as expressed in Definition 4.3.2 by

$$
\begin{aligned}
\rho_1^\star(\tilde{X}, \tilde{Y}) &= \max \left\{ \int_0^1 |\tilde{Y}_\alpha^L - \tilde{X}_\alpha^L| d\alpha, \int_0^1 |\tilde{Y}_\alpha^R - \tilde{X}_\alpha^R| d\alpha \right\} \\
&= \max \left\{ \int_0^1 |8 - 0.2\alpha| d\alpha, \int_0^1 |8 - 0.5\alpha| d\alpha \right\} \\
&= \max \left\{ 7.9, 7.75 \right\} = 7.9,
\end{aligned}
$$

$$
\rho_2^\star(\tilde{X}, \tilde{Y}) = \max \left\{ \sqrt[2]{\int_0^1 \left(\tilde{Y}_\alpha^L - \tilde{X}_\alpha^L\right)^2 d\alpha}, \sqrt[2]{\int_0^1 \left(\tilde{Y}_\alpha^R - \tilde{X}_\alpha^R\right)^2 d\alpha} \right\}
$$

$$= \max \left\{ \sqrt[2]{\int_0^1 (8 - 0.2\alpha)^2 d\alpha}, \sqrt[2]{\int_0^1 (8 - 0.5\alpha)^2 d\alpha} \right\}$$

$$= \max \left\{ 7.900211, 7.751344 \right\} = 7.900211.$$

- Finally, the Bertoluzza distance described in Definition 4.2.2 is given by

$$d_{Bertoluzza}(\tilde{X}, \tilde{Y}) = \sqrt{\int_0^1 \int_0^1 \left(\tilde{X}_\alpha^{[\lambda]} - \tilde{Y}_\alpha^{[\lambda]} \right)^2 d\gamma(\lambda) d\phi(\alpha)},$$

where $\tilde{X}_\alpha^{[\lambda]} = \lambda \tilde{X}_\alpha^R + (1 - \lambda) \tilde{X}_\alpha^L$ and $\tilde{Y}_\alpha^{[\lambda]} = \lambda \tilde{Y}_\alpha^R + (1 - \lambda) \tilde{Y}_\alpha^L$. In the same settings, this equation can be written as:

$$d_{Bertoluzza}(\tilde{X}, \tilde{Y}) = \sqrt{\int_0^1 \int_0^1 \left[(\lambda \tilde{X}_\alpha^R + (1 - \lambda) \tilde{X}_\alpha^L) - (\lambda \tilde{Y}_\alpha^R + (1 - \lambda) \tilde{Y}_\alpha^L) \right]^2 d\lambda d\alpha},$$

by replacing the α-cuts of $\tilde{X}$ and $\tilde{Y}$ using the right expressions, and by additional integration step, we get that:

$$d_{Bertoluzza}(\tilde{X}, \tilde{Y}) = 7.825812.$$

A summary of the obtained results can be found in Table 4.1. First, we can clearly see that the Propositions 4.7.4 and 4.7.5 regarding mainly the metric $d_{SGD}^{\theta^\star}$ are verified. Each of the proposed distances can be adopted for different motivations. Therefore, one can suspect that in the statistical environment, the choice between

Table 4.1 Summary of the distances between $\tilde{X}$ and $\tilde{Y}$ by different approaches—Example 4.7.1

Chosen approach	Type of distance	Distance between $\tilde{X}$ and $\tilde{Y}$
Diamond and Kloeden (1990) (Sect. 4.1)	ρ_1	7.825
	ρ_2	7.826131
Bertoluzza et al. (1995) (Sect. 4.2)	$d_{Bertoluzza}$	7.825812
	$\delta_{1,\frac{1}{2}}^\star$	7.825
Grzegorzewski (1998) (Sect. 4.3)	$\delta_{2,\frac{1}{2}}^\star$	7.826131
	$\rho_1^\star$	7.9
	$\rho_2^\star$	7.900211
Trutschnig et al. (2009) (Sect. 4.4)	$d_{mid/spr}$	7.825812
Sinova et al. (2014) (Sect. 4.5)	$d_{\phi-wabl/ldev/rdev}$	7.825377
Yao and Wu (2000) (Sect. 4.6)	d_{SGD}	-7.825
Berkachy (Sect. 4.7)	$d_{SGD}^{\theta^\star}$	7.825359
	d_{GSGD}	-7.825359

them is not innocent since their application could probably be influencing on the results of statistical analyses.

An example illustrating the concepts of the nearest trapezoidal fuzzy numbers is given as follows:

Example 4.7.2 For this example, we will first find the nearest symmetrical trapezoidal fuzzy number for $\tilde{Y}$ of Example 4.7.1, denoted by $\tilde{S} = [s_0 - 2\epsilon, s_0 - \epsilon, s_0 + \epsilon, s_0 + 2\epsilon]$ as given in Proposition 4.7.6. We have then to calculate the parameters s_0 and ϵ. This calculation can be given by

$$s_0 = d_{SGD}(\tilde{Y}, \tilde{0}) = 11.15,$$

$$\epsilon = \frac{9}{14} d_{SGD}(\tilde{Y}, \tilde{0}) - \frac{3}{7} \int_0^1 \tilde{Y}_\alpha^L (2 - \alpha) d\alpha$$

$$= \frac{9}{14} \cdot 11.15 - \frac{3}{7} \cdot 14.03333 = 7.167857 - 6.014286 = 1.153571.$$

Consequently, with respect to the metric $d_{SGD}^{\theta^*}$, the dotted trapezoidal fuzzy number $\tilde{S}$ shown in Fig. 4.2 is written by the following quadruple

$$\tilde{S} = [s_0 - 2\epsilon, \ s_0 - \epsilon, \ s_0 + \epsilon, \ s_0 + 2\epsilon] = [8.842858, \ 9.996429, \ 12.30357, \ 13.45714].$$

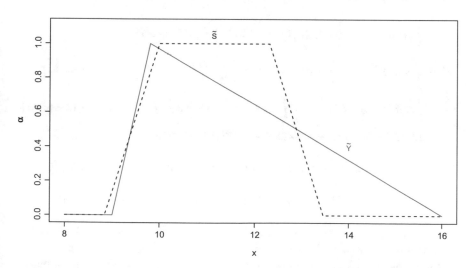

Fig. 4.2 Membership functions of $\tilde{Y}$ and its nearest symmetrical trapezoidal fuzzy number $\tilde{S}$—Example 4.7.2

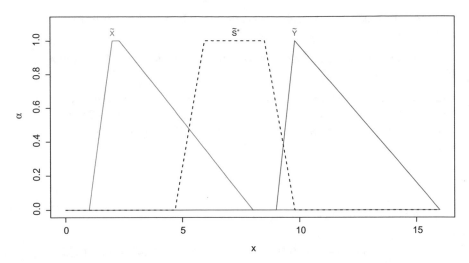

Fig. 4.3 Membership functions of $\tilde{X}$, $\tilde{Y}$ and the nearest mid-way symmetrical trapezoidal fuzzy number $\tilde{S}^\star$ between $\tilde{X}$ and $\tilde{Y}$—Example 4.7.2

From another side, we would be interested as well in calculating the nearest mid-way one between the fuzzy numbers $\tilde{X}$ and $\tilde{Y}$ of Example 4.7.1, as shown in Proposition 4.7.7. This fuzzy number is denoted by $\tilde{S}^\star = [s_0^\star - 2\epsilon^\star, s_0^\star - \epsilon^\star, s_0^\star + \epsilon^\star, s_0^\star + 2\epsilon^\star]$ such that the parameters $s_0^\star$ and $\epsilon^\star$ can be written as follows:

$$s_0^\star = \frac{1}{2}d_{SGD}(\tilde{X}, \tilde{0}) + \frac{1}{2}d_{SGD}(\tilde{Y}, \tilde{0}) = \frac{1}{2}3.325 + \frac{1}{2}11.15 = 7.2375,$$

$$\epsilon^\star = \frac{9}{28}d_{SGD}(\tilde{X}, \tilde{0}) - \frac{9}{28}d_{SGD}(\tilde{Y}, \tilde{0}) - \frac{3}{14}\Big[\int_0^1 \tilde{X}_\alpha^L(2-\alpha)d\alpha - \int_0^1 \tilde{Y}_\alpha^R(2-\alpha)d\alpha\Big]$$

$$= \frac{9}{28}3.325 - \frac{9}{28}11.15 - \frac{3}{14}\Big[\int_0^1 (1+\alpha)\cdot(2-\alpha)d\alpha - \int_0^1 (16 - 6.2\alpha)\cdot(2-\alpha)d\alpha\Big]$$

$$= 1.06875 - 3.583929 - 0.4642857 + 4.257143 = 1.277678.$$

Then, the dotted mid-way trapezoidal fuzzy number $\tilde{S}^\star$ seen in Fig. 4.3 is given by

$$\tilde{S}^\star = [4.682144, 5.959822, 8.515178, 9.792856].$$

Remark 4.7.3 (The Nearest Symmetrical Trapezoidal Fuzzy Number) By respect to the previously described distances (ρ_1, ρ_2, $d_{Bertoluzza}$, $d_{mid/spr}$, $d_{\phi-wabl/ldev/rdev}$ and some cases of $\delta_{p,q}^\star$) and following the same derivation procedure, the parameter s_0 is always given by the same equation seen in Eq. 4.34. As such, the parameter s_0 of the nearest trapezoidal fuzzy number of $\tilde{X}$ is its signed distance measured from the fuzzy origin given by

$$s_0 = d_{SGD}(\tilde{X}, \tilde{0}) = \frac{1}{2}\int_0^1 (\tilde{X}_\alpha^L + \tilde{X}_\alpha^R)d\alpha.$$

The proofs are analog to the one of Proposition 4.7.6. However, the parameter ϵ differs slightly between distances. A study on such differences could eventually be interesting in order to sort out the distances with the smallest value of ϵ, thus paving the way of a better approximation of the fuzzy number.

Appendix

This Appendix is devoted for the proofs of Chap. 4 as follows:

A Proofs of Chap. 4

Proof of Corollary 4.6.1 Using the definition of the metric ρ_1, we have that

$$\rho_1(\tilde{X}, \tilde{Y}) = \frac{1}{2}\int_0^1 |\tilde{X}_\alpha^L - \tilde{Y}_\alpha^L|d\alpha + \frac{1}{2}\int_0^1 |\tilde{X}_\alpha^R - \tilde{Y}_\alpha^R|d\alpha.$$

Then, if $|\tilde{X}_\alpha^L - \tilde{Y}_\alpha^L| = \tilde{X}_\alpha^L - \tilde{Y}_\alpha^L$ and $|\tilde{X}_\alpha^R - \tilde{Y}_\alpha^R| = \tilde{X}_\alpha^R - \tilde{Y}_\alpha^R$, we get the following

$$\begin{aligned}\rho_1(\tilde{X}, \tilde{Y}) &= \frac{1}{2}\int_0^1 (\tilde{X}_\alpha^L - \tilde{Y}_\alpha^L)d\alpha + \frac{1}{2}\int_0^1 (\tilde{X}_\alpha^R - \tilde{Y}_\alpha^R)d\alpha \\ &= \frac{1}{2}\int_0^1 (\tilde{X}_\alpha^L + \tilde{X}_\alpha^R)d\alpha - \frac{1}{2}\int_0^1 (\tilde{Y}_\alpha^L + \tilde{Y}_\alpha^R)d\alpha \\ &= d_{SGD}(\tilde{X}, \tilde{0}) - d_{SGD}(\tilde{Y}, \tilde{0}) \\ &= d_{SGD}(\tilde{X}, \tilde{Y}).\end{aligned}$$

Alternatively, if $|\tilde{X}_\alpha^L - \tilde{Y}_\alpha^L| = -(\tilde{X}_\alpha^L - \tilde{Y}_\alpha^L)$ and $|\tilde{X}_\alpha^R - \tilde{Y}_\alpha^R| = -(\tilde{X}_\alpha^R - \tilde{Y}_\alpha^R)$, we obtain that

$$d_{SGD}(\tilde{X}, \tilde{Y}) = -\rho_1(\tilde{X}, \tilde{Y}). \tag{4.46}$$

Furthermore, for all fuzzy numbers $\tilde{X}$ and $\tilde{Y} \in \mathbb{F}_c^\star(\mathbb{R})$, the following is always true

$$(\tilde{X}_\alpha^L - \tilde{Y}_\alpha^L) \le |\tilde{X}_\alpha^L - \tilde{Y}_\alpha^L| \text{ and } (\tilde{X}_\alpha^R - \tilde{Y}_\alpha^R) \le |\tilde{X}_\alpha^R - \tilde{Y}_\alpha^R|.$$

Since these expressions are integrable, we can write that

$$\frac{1}{2}\int_0^1 (\tilde{X}_\alpha^L - \tilde{Y}_\alpha^L)d\alpha + \frac{1}{2}\int_0^1 (\tilde{X}_\alpha^R - \tilde{Y}_\alpha^R)d\alpha \le \frac{1}{2}\int_0^1 |\tilde{X}_\alpha^L - \tilde{Y}_\alpha^L|d\alpha + \frac{1}{2}\int_0^1 |\tilde{X}_\alpha^R - \tilde{Y}_\alpha^R|d\alpha,$$

we have then that

$$d_{SGD}(\tilde{X}, \tilde{Y}) \le \rho_1(\tilde{X}, \tilde{Y}).$$

From another side, we know that the ρ_1 metric is symmetric and that the signed distance is directional, then from Proposition 4.6.2, we have that:

$$d_{SGD}(\tilde{Y}, \tilde{X}) \le \rho_1(\tilde{Y}, \tilde{X}) \quad \text{is equivalent to} \quad -d_{SGD}(\tilde{X}, \tilde{Y}) \le \rho_1(\tilde{X}, \tilde{Y}),$$

then, we conclude that

$$d_{SGD}(\tilde{X}, \tilde{Y}) \ge -\rho_1(\tilde{X}, \tilde{Y}).$$

$\square$

Proof of Proposition 4.7.1

1. Consider $\tilde{X}$ a fuzzy number of $\mathbb{F}_c^\star(\mathbb{R})$. From Properties 4.6.1, we know that the signed distance is reflexive, then $d_{SGD}(\tilde{X}, \tilde{X}) = 0$. The reflexivity of $d_{SGD}^{\theta^\star}$, i.e. $d_{SGD}^{\theta^\star}(\tilde{X}, \tilde{X}) = 0$, can then be proven as follows:

$$d_{SGD}^{\theta^\star}(\tilde{X}, \tilde{X}) = \sqrt{\left(d_{SGD}(\tilde{X}, \tilde{X})\right)^2 + \theta^\star \left(\int_0^1 \max\left(\mathrm{dev}^R \tilde{X} - \mathrm{dev}^L \tilde{X}, \mathrm{dev}^R \tilde{X} - \mathrm{dev}^L \tilde{X}\right)d\alpha\right)^2}$$

$$= \sqrt{0 + \theta^\star \left(\int_0^1 \left(\mathrm{dev}^R \tilde{X} - \mathrm{dev}^L \tilde{X}\right)d\alpha\right)^2},$$

with

$$\int_0^1 \left(\mathrm{dev}^R \tilde{X} - \mathrm{dev}^L \tilde{X}\right)d\alpha = \int_0^1 \left(\tilde{X}_\alpha^R - d_{SGD}(\tilde{X}, \tilde{0}) - d_{SGD}(\tilde{X}, \tilde{0}) + \tilde{X}_\alpha^L\right)d\alpha$$

$$= \int_0^1 \left[\tilde{X}_\alpha^L(\alpha) + \tilde{X}_\alpha^R(\alpha)\right]d\alpha - 2 \cdot \frac{1}{2} \int_0^1 \left[\tilde{X}_\alpha^L(\alpha) + \tilde{X}_\alpha^R(\alpha)\right]d\alpha = 0,$$

which induces that

$$d_{SGD}^{\theta^\star}(\tilde{X}, \tilde{X}) = 0.$$

2. By construction, it is clear that

$$d_{SGD}^{\theta^\star}(\tilde{X}, \tilde{Y}) \ge 0, \ \forall \tilde{X}, \tilde{Y} \in \mathbb{F}_c^\star(\mathbb{R}).$$

3. If $d_{SGD}^{\theta^\star}(\tilde{X}, \tilde{Y}) = 0$, then

$$\sqrt{\left(d_{SGD}(\tilde{X}, \tilde{Y})\right)^2 + \theta^\star \left(\int_0^1 \max\left(\mathrm{dev}^R \tilde{Y} - \mathrm{dev}^L \tilde{X}, \mathrm{dev}^R \tilde{X} - \mathrm{dev}^L \tilde{Y}\right)d\alpha\right)^2} = 0,$$

implying that

$$d_{SGD}(\tilde{X}, \tilde{Y}) = 0 \tag{4.47}$$

$$\text{and } \theta^\star \left(\int_0^1 \max \left(\text{dev}^R \tilde{Y} - \text{dev}^L \tilde{X}, \text{dev}^R \tilde{X} - \text{dev}^L \tilde{Y} \right) d\alpha \right)^2 = 0. \tag{4.48}$$

For Eq. 4.47, we know from Properties 4.6.1 that the distance d_{SGD} is non-degenerate. Then,

$$d_{SGD}(\tilde{X}, \tilde{Y}) = 0 \text{ also written as } d_{SGD}(\tilde{X}, \tilde{0}) = d_{SGD}(\tilde{Y}, \tilde{0}), \text{ is equivalent to } \tilde{X} = \tilde{Y}. \tag{4.49}$$

For Eq. 4.48, we suppose that $\theta^\star \neq 0$. If

$$max \left(\text{dev}^R \tilde{Y} - \text{dev}^L \tilde{X}, \text{dev}^R \tilde{X} - \text{dev}^L \tilde{Y} \right) = \text{dev}^R \tilde{Y} - \text{dev}^L \tilde{X},$$

then we get

$$\int_0^1 \left(\text{dev}^R \tilde{Y} - \text{dev}^L \tilde{X} \right) d\alpha = \int_0^1 \left(\tilde{Y}_\alpha^R(\alpha) - d_{SGD}(\tilde{Y}, \tilde{0}) - d_{SGD}(\tilde{X}, \tilde{0}) + \tilde{X}_\alpha^L(\alpha) \right) d\alpha$$

$$= \int_0^1 \left(\tilde{Y}_\alpha^R(\alpha) + \tilde{X}_\alpha^L(\alpha) \right) d\alpha - d_{SGD}(\tilde{Y}, \tilde{0}) - d_{SGD}(\tilde{X}, \tilde{0}).$$

Therefore, since $\left(\int_0^1 \max \left(\text{dev}^R \tilde{Y} - \text{dev}^L \tilde{X}, \text{dev}^R \tilde{X} - \text{dev}^L \tilde{Y} \right) d\alpha \right)^2 = 0$, we could write that

$$\int_0^1 \left(\tilde{Y}_\alpha^R(\alpha) + \tilde{X}_\alpha^L(\alpha) \right) d\alpha - d_{SGD}(\tilde{Y}, \tilde{0}) - d_{SGD}(\tilde{X}, \tilde{0}) = 0.$$

If we would like now to verify Eq. 4.48 by the inclusive results of Eq. 4.49, we could replace $d_{SGD}(\tilde{Y}, \tilde{0})$ by $d_{SGD}(\tilde{X}, \tilde{0})$. We get that

$$\int_0^1 \left(\tilde{Y}_\alpha^R(\alpha) + \tilde{X}_\alpha^L(\alpha) \right) d\alpha - d_{SGD}(\tilde{Y}, \tilde{0}) - d_{SGD}(\tilde{X}, \tilde{0})$$

$$= \int_0^1 \left(\tilde{Y}_\alpha^R(\alpha) + \tilde{X}_\alpha^L(\alpha) \right) d\alpha - 2 \cdot d_{SGD}(\tilde{X}, \tilde{0})$$

$$= \int_0^1 \left(\tilde{Y}_\alpha^R(\alpha) + \tilde{X}_\alpha^L(\alpha) \right) d\alpha - \int_0^1 \left(\tilde{X}_\alpha^L(\alpha) + \tilde{X}_\alpha^R(\alpha) \right) d\alpha$$

$$= \int_0^1 \left(\tilde{Y}_\alpha^R(\alpha) - \tilde{X}_\alpha^R(\alpha) \right) d\alpha,$$

where $\tilde{Y}_\alpha^R(\alpha) - \tilde{X}_\alpha^R(\alpha)$ is integrable and non-negative. From Eq. 4.48, since $\tilde{X} = \tilde{Y}$, then $\tilde{X}_\alpha^R(\alpha) = \tilde{Y}_\alpha^R(\alpha)$. We deduce that

$$\int_0^1 \left(\tilde{Y}_\alpha^R(\alpha) - \tilde{X}_\alpha^R(\alpha) \right) d\alpha = 0.$$

For the left hand sides of $\tilde{X}$ and $\tilde{Y}$, we replace this time $d_{SGD}(\tilde{X}, \tilde{0})$ by $d_{SGD}(\tilde{Y}, \tilde{0})$. We get that

$$\int_0^1 \left(\tilde{Y}_\alpha^R(\alpha) + \tilde{X}_\alpha^L(\alpha) \right) d\alpha - d_{SGD}(\tilde{Y}, \tilde{0}) - d_{SGD}(\tilde{X}, \tilde{0})$$

$$= \int_0^1 \left(\tilde{Y}_\alpha^R(\alpha) + \tilde{X}_\alpha^L(\alpha) \right) d\alpha - 2 \cdot d_{SGD}(\tilde{Y}, \tilde{0})$$

$$= \int_0^1 \left(\tilde{Y}_\alpha^R(\alpha) + \tilde{X}_\alpha^L(\alpha) \right) d\alpha - \int_0^1 \left(\tilde{Y}_\alpha^L(\alpha) + \tilde{Y}_\alpha^R(\alpha) \right) d\alpha$$

$$= \int_0^1 \left(\tilde{X}_\alpha^L(\alpha) - \tilde{Y}_\alpha^L(\alpha) \right) d\alpha = 0,$$

since $\tilde{X} = \tilde{Y}$ and consequently $\tilde{X}_\alpha^L(\alpha) = \tilde{Y}_\alpha^L(\alpha)$. Note that $\tilde{X}_\alpha^L(\alpha) - \tilde{Y}_\alpha^L(\alpha)$ is integrable and non-negative.

For the case where $max\left(\text{dev}^R \tilde{Y} - \text{dev}^L \tilde{X}, \text{dev}^R \tilde{X} - \text{dev}^L \tilde{Y}\right) = \text{dev}^R \tilde{X} - \text{dev}^L \tilde{Y}$, we get exactly the same result.

On the other hand, in order to confirm the property-based equivalence, we have to prove now that if $\tilde{X} = \tilde{Y}$, then $d_{SGD}^{\theta^\star}(\tilde{X}, \tilde{Y}) = 0$. It can be done by means of the following:

We know that if $\tilde{X} = \tilde{Y}$, then

$$\text{dev}^R \tilde{Y} - \text{dev}^L \tilde{X} = \text{dev}^R \tilde{X} - \text{dev}^L \tilde{Y} = \text{dev}^R \tilde{X} - \text{dev}^L \tilde{X}$$

$$= \tilde{X}_\alpha^R - d_{SGD}(\tilde{X}, \tilde{0}) - d_{SGD}(\tilde{X}, \tilde{0}) - \tilde{X}_\alpha^L$$

$$= \tilde{X}_\alpha^R - 2d_{SGD}(\tilde{X}, \tilde{0}) - \tilde{X}_\alpha^L.$$

Furthermore, we get that

$$\int_0^1 max\left(\text{dev}^R \tilde{Y} - \text{dev}^L \tilde{X}, \text{dev}^R \tilde{X} - \text{dev}^L \tilde{Y}\right) d\alpha = \int_0^1 \left(\tilde{X}_\alpha^R - 2d_{SGD}(\tilde{X}, \tilde{0}) - \tilde{X}_\alpha^L \right) d\alpha$$

$$= 2d_{SGD}(\tilde{X}, \tilde{0}) - 2d_{SGD}(\tilde{X}, \tilde{0}) \int_0^1 d\alpha = 0.$$

4. For the fuzzy numbers $\tilde{X}$ and $\tilde{Y}$, we know that d_{SGD} is partially symmetric as seen in Proposition 4.6.2, then we have that

$$d_{SGD}^{\theta^\star}(\tilde{X}, \tilde{Y}) = \sqrt{\left(d_{SGD}(\tilde{X}, \tilde{Y})\right)^2 + \theta^\star \left(\int_0^1 \max \left(\mathrm{dev}^R \tilde{Y} - \mathrm{dev}^L \tilde{X}, \mathrm{dev}^R \tilde{X} - \mathrm{dev}^L \tilde{Y} \right) d\alpha \right)^2}$$

$$= \sqrt{\left(d_{SGD}(\tilde{Y}, \tilde{X})\right)^2 + \theta^\star \left(\int_0^1 \max \left(\mathrm{dev}^R \tilde{X} - \mathrm{dev}^L \tilde{Y}, \mathrm{dev}^R \tilde{Y} - \mathrm{dev}^L \tilde{X} \right) d\alpha \right)^2}$$

$$= d_{SGD}^{\theta^\star}(\tilde{Y}, \tilde{X}).$$

5. Let $\tilde{X}$, $\tilde{Y}$, and $\tilde{Z}$ be fuzzy numbers. We want to prove the following triangular inequality

$$d_{SGD}^{\theta^\star}(\tilde{X}, \tilde{Y}) \leq d_{SGD}^{\theta^\star}(\tilde{X}, \tilde{Z}) + d_{SGD}^{\theta^\star}(\tilde{Z}, \tilde{Y}).$$

We know that

$$\left(d_{SGD}^{\theta^\star}(\tilde{X}, \tilde{Y})\right)^2 = \left(d_{SGD}(\tilde{X}, \tilde{Y})\right)^2 + \theta^\star \left(\int_0^1 \max \left(\mathrm{dev}^R \tilde{Y} - \mathrm{dev}^L \tilde{X}, \mathrm{dev}^R \tilde{X} - \mathrm{dev}^L \tilde{Y} \right) d\alpha \right)^2.$$

Let us decompose the previous expression. From the properties of the signed distance, we have that

$$d_{SGD}(\tilde{X}, \tilde{Y}) = d_{SGD}(\tilde{X}, \tilde{Z}) + d_{SGD}(\tilde{Z}, \tilde{Y}).$$

Then, we get

$$\left(d_{SGD}(\tilde{X}, \tilde{Y})\right)^2 \leq \left(d_{SGD}(\tilde{X}, \tilde{Z})\right)^2 + \left(d_{SGD}(\tilde{Z}, \tilde{Y})\right)^2,$$

since $d_{SGD}(\tilde{X}, \tilde{Z}) \cdot d_{SGD}(\tilde{Z}, \tilde{Y}) \geq 0$.
For the second part, suppose that

$$\max \left(\mathrm{dev}^R \tilde{X} - \mathrm{dev}^L \tilde{Y}, \mathrm{dev}^R \tilde{Y} - \mathrm{dev}^L \tilde{X} \right) = \left(\mathrm{dev}^R \tilde{Y} - \mathrm{dev}^L \tilde{X} \right),$$

and with $\int_0^1 \left(\mathrm{dev}^L \tilde{Z} - \mathrm{dev}^R \tilde{Z} \right) d\alpha = 0$, we have that:

$$\left(\int_0^1 \max \left(\mathrm{dev}^R \tilde{X} - \mathrm{dev}^L \tilde{Y}, \mathrm{dev}^R \tilde{Y} - \mathrm{dev}^L \tilde{X} \right) d\alpha \right)^2$$

$$= \left(\int_0^1 \left(\mathrm{dev}^R \tilde{Y} - \mathrm{dev}^L \tilde{X} \right) d\alpha \right)^2$$

$$= \left(\int_0^1 \left(\mathrm{dev}^R \tilde{Y} - \mathrm{dev}^L \tilde{X} \right) d\alpha - \int_0^1 \left(\mathrm{dev}^L \tilde{Z} - \mathrm{dev}^R \tilde{Z} \right) d\alpha \right)^2$$

$$= \left(\int_0^1 (\mathrm{dev}^R \tilde{Z} - \mathrm{dev}^L \tilde{X}) d\alpha - \int_0^1 (\mathrm{dev}^R \tilde{Y} - \mathrm{dev}^L \tilde{Z}) d\alpha \right)^2$$

$$\leq \left(\int_0^1 (\mathrm{dev}^R \tilde{Z} - \mathrm{dev}^L \tilde{X}) d\alpha \right)^2 + \left(\int_0^1 (\mathrm{dev}^R \tilde{Y} - \mathrm{dev}^L \tilde{Z}) d\alpha \right)^2. \quad (4.50)$$

Then, since the fuzzy number $\tilde{Z}$ is supposed to be an intermediate number between $\tilde{X}$ and $\tilde{Y}$, then we have that

$$\max \left(\mathrm{dev}^R \tilde{Z} - \mathrm{dev}^L \tilde{X}, \mathrm{dev}^R \tilde{X} - \mathrm{dev}^L \tilde{Z} \right) = \mathrm{dev}^R \tilde{Z} - \mathrm{dev}^L \tilde{X}$$

$$\max \left(\mathrm{dev}^R \tilde{Z} - \mathrm{dev}^L \tilde{Y}, \mathrm{dev}^R \tilde{Y} - \mathrm{dev}^L \tilde{Z} \right) = \mathrm{dev}^R \tilde{Y} - \mathrm{dev}^L \tilde{Z}.$$

From Eq. 4.50, we get the following equality:

$$\left(\int_0^1 (\mathrm{dev}^R \tilde{Z} - \mathrm{dev}^L \tilde{X}) d\alpha \right)^2 + \left(\int_0^1 (\mathrm{dev}^R \tilde{Y} - \mathrm{dev}^L \tilde{Z}) d\alpha \right)^2 \quad (4.51)$$

$$= \left(\int_0^1 \max (\mathrm{dev}^R \tilde{Z} - \mathrm{dev}^L \tilde{X}, \mathrm{dev}^R \tilde{X} - \mathrm{dev}^L \tilde{Z}) d\alpha \right)^2$$

$$+ \left(\int_0^1 \max (\mathrm{dev}^R \tilde{Y} - \mathrm{dev}^L \tilde{Z}, \mathrm{dev}^R \tilde{Z} - \mathrm{dev}^L \tilde{Y}) d\alpha \right)^2. \quad (4.52)$$

We finally deduce that for $0 \leq \theta^\star \leq 1$,

$$(d_{SGD}(\tilde{X}, \tilde{Y}))^2 + \theta^\star \left(\int_0^1 \max (\mathrm{dev}^R \tilde{X} - \mathrm{dev}^L \tilde{Y}, \mathrm{dev}^R \tilde{Y} - \mathrm{dev}^L \tilde{X}) d\alpha \right)^2$$

$$\leq (d_{SGD}(\tilde{X}, \tilde{Z}))^2 + \theta^\star \left(\int_0^1 \max (\mathrm{dev}^R \tilde{Z} - \mathrm{dev}^L \tilde{X}, \mathrm{dev}^R \tilde{X} - \mathrm{dev}^L \tilde{Z}) d\alpha \right)^2$$

$$+ (d_{SGD}(\tilde{Z}, \tilde{Y}))^2 + \theta^\star \left(\int_0^1 \max (\mathrm{dev}^R \tilde{Y} - \mathrm{dev}^L \tilde{Z}, \mathrm{dev}^R \tilde{Z} - \mathrm{dev}^L \tilde{Y}) d\alpha \right)^2$$

$$\Leftrightarrow d_{SGD}^{\theta^\star}(\tilde{X}, \tilde{Y}) \leq d_{SGD}^{\theta^\star}(\tilde{X}, \tilde{Z}) + d_{SGD}^{\theta^\star}(\tilde{Z}, \tilde{Y}), \quad (4.53)$$

since $d_{SGD}^{\theta^\star}(\tilde{X}, \tilde{Y})$, $d_{SGD}^{\theta^\star}(\tilde{X}, \tilde{Z})$, and $d_{SGD}^{\theta^\star}(\tilde{Z}, \tilde{Y})$ are positive.   □

Proof of Proposition 4.7.2

- For the translation invariance, consider the fuzzy numbers $\tilde{X}$ and $\tilde{Y} \in \mathbb{F}_c^\star(\mathbb{R})$, and $k \in \mathbb{R}$, by applying $d_{SGD}^{\theta^\star}$ on $\tilde{X} + k$ and $\tilde{Y} + k$, we have that

$$d_{SGD}^{\theta^\star} \left((\tilde{X} + k), (\tilde{Y} + k) \right) = \left[(d_{SGD}((\tilde{X} + k), (\tilde{Y} + k)))^2 \right.$$

$$\left. + \theta^\star \left(\int_0^1 \max (\mathrm{dev}^R(\tilde{Y} + k) - \mathrm{dev}^L(\tilde{X} + k), \mathrm{dev}^R(\tilde{X} + k) - \mathrm{dev}^L(\tilde{Y} + k)) d\alpha \right)^2 \right]^{\frac{1}{2}}.$$

From Properties 4.6.1, since d_{SGD} is translation invariant, i.e. $d_{SGD}\big((\tilde{X} + k), (\tilde{Y} + k)\big) = d_{SGD}\big(\tilde{X}, \tilde{Y}\big)$, the deviations in the shapes of $\tilde{X} + k$ and $\tilde{Y} + k$ are then

$$\mathrm{dev}^L(\tilde{X} + k) = d_{SGD}(\tilde{X} + k, \tilde{0}) - (\tilde{X} + k)_\alpha^L = d_{SGD}(\tilde{X}, \tilde{0}) + k - \tilde{X}_\alpha^L - k = \mathrm{dev}^L \tilde{X},$$

$$\mathrm{dev}^R(\tilde{X} + k) = (\tilde{X} + k)_\alpha^R - d_{SGD}(\tilde{X} + k, \tilde{0}) = \tilde{X}_\alpha^R + k - d_{SGD}(\tilde{X}, \tilde{0}) - k = \mathrm{dev}^R \tilde{X}.$$

It is the same for $\mathrm{dev}^L(\tilde{Y} + k)$ and $\mathrm{dev}^R(\tilde{Y} + k)$. Then, we directly get that

$$d_{SGD}^{\theta^\star}\big((\tilde{X} + k), (\tilde{Y} + k)\big) = d_{SGD}^{\theta^\star}\big(\tilde{X}, \tilde{Y}\big).$$

- For the scale invariance, let $\tilde{X}$ and $\tilde{Y}$ be two fuzzy numbers in $\mathbb{F}_c^\star(\mathbb{R})$, and $k \in \mathbb{R}$. The $d_{SGD}^{\theta^\star}$ distance between $(k \cdot \tilde{X})$ and $(k \cdot \tilde{Y})$ is written as

$$d_{SGD}^{\theta^\star} \quad \big((k \cdot \tilde{X}), (k \cdot \tilde{Y})\big) = \Big[\big(d_{SGD}((k \cdot \tilde{X}), (k \cdot \tilde{Y}))\big)^2$$
$$+ \theta^\star \Big(\int_0^1 \max \big(\mathrm{dev}^R(k \cdot \tilde{Y}) - \mathrm{dev}^L(k \cdot \tilde{X}), \mathrm{dev}^R(k \cdot \tilde{X}) - \mathrm{dev}^L(k \cdot \tilde{Y})\big) d\alpha\Big)^2\Big]^{\frac{1}{2}}.$$

We have that if $k \geq 0$,

$$\mathrm{dev}^L(k \cdot \tilde{X}) = d_{SGD}(k \cdot \tilde{X}, \tilde{0}) - (k \cdot \tilde{X})_\alpha^L = k \cdot d_{SGD}(\tilde{X}, \tilde{0}) - k \cdot \tilde{X}_\alpha^L = k \cdot \mathrm{dev}^L \tilde{X}, \quad (4.54)$$

$$\mathrm{dev}^R(k \cdot \tilde{X}) = (k \cdot \tilde{X})_\alpha^R - d_{SGD}(k \cdot \tilde{X}, \tilde{0}) = k \cdot \tilde{X}_\alpha^R - k \cdot d_{SGD}(\tilde{X}, \tilde{0}) = k \cdot \mathrm{dev}^R \tilde{X}. \quad (4.55)$$

If $k < 0$, the case is much more complicated. According to the negative value of k, a rotation between the left and right sides of the fuzzy number arises. A subtle consequence related to the α-cuts of the obtained fuzzy number is as follows:

$$(k \cdot \tilde{X})_\alpha^L = k \cdot \tilde{X}_\alpha^R \quad \text{and} \quad (k \cdot \tilde{X})_\alpha^R = k \cdot \tilde{X}_\alpha^L.$$

Then, we get that:

$$\mathrm{dev}^L(k \cdot \tilde{X}) = d_{SGD}(k \cdot \tilde{X}, \tilde{0}) - (k \cdot \tilde{X})_\alpha^L = k \cdot d_{SGD}(\tilde{X}, \tilde{0}) - k \cdot \tilde{X}_\alpha^R = -k \cdot \mathrm{dev}^R \tilde{X}, \quad (4.56)$$

$$\mathrm{dev}^R(k \cdot \tilde{X}) = (k \cdot \tilde{X})_\alpha^R - d_{SGD}(k \cdot \tilde{X}, \tilde{0}) = k \cdot \tilde{X}_\alpha^L - k \cdot d_{SGD}(\tilde{X}, \tilde{0}) = -k \cdot \mathrm{dev}^L \tilde{X}. \quad (4.57)$$

According to both conditions on k, i.e. $k \geq 0$ and $k < 0$, the scale invariance property written by

$$d_{SGD}^{\theta^\star}\big((k \cdot \tilde{X}), (k \cdot \tilde{Y})\big) = |k| \cdot d_{SGD}^{\theta^\star}\big(\tilde{X}, \tilde{Y}\big), \; \forall k \in \mathbb{R},$$

can then be easily deduced.

$\square$

Proof of Proposition 4.7.3 For $\theta^\star \neq 0$, we have that

$$d^{\theta^\star}_{SGD}(\tilde{X}, \tilde{Y}) = |d_{SGD}(\tilde{X}, \tilde{Y})| \Leftrightarrow \int_0^1 \max\left(\mathrm{dev}^R \tilde{Y} - \mathrm{dev}^L \tilde{X}, \mathrm{dev}^R \tilde{X} - \mathrm{dev}^L \tilde{Y}\right) d\alpha = 0$$

$$\Leftrightarrow \int_0^1 \left(\mathrm{dev}^R \tilde{Y} - \mathrm{dev}^L \tilde{X}\right) d\alpha = 0$$

$$\text{or } \int_0^1 \left(\mathrm{dev}^R \tilde{X} - \mathrm{dev}^L \tilde{Y}\right) d\alpha = 0.$$

Suppose the case for which $\int_0^1 \left(\mathrm{dev}^R \tilde{Y} - \mathrm{dev}^L \tilde{X}\right) d\alpha = 0$. Then, we get

$$\int_0^1 \left(\mathrm{dev}^R \tilde{Y} - \mathrm{dev}^L \tilde{X}\right) d\alpha = 0 \Leftrightarrow \int_0^1 \left(\tilde{Y}^R_\alpha - d_{SGD}(\tilde{Y}, \tilde{0}) - d_{SGD}(\tilde{X}, \tilde{0}) + \tilde{X}^L_\alpha\right) d\alpha = 0$$

$$\Leftrightarrow \frac{1}{2} \int_0^1 \left(\tilde{X}^L_\alpha - \tilde{X}^R_\alpha\right) d\alpha = \frac{1}{2} \int_0^1 \left(\tilde{Y}^L_\alpha - \tilde{Y}^R_\alpha\right) d\alpha$$

$$\Leftrightarrow spr(\tilde{X}) = spr(\tilde{Y}).$$

The result is the same for the case $\int_0^1 \left(\mathrm{dev}^R \tilde{X} - \mathrm{dev}^L \tilde{Y}\right) d\alpha = 0$.
We have now to prove that if $spr(\tilde{X}) = spr(\tilde{Y})$, then $d^{\theta^\star}_{SGD}(\tilde{X}, \tilde{Y}) = |d_{SGD}(\tilde{X}, \tilde{Y})|$.
In other words, we would like to prove that if $spr(\tilde{X}) = spr(\tilde{Y})$, then $\int_0^1 \max\left(\mathrm{dev}^R \tilde{Y} - \mathrm{dev}^L \tilde{X}, \mathrm{dev}^R \tilde{X} - \mathrm{dev}^L \tilde{Y}\right) d\alpha = 0$.

$$spr(\tilde{X}) = spr(\tilde{Y}) \Leftrightarrow \frac{1}{2} \int_0^1 \left(\tilde{X}^L_\alpha - \tilde{X}^R_\alpha\right) d\alpha = \frac{1}{2} \int_0^1 \left(\tilde{Y}^L_\alpha - \tilde{Y}^R_\alpha\right) d\alpha$$

$$\Leftrightarrow \frac{1}{2} \int_0^1 \left(\tilde{X}^L_\alpha - \tilde{X}^R_\alpha\right) d\alpha - \frac{1}{2} \int_0^1 \tilde{X}^L_\alpha d\alpha + \frac{1}{2} \int_0^1 \tilde{X}^L_\alpha d\alpha$$

$$- \frac{1}{2} \int_0^1 \left(\tilde{Y}^L_\alpha - \tilde{Y}^R_\alpha\right) d\alpha + \frac{1}{2} \int_0^1 \tilde{Y}^R_\alpha d\alpha - \frac{1}{2} \int_0^1 \tilde{Y}^R_\alpha d\alpha = 0$$

$$\Leftrightarrow \frac{1}{2} \int_0^1 \tilde{X}^L_\alpha d\alpha - \frac{1}{2} \int_0^1 \tilde{X}^R_\alpha d\alpha - \frac{1}{2} \int_0^1 \tilde{X}^L_\alpha d\alpha + \frac{1}{2} \int_0^1 \tilde{X}^R_\alpha d\alpha$$

$$- \frac{1}{2} \int_0^1 \tilde{Y}^L_\alpha d\alpha + \frac{1}{2} \int_0^1 \tilde{Y}^R_\alpha d\alpha + \frac{1}{2} \int_0^1 \tilde{Y}^R_\alpha d\alpha - \frac{1}{2} \int_0^1 \tilde{Y}^R_\alpha d\alpha = 0$$

$$\Leftrightarrow \int_0^1 \tilde{X}^L_\alpha d\alpha - \frac{1}{2} \int_0^1 \left(\tilde{X}^L_\alpha + \tilde{X}^R_\alpha\right) d\alpha - \frac{1}{2} \int_0^1 \left(\tilde{Y}^L_\alpha + \tilde{Y}^R_\alpha\right) d\alpha + \int_0^1 \tilde{Y}^R_\alpha d\alpha = 0$$

$$\Leftrightarrow \int_0^1 \left(\tilde{X}^L_\alpha - d_{SGD}(\tilde{X}, \tilde{0}) - d_{SGD}(\tilde{Y}, \tilde{0}) + \tilde{Y}^R_\alpha\right) d\alpha = 0$$

$$\Leftrightarrow \int_0^1 (\mathrm{dev}^R \tilde{Y} - \mathrm{dev}^L \tilde{X})\mathrm{d}\alpha = 0$$

$$\Leftrightarrow \int_0^1 \max (\mathrm{dev}^R \tilde{Y} - \mathrm{dev}^L \tilde{X}, \mathrm{dev}^R \tilde{X} - \mathrm{dev}^L \tilde{Y})\mathrm{d}\alpha = 0,$$

with $|d_{SGD}(\tilde{X}, \tilde{0})| \leq |d_{SGD}(\tilde{Y}, \tilde{0})|$.

For the case where $\int_0^1 \max (\mathrm{dev}^R \tilde{Y} - \mathrm{dev}^L \tilde{X}, \mathrm{dev}^R \tilde{X} - \mathrm{dev}^L \tilde{Y})\mathrm{d}\alpha = \mathrm{dev}^R \tilde{X} - \mathrm{dev}^L \tilde{Y}$, the result can be extended. $\qquad\qquad\qquad\qquad\qquad\qquad\qquad\square$

Proof of Proposition 4.7.4 By applying the Jensen's inequality in the following expression, we have that:

$$\big(d_{SGD}(\tilde{X}, \tilde{Y})\big)^2 = \big(d_{SGD}(\tilde{X}, \tilde{0}) - d_{SGD}(\tilde{Y}, \tilde{0})\big)^2$$

$$= \Big(\frac{1}{2}\int_0^1 (\tilde{X}_\alpha^L + \tilde{X}_\alpha^R)\mathrm{d}\alpha - \frac{1}{2}\int_0^1 (\tilde{Y}_\alpha^L + \tilde{Y}_\alpha^R)\mathrm{d}\alpha\Big)^2$$

$$= \Big(\frac{1}{2}\int_0^1 (\tilde{X}_\alpha^L - \tilde{Y}_\alpha^L)\mathrm{d}\alpha + \frac{1}{2}\int_0^1 (\tilde{X}_\alpha^R - \tilde{Y}_\alpha^R)\mathrm{d}\alpha\Big)^2$$

$$\leq \frac{1}{2}\Big(\int_0^1 \Big(\frac{1}{2}(\tilde{X}_\alpha^L - \tilde{Y}_\alpha^L)^2 + \frac{1}{2}(\tilde{X}_\alpha^R - \tilde{Y}_\alpha^R)^2\Big)\mathrm{d}\alpha\Big) = \frac{1}{2}\big(\rho_2(\tilde{X}, \tilde{Y})\big)^2.$$

Suppose now that

$$\int_0^1 \max (\mathrm{dev}^R \tilde{Y} - \mathrm{dev}^L \tilde{X}, \mathrm{dev}^R \tilde{X} - \mathrm{dev}^L \tilde{Y})\mathrm{d}\alpha = \int_0^1 (\mathrm{dev}^R \tilde{Y} - \mathrm{dev}^L \tilde{X})\mathrm{d}\alpha.$$

Then, we have that

$$\int_0^1 \max (\mathrm{dev}^R \tilde{Y} - \mathrm{dev}^L \tilde{X}, \mathrm{dev}^R \tilde{X} - \mathrm{dev}^L \tilde{Y})\mathrm{d}\alpha = \int_0^1 (\tilde{Y}_\alpha^R - d_{SGD}(\tilde{Y}, \tilde{0}) - d_{SGD}(\tilde{X}, \tilde{0}) + \tilde{X}_\alpha^L)\mathrm{d}\alpha$$

$$= \frac{1}{2}\int_0^1 (\tilde{X}_\alpha^L - \tilde{Y}_\alpha^L)\mathrm{d}\alpha - \frac{1}{2}\int_0^1 (\tilde{X}_\alpha^R - \tilde{Y}_\alpha^R)\mathrm{d}\alpha.$$

It follows that

$$\Big(\int_0^1 (\mathrm{dev}^R \tilde{Y} - \mathrm{dev}^L \tilde{X})\mathrm{d}\alpha\Big)^2 = \Big(\frac{1}{2}\int_0^1 (\tilde{X}_\alpha^L - \tilde{Y}_\alpha^L)\mathrm{d}\alpha - \frac{1}{2}\int_0^1 (\tilde{X}_\alpha^R - \tilde{Y}_\alpha^R)\mathrm{d}\alpha\Big)^2$$

$$\leq \frac{1}{2}\Big(\frac{1}{2}\int_0^1 (\tilde{X}_\alpha^L - \tilde{Y}_\alpha^L)^2\mathrm{d}\alpha + \frac{1}{2}\int_0^1 (\tilde{X}_\alpha^R - \tilde{Y}_\alpha^R)^2\mathrm{d}\alpha\Big)$$

$$= \frac{1}{2}\big(\rho_2(\tilde{X}, \tilde{Y})\big)^2.$$

Then, we get the following inequality:

$$d_{SGD}^{\theta^\star}(\tilde{X}, \tilde{Y}) = \sqrt{\left(d_{SGD}(\tilde{X}, \tilde{Y})\right)^2 + \theta^\star \left(\int_0^1 \max\left(\mathrm{dev}^R \tilde{Y} - \mathrm{dev}^L \tilde{X}, \mathrm{dev}^R \tilde{X} - \mathrm{dev}^L \tilde{Y}\right)\mathrm{d}\alpha\right)^2}$$

$$\leq \sqrt{\frac{1}{2}\left(\rho_2(\tilde{X}, \tilde{Y})\right)^2 + \frac{1}{2}\theta^\star \left(\rho_2(\tilde{X}, \tilde{Y})\right)^2}$$

$$\leq \sqrt{\frac{1}{2} + \frac{\theta^\star}{2}} \rho_2(\tilde{X}, \tilde{Y}) \leq \rho_2(\tilde{X}, \tilde{Y}),$$

since $\sqrt{\frac{1}{2} + \frac{\theta^\star}{2}} \leq 1$ for $\theta^\star \leq 1$.

It is the same for the case $\int_0^1 \max\left(\mathrm{dev}^R \tilde{Y} - \mathrm{dev}^L \tilde{X}, \mathrm{dev}^R \tilde{X} - \mathrm{dev}^L \tilde{Y}\right)\mathrm{d}\alpha = \int_0^1 \left(\mathrm{dev}^R \tilde{X} - \mathrm{dev}^L \tilde{Y}\right)\mathrm{d}\alpha$.

For the lower bound of the inequality, we have from Corollary 4.6.1 that $d_{SGD}(\tilde{X}, \tilde{Y}) \geq -\rho_1(\tilde{X}, \tilde{Y})$. Then,

$$d_{SGD}^{\theta^\star}(\tilde{X}, \tilde{Y}) = \sqrt{\left(d_{SGD}(\tilde{X}, \tilde{Y})\right)^2 + \theta^\star \left(\int_0^1 \max\left(\mathrm{dev}^R \tilde{Y} - \mathrm{dev}^L \tilde{X}, \mathrm{dev}^R \tilde{X} - \mathrm{dev}^L \tilde{Y}\right)\mathrm{d}\alpha\right)^2}$$

$$\geq \sqrt{\left(-\rho_1(\tilde{X}, \tilde{Y})\right)^2 + \theta^\star \left(\int_0^1 \max\left(\mathrm{dev}^R \tilde{Y} - \mathrm{dev}^L \tilde{X}, \mathrm{dev}^R \tilde{X} - \mathrm{dev}^L \tilde{Y}\right)\mathrm{d}\alpha\right)^2}$$

$$\geq \rho_1(\tilde{X}, \tilde{Y}), \tag{4.58}$$

since $\theta^\star \left(\int_0^1 \max\left(\mathrm{dev}^R \tilde{Y} - \mathrm{dev}^L \tilde{X}, \mathrm{dev}^R \tilde{X} - \mathrm{dev}^L \tilde{Y}\right)\mathrm{d}\alpha\right)^2 \geq 0$ and $\rho_1(\tilde{X}, \tilde{Y}) \geq 0$.
□

Proof of Proposition 4.7.5

1. It is sufficient to prove that $d_{SGD}^{\theta^\star} \leq d_{mid/spr}$ to directly conclude that $d_{SGD}^{\theta^\star} \leq d_{Bertoluzza}$. Thus, we will subsequently prove the first inequality only.

   Since the signed distance measured from the fuzzy origin coincides with the mid measure as seen in Remark 4.6.1, and according to Definition 4.5.1, the idea is to compare

   $$\int_0^1 \left(\mathrm{spr}\tilde{X} - \mathrm{spr}\tilde{Y}\right)^2 \mathrm{d}\alpha \text{ to } \theta^\star \left(\int_0^1 \max\left(\mathrm{dev}^R \tilde{Y} - \mathrm{dev}^L \tilde{X}, \mathrm{dev}^R \tilde{X} - \mathrm{dev}^L \tilde{Y}\right)\mathrm{d}\alpha\right)^2 \text{ with } 0 \leq \theta^\star \leq 1.$$

   Suppose that

   $$\theta^\star \left(\int_0^1 \max\left(\mathrm{dev}^R \tilde{Y} - \mathrm{dev}^L \tilde{X}, \mathrm{dev}^R \tilde{X} - \mathrm{dev}^L \tilde{Y}\right)\mathrm{d}\alpha\right)^2 = \theta^\star \left(\int_0^1 \left(\mathrm{dev}^R \tilde{Y} - \mathrm{dev}^L \tilde{X}\right)\mathrm{d}\alpha\right)^2.$$

Accordingly, the proof can be done as follows:

$$\theta^{\star}\left(\int_0^1 (\mathrm{dev}^R \tilde{Y} - \mathrm{dev}^L \tilde{X})\mathrm{d}\alpha\right)^2 = \theta^{\star}\left(\int_0^1 (\tilde{Y}_{\alpha}^R - d_{SGD}(\tilde{Y},\tilde{0}) - d_{SGD}(\tilde{X},\tilde{0}) + \tilde{X}_{\alpha}^L)\mathrm{d}\alpha\right)^2$$

$$= \theta^{\star}\left(\int_0^1 (\frac{1}{2}(\tilde{Y}_{\alpha}^R - \tilde{Y}_{\alpha}^L) - \frac{1}{2}(\tilde{X}_{\alpha}^R - \tilde{X}_{\alpha}^L))\mathrm{d}\alpha\right)^2$$

$$\leq \theta^{\star}\int_0^1 \left(\frac{1}{2}(\tilde{Y}_{\alpha}^R - \tilde{Y}_{\alpha}^L) - \frac{1}{2}(\tilde{X}_{\alpha}^R - \tilde{X}_{\alpha}^L)\right)^2 \mathrm{d}\alpha$$

$$\leq \theta^{\star}\int_0^1 (\mathrm{spr}\tilde{Y} - \mathrm{spr}\tilde{X})^2 \mathrm{d}\alpha$$

$$\leq \int_0^1 (\mathrm{spr}\tilde{X} - \mathrm{spr}\tilde{Y})^2 \mathrm{d}\alpha.$$

We then conclude that

$$d_{SGD}^{\theta^{\star}}(\tilde{X}, \tilde{Y}) \leq d_{mid/spr}(\tilde{X}, \tilde{Y}).$$

2. We would like to compare now $\int_0^1 \left(\frac{1}{2}(\mathrm{ldev}^{\phi}\tilde{X} - \mathrm{ldev}^{\phi}\tilde{Y})^2 + \frac{1}{2}(\mathrm{rdev}^{\phi}\tilde{X} - \mathrm{rdev}^{\phi}\tilde{Y})^2\right)\mathrm{d}\alpha$ to $\left(\int_0^1 \max(\mathrm{dev}^R\tilde{Y} - \mathrm{dev}^L\tilde{X}, \mathrm{dev}^R\tilde{X} - \mathrm{dev}^L\tilde{Y})\mathrm{d}\alpha\right)^2$, when ϕ is Lebesgue measured. Suppose that $\left(\int_0^1 \max(\mathrm{dev}^R\tilde{Y} - \mathrm{dev}^L\tilde{X}, \mathrm{dev}^R\tilde{X} - \mathrm{dev}^L\tilde{Y})\mathrm{d}\alpha\right)^2 = \left(\int_0^1 (\mathrm{dev}^R\tilde{Y} - \mathrm{dev}^L\tilde{X})\mathrm{d}\alpha\right)^2$. We note that by decomposing $d_{SGD}(\tilde{X}, \tilde{0})$, we can easily prove the following

$$\int_0^1 \left(d_{SGD}(\tilde{X},\tilde{0}) - d_{SGD}(\tilde{X},\tilde{0})\right)\mathrm{d}\alpha = \int_0^1 \left(\frac{1}{2}\mathrm{dev}^L\tilde{X} - \frac{1}{2}\mathrm{dev}^R\tilde{X}\right)\mathrm{d}\alpha = 0.$$

Then, from Remark 4.7.1, we have that:

$$\left(\int_0^1 (\mathrm{dev}^R\tilde{Y} - \mathrm{dev}^L\tilde{X})\mathrm{d}\alpha\right)^2 = \left(\int_0^1 (\mathrm{dev}^R\tilde{Y} + d_{SGD}(\tilde{Y},\tilde{0}) - d_{SGD}(\tilde{Y},\tilde{0}) - \mathrm{dev}^L\tilde{X}\right.$$

$$\left. + d_{SGD}(\tilde{X},\tilde{0}) - d_{SGD}(\tilde{X},\tilde{0}))\mathrm{d}\alpha\right)^2$$

$$= \left(\int_0^1 (\mathrm{dev}^R\tilde{Y} + \frac{1}{2}\mathrm{dev}^L\tilde{Y} - \frac{1}{2}\mathrm{dev}^R\tilde{Y} - \mathrm{dev}^L\tilde{X} + \frac{1}{2}\mathrm{dev}^L\tilde{X} - \frac{1}{2}\mathrm{dev}^R\tilde{X})\mathrm{d}\alpha\right)^2$$

$$= \frac{1}{4}\left(\int_0^1 (\mathrm{dev}^L\tilde{Y} - \mathrm{dev}^L\tilde{X}) + (\mathrm{dev}^R\tilde{Y} - \mathrm{dev}^R\tilde{X})\mathrm{d}\alpha\right)^2$$

$$\leq \frac{1}{2}\int_0^1 \left(\frac{1}{2}(\mathrm{dev}^L\tilde{Y} - \mathrm{dev}^L\tilde{X})^2 + \frac{1}{2}(\mathrm{dev}^R\tilde{Y} - \mathrm{dev}^R\tilde{X})^2\right)\mathrm{d}\alpha$$

$$= \frac{1}{2}\int_0^1 \left(\frac{1}{2}(\mathrm{ldev}^{\phi}\tilde{X} - \mathrm{ldev}^{\phi}\tilde{Y})^2 + \frac{1}{2}(\mathrm{rdev}^{\phi}\tilde{X} - \mathrm{rdev}^{\phi}\tilde{Y})^2\right)\mathrm{d}\alpha.$$

Finally, since

$$\left[d_{SGD}(\tilde{X}, \tilde{0}) - d_{SGD}(\tilde{Y}, \tilde{0}) \right]^2 = \left[\text{wabl}^\phi(\tilde{X}) - \text{wabl}^\phi(\tilde{Y}) \right]^2,$$

we can deduce that

$$d_{SGD}^{\theta^\star}(\tilde{X}, \tilde{Y}) \leq d_{\phi-wabl//ldev/rdev}(\tilde{X}, \tilde{Y}).$$

$\square$

Proof of Proposition 4.7.6 The nearest trapezoidal fuzzy number to a given fuzzy number $\tilde{X}$ denoted by $\tilde{S}$ is given by $[s_0 - 2\epsilon, s_0 - \epsilon, s_0 + \epsilon, s_0 + 2\epsilon]$. Its α-cuts are written as $\tilde{S}_\alpha = [s_0 - 2\epsilon + \epsilon\alpha, \ s_0 + 2\epsilon - \epsilon\alpha]$. Its signed distance is $d_{SGD}(\tilde{S}, \tilde{0}) = s_0$. In addition, the deviations in the shape are as follows:

$$\text{dev}^L \tilde{S} = d_{SGD}(\tilde{S}, \tilde{0}) - \tilde{S}_\alpha^L = 2\epsilon - \epsilon\alpha,$$
$$\text{dev}^R \tilde{S} = \tilde{S}_\alpha^R - d_{SGD}(\tilde{S}, \tilde{0}) = 2\epsilon - \epsilon\alpha.$$

Suppose now that

$$\max\left(\text{dev}^R \tilde{S} - \text{dev}^L \tilde{X}, \text{dev}^R \tilde{X} - \text{dev}^L \tilde{S} \right) = \text{dev}^R \tilde{S} - \text{dev}^L \tilde{X}.$$

By respect to the metric $d_{SGD}^{\theta^\star}$, the parameters s_0 and ϵ can be found as follows:

$$d_{SGD}^{\theta^\star}(\tilde{X}, \tilde{S}) = \sqrt{ \left(d_{SGD}(\tilde{X}, \tilde{S}) \right)^2 + \theta^\star \left(\int_0^1 \max\left(\text{dev}^R \tilde{S} - \text{dev}^L \tilde{X}, \text{dev}^R \tilde{X} - \text{dev}^L \tilde{S} \right) d\alpha \right)^2 }$$

$$= \sqrt{ \left(d_{SGD}(\tilde{X}, \tilde{S}) \right)^2 + \theta^\star \left(\int_0^1 \left(\text{dev}^R \tilde{S} - \text{dev}^L \tilde{X} \right) d\alpha \right)^2 }$$

$$= \sqrt{ \left(d_{SGD}(\tilde{X}, \tilde{0}) - s_0 \right)^2 + \theta^\star \left(\int_0^1 \left(2\epsilon - \epsilon\alpha - d_{SGD}(\tilde{X}, \tilde{0}) + \tilde{X}_\alpha^L \right) d\alpha \right)^2 }.$$

Finding the nearest fuzzy number means to find the optimum of $d_{SGD}^{\theta^\star}$ for s_0 and ϵ. They can be calculated by solving the equations based on the derivatives of $d_{SGD}^{\theta^\star}(\tilde{X}, \tilde{S})$ over s_0 and ϵ. We get

$$\frac{\partial d_{SGD}^{\theta^\star}(\tilde{X}, \tilde{S})}{\partial s_0} = 0 \Leftrightarrow \frac{1}{2}(d_{SGD}^{\theta^\star}(\tilde{X}, \tilde{S}))^{\frac{1}{2}-1} \left[-2 \cdot \left(d_{SGD}(\tilde{X}, \tilde{0}) - s_0 \right) \right] = 0$$

$$\Leftrightarrow s_0 = d_{SGD}(\tilde{X}, \tilde{0}).$$

Similarly, for $\theta^\star \neq 0$ and $d_{SGD}^{\theta^\star}(\tilde{X}, \tilde{S}) \neq 0$, we have

$$\frac{\partial d_{SGD}^{\theta^\star}(\tilde{X}, \tilde{S})}{\partial \epsilon} = 0 \Leftrightarrow \frac{1}{2}(d_{SGD}^{\theta^\star}(\tilde{X}, \tilde{S}))^{\frac{1}{2}-1}\theta^\star\left[2\int_0^1 (2-\alpha)\cdot(2\epsilon - \epsilon\alpha - d_{SGD}(\tilde{X}, \tilde{0}) + \tilde{X}_\alpha^L)d\alpha\right] = 0$$

$$\Leftrightarrow \epsilon = \frac{9}{14}d_{SGD}(\tilde{X}, \tilde{0}) - \frac{3}{7}\int_0^1 \tilde{X}_\alpha^L(2-\alpha)d\alpha.$$

For $\max\left(\text{dev}^R\tilde{S} - \text{dev}^L\tilde{X}, \text{dev}^R\tilde{X} - \text{dev}^L\tilde{S}\right) = \text{dev}^R\tilde{X} - \text{dev}^L\tilde{S}$, the same result can be verified. $\qquad\square$

Proof of Proposition 4.7.7 Consider now two fuzzy numbers $\tilde{X}$ and $\tilde{Y}$. The aim is to find the nearest mid-way symmetrical trapezoidal fuzzy number by respect to the $d_{SGD}^{\theta^\star}$. This fuzzy number is designated by $\tilde{S}$ such that $\tilde{S} = [s_0 - 2\epsilon, s_0 - \epsilon, s_0 + \epsilon, s_0 + 2\epsilon]$, with its α-cuts such that $\tilde{S}_\alpha = [s_0 - 2\epsilon + \epsilon\alpha, s_0 + 2\epsilon - \epsilon\alpha]$. The signed distance and the deviations in the shape of $\tilde{S}$ are given in the previous proof.

The parameters s_0 and ϵ can be found by solving the derivatives of the distances over these parameters. These latter can be found as follows:

$$d_{SGD}^{\theta^\star}(\tilde{X}, \tilde{Y}) = d_{SGD}^{\theta^\star}(\tilde{X}, \tilde{S}) + d_{SGD}^{\theta^\star}(\tilde{S}, \tilde{Y})$$

$$= \sqrt{\left(d_{SGD}(\tilde{X}, \tilde{S})\right)^2 + \theta^\star\left(\int_0^1 \max\left(\text{dev}^R\tilde{S} - \text{dev}^L\tilde{X}, \text{dev}^R\tilde{X} - \text{dev}^L\tilde{S}\right)d\alpha\right)^2}$$

$$+ \sqrt{\left(d_{SGD}(\tilde{S}, \tilde{Y})\right)^2 + \theta^\star\left(\int_0^1 \max\left(\text{dev}^R\tilde{Y} - \text{dev}^L\tilde{S}, \text{dev}^R\tilde{S} - \text{dev}^L\tilde{Y}\right)d\alpha\right)^2}$$

$$= \sqrt{\left(d_{SGD}(\tilde{X}, \tilde{S})\right)^2 + \theta^\star\left(\int_0^1 (\text{dev}^R\tilde{S} - \text{dev}^L\tilde{X})d\alpha\right)^2}$$

$$+ \sqrt{\left(d_{SGD}(\tilde{S}, \tilde{Y})\right)^2 + \theta^\star\left(\int_0^1 (\text{dev}^R\tilde{Y} - \text{dev}^L\tilde{S})d\alpha\right)^2},$$

with

$$\max\left(\text{dev}^R\tilde{S} - \text{dev}^L\tilde{X}, \text{dev}^R\tilde{X} - \text{dev}^L\tilde{S}\right) = \text{dev}^R\tilde{S} - \text{dev}^L\tilde{X}$$

$$\text{and } \max\left(\text{dev}^R\tilde{S} - \text{dev}^L\tilde{Y}, \text{dev}^R\tilde{Y} - \text{dev}^L\tilde{S}\right) = \text{dev}^R\tilde{Y} - \text{dev}^L\tilde{S},$$

since $\tilde{S}$ is seen as an intermediate fuzzy number between $\tilde{X}$ and $\tilde{Y}$. Furthermore, the fuzzy number $\tilde{S}$ is in the mid-way between $\tilde{X}$ and $\tilde{Y}$. It follows that

$$d_{SGD}^{\theta^\star}(\tilde{X}, \tilde{S}) = d_{SGD}^{\theta^\star}(\tilde{S}, \tilde{Y}).$$

Suppose that the same weight $\theta^\star$ is considered. Then, we have that:

$$\frac{\partial d_{SGD}^{\theta^\star}(\tilde{X}, \tilde{Y})}{\partial s_0} = 0 \Leftrightarrow \frac{\partial d_{SGD}^{\theta^\star}(\tilde{X}, \tilde{S})}{\partial s_0} + \frac{\partial d_{SGD}^{\theta^\star}(\tilde{Y}, \tilde{S})}{\partial s_0} = 0$$

$$\Leftrightarrow \frac{1}{2} \cdot (d_{SGD}^{\theta^\star}(\tilde{X}, \tilde{S}))^{\frac{1}{2}-1} \cdot \left[-2 \cdot \left(d_{SGD}(\tilde{X}, \tilde{0}) - s_0 \right) \right]$$

$$+ \frac{1}{2} \cdot (d_{SGD}^{\theta^\star}(\tilde{Y}, \tilde{S}))^{\frac{1}{2}-1} \cdot \left[-2 \cdot \left(d_{SGD}(\tilde{Y}, \tilde{0}) - s_0 \right) \right] = 0$$

$$\Leftrightarrow s_0 = \frac{1}{2} d_{SGD}(\tilde{X}, \tilde{0}) + \frac{1}{2} d_{SGD}(\tilde{Y}, \tilde{0}).$$

For the parameter ϵ, we have that:

$$\frac{\partial d_{SGD}^{\theta^\star}(\tilde{X}, \tilde{S})}{\partial \epsilon} = 0 \Leftrightarrow \frac{\partial d_{SGD}^{\theta^\star}(\tilde{X}, \tilde{S})}{\partial \epsilon} + \frac{\partial d_{SGD}^{\theta^\star}(\tilde{Y}, \tilde{S})}{\partial \epsilon} = 0$$

$$\Leftrightarrow \frac{1}{2}(d_{SGD}^{\theta^\star}(\tilde{X}, \tilde{S}))^{\frac{1}{2}-1}\theta^\star \left[2 \int_0^1 (2 - \alpha) \cdot \left(2\epsilon - \epsilon\alpha - d_{SGD}(\tilde{X}, \tilde{0}) + \tilde{X}_\alpha^L \right) d\alpha \right]$$

$$+ \frac{1}{2}(d_{SGD}^{\theta^\star}(\tilde{Y}, \tilde{S}))^{\frac{1}{2}-1}\theta^\star \left[2 \int_0^1 (-2 + \alpha) \cdot \left(\tilde{Y}_\alpha^R - d_{SGD}(\tilde{Y}, \tilde{0}) - 2\epsilon + \epsilon\alpha \right) d\alpha \right] = 0$$

$$\Leftrightarrow \epsilon = \frac{9}{28} d_{SGD}\left(\tilde{X}, \tilde{0} \right) - \frac{9}{28} d_{SGD}\left(\tilde{Y}, \tilde{0} \right) - \frac{3}{14}\left[\int_0^1 \tilde{X}_\alpha^L (2 - \alpha) d\alpha - \int_0^1 \tilde{Y}_\alpha^R (2 - \alpha) d\alpha \right].$$

$\square$

References

Abbasbandy, S., & Asady, B. (2006). Ranking of fuzzy numbers by sign distance. *Information Sciences, 176*(16), 2405. issn: 0020-0255. https://doi.org/10.1016/j.ins.2005.03.013.

Bertoluzza, C., Corral, N., & Salas, A. (1995). On a new class of distances between fuzzy numbers. *Mathware & Soft Computing, 2,* 71–84.

Diamond, P., & Kloeden, P. (1990). Metric spaces of fuzzy sets. *Fuzzy Sets and Systems, 35*(2), 241–249. issn: 0165-0114. https://doi.org/10.1016/0165-0114(90)90197-E.

Dubois, D., & Prade, H. (1987). The mean value of a fuzzy number. *Fuzzy Sets and Systems, 24*(3), 279–300. issn: 0165-0114. https://doi.org/10.1016/0165-0114(87)90028-5. http://www.sciencedirect.com/science/article/pii/0165011487900285

Dubois, D., & Prade, H. (2008). Gradual elements in a fuzzy set. *Soft Computing, 12*(2), 165–175. issn: 1433-7479. https://doi.org/10.1007/s00500-007-0187-6.

George, A., & Veeramani, P. (1994). On some results in fuzzy metric spaces. *Fuzzy Sets and Systems, 64*(3), 395–399. issn: 0165-0114. https://doi.org/10.1016/0165-0114(94)90162-7. http://www.sciencedirect.com/science/article/pii/0165011494901627

Grzegorzewski, P. (1998). Metrics and orders in space of fuzzy numbers. *Fuzzy Sets and Systems, 97*(1), 83–94. issn: 0165-0114. https://doi.org/10.1016/S0165-0114(96)00322-3.

Heilpern, S. (1992). The expected value of a fuzzy number. *Fuzzy Sets and Systems, 47*(1), 81–86. issn: 0165-0114. https://doi.org/10.1016/0165-0114(92)90062-9. http://www.sciencedirect.com/science/article/pii/0165011492900629.

Kramosil, I., & Michalek, J. (1975). Fuzzy metric and statistical metric spaces. *Kybernetika, 11,* 336–344.

Kaleva, O. (1987). Fuzzy differential equations. *Fuzzy Sets and Systems, 24*(3), 301–317. issn: 0165-0114. https://doi.org/10.1016/0165-0114(87)90029-7. http://www.sciencedirect.com/science/article/pii/0165011487900297

Klement, E. P., Puri, M. L., & Ralescu, D. A. (1986). Limit theorems for fuzzy random variables. *Proceedings of the Royal Society of London. A. Mathematical and Physical Sciences, 407*(1832), 171–182. https://doi.org/10.1098/rspa.1986.0091. https://royalsocietypublishing.org/doi/abs/10.1098/rspa.1986.0091

McCulloch, J. C. (2016). *Novel methods of measuring the similarity and distance between complex fuzzy sets.* Ph.D. Thesis. University of Nottingham. http://eprints.nottingham.ac.uk/33401/1/thesis.pdf

Puri, M. L., & Ralescu, D. A. (1985). The concept of normality for fuzzy random variables. *The Annals of Probability, 13*(4), 1373–1379.

Sinova, B., Gila, M. Á., Lõpeza, M. T., & Van Aelstbc, S. (2014). A parameterized L2 metric between fuzzy numbers and its parameter interpretation. *Fuzzy Sets and Systems, 245,* 101–115.

Trutschnig, W., González-Rodríguez, G., Colubi, A., & Ángeles Gil, M. (2009). A new family of metrics for compact, convex (fuzzy) sets based on a generalized concept of mid and spread. *Information Sciences, 179*(23), 3964–3972. ISSN: 0020-0255. https://doi.org/10.1016/j.ins.2009.06.023

Yao, J.-S., & Wu, K. (2000). Ranking fuzzy numbers based on decomposition principle and signed distance. *Fuzzy Sets and Systems, 116*(2), 275–288.

Chapter 5
Fuzzy Random Variables and Fuzzy Distributions

Data sets collected from real experiments are often affected by the randomness, while this information could have an imprecise nature. Therefore, in such random data sets, the probability theory could not always be sufficient for modelling this uncertainty. This latter requires complementary tools. Thus, a combination between probability and fuzzy theories should be considered. It is important to introduce the concept of a random variable in the fuzzy context. This concept will be extensively used later on in developing statistical methods based on fuzzy data.

Two out of many approaches defining fuzzy random variables are widely used in the literature: the *Kwakernaak–Kruse and Meyer* and the *Féron–Puri and Ralescu* approaches; namely, Kwakernaak (1978, 1979) gave a definition of the fuzzy random variable (FRV). It is defined as the fuzzy perception of an unobservable real-valued random variable. It is actually a description of the fuzziness contained in a traditional random variable often called the *"original."* The Kwakernaak ideas were after elaborated by Kruse and Meyer (1987) in a mathematical framework.

From another perspective, in a series of papers, Féron (1976a, 1976b, 1979), etc., the author proposed the concept of the random fuzzy variable (RFV) in the context of modelling a given random experiment which produces fuzzy sets of a metric space. This random fuzzy set is defined as the "levelwise extension" of the notion of random sets in real-based spaces or as random elements defined in the spaces of fuzzy sets. These spaces are usually provided with a given Borel σ-algebra. Puri and Ralescu (1986) asserted that some metrics are missing in Féron (1976a)'s contribution. Given that, they revisited Féron (1976a)'s concepts and gave an expanded version of them. They added, for instance, the notion of expectation. The *Féron–Puri and Ralescu* definition corresponds to the ontic case since it models uncertain random variates, often associated with the human perception of things. In the *Féron–Puri and Ralescu* sense, the produced data are fuzzy. The theory of random fuzzy sets is considered as the extension of the theory of random sets and random variables. It is seen as a way to preserve the settings of the real space and extend them conveniently to statistical procedures with fuzzy entities.

R. Berkachy, *The Signed Distance Measure in Fuzzy Statistical Analysis*,
Fuzzy Management Methods, https://doi.org/10.1007/978-3-030-76916-1_5

However, Puri and Ralescu renounced the use of the terminology RFV and often kept using the FRV one. Consequently, confusion has been probably induced between both approaches, but stronger conceptual differences exist between both. Several references have deeply discussed about the differences between these approaches. We note, for example, Gil et al. (2014), Gil (2018), etc. In addition, in compliance with the notions of epistemic and ontic fuzzy values as we explained, one could easily see that the *Kwakernaak–Kruse and Meyer* approach is rather used in the case of epistemic fuzzy entities, while the ontic could be referred to the *Féron–Puri and Ralescu* one. Therefore, the proposed definition of an FRV is well confirmed as a fuzzy perception of these entities. In the next sections, we will ourselves briefly review these two approaches in order to propose related fuzzy statistical approaches in the upcoming chapters.

We will first open the chapter in Sect. 5.1 with the concept of fuzzy events and their probabilities, as defended by Zadeh (1968). The purpose is then to be able to propose the approaches defining the FRV (eventually the RFV) in Sect. 5.2. Moreover, we will not only define a random variable in the fuzzy context but also show in Sect. 5.3 the distribution function associated with it. After that we give in Sect. 5.4 the expectation and the variance related to both of the considered approaches. Finally, Sect. 5.5 aims to propose the concepts of empirical estimators of the distribution parameters.

5.1 Fuzzy Events and Their Probabilities

Zadeh (1968) proposed a definition of the concept of a fuzzy event over $\mathbb{R}^n$ and the probability of this event. Note that the subsequent notions will be defined on $\mathbb{R}^n$ in coherence with the Zadeh (1968) ideas, but one could easily deduce the case where $n = 1$. These notions are the combination of the probability theory based on measuring the occurrence of a given event from one side and the fuzziness mainly due to its nature. The concepts related to the fuzzy events are given by the following:

Definition 5.1.1 (Fuzzy Event) Let $(\mathbb{R}^n, \mathcal{A}, \mathbb{P})$ be a probability space. A fuzzy subset $\tilde{X}$ of $\mathbb{R}^n$ is said to be a fuzzy event of $\mathbb{R}^n$ such that it is associated with an MF $\mu_{\tilde{X}}$ measurable in the Borel space. The fuzzy event is defined such that each α-level set belongs to the σ-field $\mathcal{A}$.

Zadeh (1968) asserted that his notion of the probability of a fuzzy event is considered to be a crisp value, defined in terms of the measure $\mathbb{P}$ as follows:

Definition 5.1.2 (Probability of a Fuzzy Event) Consider a fuzzy event $\tilde{X}$ of the probability space $(\mathbb{R}^n, \mathcal{A}, \mathbb{P})$. The probability of a fuzzy event $\tilde{X}$ is the expectation $\mathbb{E}(\mu_{\tilde{X}})$ of its MF $\mu_{\tilde{X}}$ with respect to the probability measure $\mathbb{P}$. It is given by

$$P(\tilde{X}) = \int_{\mathbb{R}^n} \mu_{\tilde{X}}(x)\mathrm{d}\mathbb{P}(x) = \mathbb{E}(\mu_{\tilde{X}}). \tag{5.1}$$

Accordingly, we define the conditional probability of two fuzzy events $\tilde{X}_1$ and $\tilde{X}_2$, as follows:

Definition 5.1.3 (Conditional Probability of Fuzzy Events) Let $\tilde{X}_1$ and $\tilde{X}_2$ be two fuzzy events on the probability space $(\mathbb{R}^n, \mathcal{A}, \mathbb{P})$. Their membership functions are, respectively, denoted by $\mu_{\tilde{X}_1}$ and $\mu_{\tilde{X}_2}$. Consider a fuzzy subset of $\mathbb{R}^n$ denoted by $\tilde{Y}$, which expresses the product of $\tilde{X}_1$ and $\tilde{X}_2$, and its MF $\mu_{\tilde{Y}}$. The conditional probability of $\tilde{X}_1$ given $\tilde{X}_2$ is written as

$$P(\tilde{X}_1 \mid \tilde{X}_2) = \frac{P(\tilde{Y})}{P(\tilde{X}_2)} = \frac{\mathbb{E}(\mu_{\tilde{Y}})}{\mathbb{E}(\mu_{\tilde{X}_2})}, \tag{5.2}$$

such that $P(\tilde{X}_2)$ is strictly positive.

We remind that in the classical approach, the conditional probability of two given events X_1 and X_2 can be given by calculating the probability of the intersection between both events divided by the probability of X_2, as given in the following equation:

$$P(X_1 \mid X_2) = \frac{P(X_1 \cap X_2)}{P(X_2)}. \tag{5.3}$$

It is natural to propose the definition of independent fuzzy events. It is given by the following definition.

Definition 5.1.4 (Independent Fuzzy Events) Consider two fuzzy events $\tilde{X}_1$ and $\tilde{X}_2$ defined on the probability space $(\mathbb{R}^n, \mathcal{A}, \mathbb{P})$. Let $\tilde{Y} = \tilde{X}_1 \tilde{X}_2$ be a fuzzy subset of $\mathbb{R}^n$ with its MF $\mu_{\tilde{X}_1 \tilde{X}_2}(y)$ given by $\mu_{\tilde{X}_1}(y) \cdot \mu_{\tilde{X}_2}(y)$, $\forall y \in \mathbb{R}^n$. The fuzzy events $\tilde{X}_1$ and $\tilde{X}_2$ are said to be independent if and only if the following is verified:

$$P(\tilde{X}_1 \tilde{X}_2) = P(\tilde{X}_1)P(\tilde{X}_2). \tag{5.4}$$

The conditional probability of $\tilde{X}_1$ given $\tilde{X}_2$ can then be written as

$$P(\tilde{X}_1 \mid \tilde{X}_2) = P(\tilde{X}_1). \tag{5.5}$$

Consequently, the definition of the joint membership function of n independent fuzzy events is expressed as follows:

Definition 5.1.5 Consider a collection $\underline{\tilde{X}}$ of n mutually independent fuzzy events $\tilde{X}_i$, $i = 1, \ldots, n$, and the joint membership function $\mu_{\underline{\tilde{X}}}$ can be written as

$$\mu_{\underline{\tilde{X}}}(\underline{x}) = \mu_{\tilde{X}_1}(x_1) \times \ldots \times \mu_{\tilde{X}_n}(x_n). \tag{5.6}$$

Note that in our case, we used the product t-norm in the purpose of facilitating the calculation of further integrations.

Finally, an interesting definition is provided by Yager (1979) who introduced the concept of the fuzzy probability of fuzzy events. The main difference in his approach is the fuzzy nature of this probability rather a scalar one as proposed in Zadeh (1968). The fuzzy probability of a fuzzy event is then defined as follows:

Definition 5.1.6 (Fuzzy Probability of Fuzzy Event)   Consider a fuzzy event $\tilde{X}$ of the probability space $(\mathbb{R}^n, \mathcal{A}, \mathbb{P})$. The fuzzy probability of a fuzzy event $\tilde{X}$ is defined as the fuzzy set $\tilde{P}(\tilde{X})$ written as

$$\tilde{P}(\tilde{X}) = \bigcup_{0 \leq \alpha \leq 1} \alpha \, P[\tilde{X}_\alpha]. \tag{5.7}$$

In other terms, Yager (1979) defined his concept of probability of a fuzzy event as a fuzzy set defined on the interval $[0; 1]$ for which the fuzzy probability is found using the probabilities of the α-cuts of the concerned fuzzy event. Additional properties on this probability can be found in Yager (1979).

5.2   Fuzzy Random Variables

As previously mentioned, we expose mainly two approaches defining a random variable in the fuzzy context: the Kwakernaak–Kruse and Meyer and the Féron–Puri and Ralescu approaches. We highlight that in particular situations, a coincidence between both of them is expected.

We note that all the proposed approaches of fuzzy random variables of this chapter are stated as usual in a probability space denoted by $(\Omega, \mathcal{A}, \mathbb{P})$ and intended to model a given random experiment, such that Ω is the universe of all possible results of the experiment, $\mathcal{A}$ is a σ-algebra of the subsets of the universe Ω, and $\mathbb{P}$ is a probability measure defined on $\mathcal{A}$. We open the section by the Kwakernaak–Kruse and Meyer definition.

5.2.1   The Kwakernaak–Kruse and Meyer Approach

In his first paper of the series, Kwakernaak (1978) stated that "Fuzzy random variables are random variables whose values are not real, but fuzzy numbers" (pp.1–1.2-3). In other terms, the approach by Kwakernaak (1978, 1979) and revisited by Kruse and Meyer (1987), both asserted that an FRV $\tilde{X}$ is not only based on a real "crisp" random variable but also given by an uncertain cognition of the unobservable random variable $X : \Omega \rightarrow \mathbb{R}$. They assumed that this FRV is obtained from the composition of the fuzzy perception written as a mapping from $\mathbb{R}$ to $\mathbb{F}(\mathbb{R})$ and the so-called original X. Mention that the original X is contained in the set of originals denoted by $\mathcal{X}$ and constituting a proper universe.

Let us first define the FRV à la Kwakernaak–Kruse and Meyer, as follows:

Definition 5.2.1 (Fuzzy Random Variable) Consider the probability space $(\Omega, \mathcal{A}, \mathbb{P})$ and the set of all fuzzy numbers $\mathbb{F}(\mathbb{R})$. A fuzzy random variable (FRV) $\tilde{X}$ is the mapping

$$\tilde{X} : \quad \Omega \quad \rightarrow \quad \mathbb{F}(\mathbb{R}),$$

such that for all $\omega \in \Omega$ and $\alpha \in [0; 1]$, the system $\left\{ \tilde{X}_\alpha(\omega) \mid \omega \in \Omega, \, \alpha \in [0; 1] \right\}$ is a measurable representation set of $\tilde{X}$ where the real-valued applications $\tilde{X}_\alpha^L$ and $\tilde{X}_\alpha^R$ given by

$$\tilde{X}_\alpha^L : \quad \Omega \quad \rightarrow \quad \mathbb{R} \quad \text{and} \quad \tilde{X}_\alpha^R : \quad \Omega \quad \rightarrow \quad \mathbb{R}$$

are, for each $\alpha \in [0; 1]$, measurable random variables. Thus, we have that $\tilde{X}_\alpha$ is an interval-valued application levelwise with

$$\tilde{X}_\alpha(\omega) = \left(\tilde{X}(\omega) \right)_\alpha = \left[\left(\tilde{X}(\omega) \right)_\alpha^L ; \left(\tilde{X}(\omega) \right)_\alpha^R \right], \quad \forall \omega \in \Omega.$$

An example illustrating the approach of Kwakernaak (1978) is given by the following:

Example 5.2.1 (The `weather` *Example)* We would like to ask a given population about the `weather` in their city. Thus, they are required to give a linguistic term between very bad, bad, moderate, good, and very good. Each person has to choose one linguistic only. Choosing precisely one linguistic only is theoretically possible. However, since human feelings often wave between two or both linguistics, the choice of one answer is perceived as fuzzy. For instance, the perception of the weather in a given city could be between very bad and bad, according to the circumstances and most probably according to the vision and preference of every citizen. Consider then the random variable

$$W = \left\{ \text{ the perception of the weather in a given city } \omega \right\}.$$

We could model the situation by the Kwakernaak–Kruse and Meyer approach, and we define the FRV $\tilde{W}$, for which, the universe Ω will be the set of considered cities and $\mathcal{A}$ its power set. The FRV can then take on the following values:

$$\tilde{W}_1 = \widetilde{\text{very bad}}, \quad \tilde{W}_2 = \widetilde{\text{bad}}, \quad \tilde{W}_3 = \widetilde{\text{moderate}}, \quad \tilde{W}_4 = \widetilde{\text{good}}, \quad \tilde{W}_5 = \widetilde{\text{very good}},$$

such that they could be modeled by common fuzzy numbers as given in Sect. 2.2. In addition, we associate with this FRV an original $W_0 : \Omega \rightarrow \mathbb{R}$ given as

$$\tilde{W}_0(\omega) = \left\{ \text{ the true weather in the city } \omega \right\}.$$

The FRV $\tilde{W}_0$ is then seen as the fuzzy perception of the weather described by the original W_0.

5.2.2   The Féron–Puri and Ralescu Approach

From their side and contrary to the Kwakernaak–Kruse and Meyer approach, Féron–Puri and Ralescu supposed that an RFV is rather a tool to model random imprecise-valued components and obviously not a fuzzy perception of real-valued ones. The definition given by Puri and Ralescu (1986) is then the following:

Definition 5.2.2 (Random Fuzzy Variable) Consider the probability space $(\Omega, \mathcal{A}, \mathbb{P})$ and $\mathbb{F}(\mathbb{R})$ the family of fuzzy numbers on the real space. Let $\mathcal{B}$ be the set of the Borel-subsets of $\mathbb{R}$. A random fuzzy variable RFV on the space $\mathcal{A}$ is a measurable function in a Borel sense given by

$$\tilde{X}: \quad \Omega \quad \rightarrow \quad \mathbb{F}(\mathbb{R})$$

such that for every $\alpha \in [0; 1]$, we have that

$$\left\{ (\omega, x) \mid \omega \in \Omega, x \in \tilde{X}_\alpha(\omega) \right\} \in \mathcal{A} \times \mathcal{B}. \tag{5.8}$$

Furthermore, one can say that for every $\alpha \in [0; 1]$, the α-level set represented by $\tilde{X}_\alpha$ such that $\tilde{X}_\alpha(\omega) = \left(\tilde{X}(\omega) \right)_\alpha$ is a random compact set.

By that, we could say that in the sense of Puri and Ralescu, the RFV $\tilde{X}$ is the direct evaluation of fuzziness contained in a given experimental output. We illustrate this approach by the following example:

Example 5.2.2 (The Work-Home Distance) We are interested again in the work-home distance of a company population given in Example 2.2.1. For this purpose, we asked a number of people about the approximate distance between their home and their work place. Consider then the random variable

$$H = \left\{ \text{ the work-home distance of a given employee of a company} \right\}.$$

Many of the respondents asserted that they are giving uncertain answers since they do not have a precise value for the distance. Then, this evaluation is seen to be naturally fuzzy. In addition, the distance and the variability in the answer vary from one person to another. If we suggest to a person to express his/her feeling by drawing a closed convex fuzzy number for which he/she has to decide where to place his/her α-level sets, and the shape of the curve interpolating the different levels chosen, this information can be viewed as an RFV by the Puri and Ralescu (1986) approach. This RFV will then be denoted by $\tilde{H}$ and describing the approximate work-home distance of the employee. In this case, Ω is the universe of all the employees of the company.

5.2.3 Synthesis

To sum up, despite that both approaches mentioned in this chapter are interesting tools for modelling fuzzy samples, their motivations are quite different. The difference is mainly in the modelling situations themselves. As an instance, the random fuzzy variables RFV by the Féron–Puri and Ralescu approach are appropriate to model random variables taking on fuzzy values—the ontic perspective—uncertain human response, for example. However, the FRV of Kwakernaak–Kruse and Meyer approach is used to model a different mechanism, where in a given random experiment, the response is obtained from the real unobservable original random variable. Instead, this random variable will be fuzzily assessed—the epistemic perspective. Thereupon, in a one-dimensional setting, it is easy to see that from a theoretical point of view, both approaches match. The idea of the following proposition seen in Gil et al. (2006) and Blanco-Fernandez et al. (2014) and relating both approaches is interesting to display.

Proposition 5.2.1 *Let* $(\Omega, \mathcal{A}, \mathbb{P})$ *be the probability space of a given random experiment; then,* $\tilde{X}$ *is an FRV by the Kwakernaak–Kruse and Meyer approach if and only if it is an RFV by the Féron–Puri and Ralescu one.*

By consequence, if we intend to consider the notion of randomness contained in the variables of a given fuzzy analysis, we should be aware of the choice of the approach. As an instance, the concept of Kwakernaak–Kruse and Meyer is related to the distribution of the originals of the considered random experiment and to the fuzzy perception of them, while the concept of the Féron–Puri and Ralescu approach is based on the randomness of the fuzzy numbers themselves. However, because of this "equivalence" of both approaches, Blanco-Fernandez et al. (2014) stated that the Féron–Puri and Ralescu reasoning can be even used in scenarios where fuzzy perceptions are considered. The authors highlighted that "conclusions could be 'technically' drawn" (pp. 1493—1.39) for such situations.

In our case, the presented methods are mainly defined in a one-dimensional setting, i.e. $n = 1$, where fuzzy convex cases are only adopted. Thus, both described approaches can be simultaneously used. It is then interesting and useful to investigate both approaches in several setups. For this purpose, we will subsequently propose different statistical contexts where we juggle with both approaches in order to better understand their use and fit.

5.3 Fuzzy Probability Distributions of Fuzzy Random Variables

It is natural to associate the fuzzy random variables with their corresponding probability density and distribution functions. These latter are often seen as fuzzy, since they model fuzzy entities. However, the definitions of distribution functions differ theoretically between approaches. As such, Viertl (2011) stated in Definition 8.1 of Chapter 8 the following general concept of a fuzzy probability density.

Definition 5.3.1 (Fuzzy Probability Density) Consider a probability space $(\Omega, \mathcal{A}, \mathbb{P})$. A fuzzy probability density denoted by $\tilde{f}$ is a fuzzy valued function $\tilde{f} : \Omega \rightarrow \mathbb{F}_c^{\star}(\mathbb{R})$, represented by its α-cuts $\left[\tilde{f}_\alpha^L(x); \tilde{f}_\alpha^R(x) \right]$, verifying the following conditions:

- The integrals $\int_\Omega \tilde{f}_\alpha^L(x)\mathrm{d}\mathbb{P}(x)$ and $\int_\Omega \tilde{f}_\alpha^R(x)\mathrm{d}\mathbb{P}(x)$ exist.
- The value 1 belongs to the kernel of the fuzzy interval given by its left and right α-level sets

$$\left[\int_\Omega \tilde{f}_\alpha^L(x)\mathrm{d}\mathbb{P}(x); \int_\Omega \tilde{f}_\alpha^R(x)\mathrm{d}\mathbb{P}(x) \right], \quad \forall \alpha \in [0; 1],$$

  and its support is included in $\mathbb{R}^+$.
- A crisp probability density $f : \Omega \rightarrow \mathbb{R}^+$ exists such that for the highest grade of α, i.e. $\alpha = 1$, we have that

$$\tilde{f}_\alpha^L(x) \le f(x) \le \tilde{f}_\alpha^R(x), \quad \forall x \in \Omega.$$

5.3.1   The Kwakernaak–Kruse and Meyer Approach

For the Kwakernaak (1978)—FRV, the associated distribution function is considered to be derived from the distribution of the original since it is seen as its fuzzy perception. The author asserted that this derivation relies basically on the extension principle of Zadeh. We can then clearly see that the intended distribution is naturally fuzzy. We express the distribution function of an FRV denoted by $\tilde{X}$ in the following way:

Definition 5.3.2 (Fuzzy Distribution Function of a Fuzzy Random Variable) Consider the probability space $(\Omega, \mathcal{A}, \mathbb{P})$ of a given random experiment. Let $\tilde{X}$ be an FRV in the Kwakernaak (1978) sense, with its corresponding set of originals denoted by $\mathcal{X}$. The fuzzy distribution function of the FRV $\tilde{X}$ is the mapping $\tilde{F}_{\tilde{X}} : \mathbb{R} \rightarrow \mathbb{F}(\mathbb{R})$, such that

$$\left(\tilde{F}_{\tilde{X}}(x) \right)(u) = \begin{cases} \sup_{X \in \mathcal{X} : F_X(x) = u} \ \inf_{\omega \in \Omega} \ \tilde{X}\left(X(\omega) \right) & \text{if} \quad u \in [0; 1], \\ 0 & \text{otherwise,} \end{cases}$$

$$(5.9)$$

where $F_X(x)$ is the distribution function of the original $X \in \mathcal{X}$ on the considered probability space, for a given level u and $\forall x \in \mathbb{R}$.

The concept of identically and independently distributed FRV are important to show. They are defined by the following definition.

Definition 5.3.3 (Identically Distributed Fuzzy Random Variables) Two fuzzy random variables $\tilde{X}$ and $\tilde{Y}$ are identically distributed on the probability space $(\Omega, \mathcal{A}, \mathbb{P})$ if and only if for $\forall \alpha \in [0; 1]$,

- the random variables $\tilde{X}_\alpha^L$ and $\tilde{Y}_\alpha^L$ are identically distributed and
- the random variables $\tilde{X}_\alpha^R$ and $\tilde{Y}_\alpha^R$ are identically distributed.

As a consequence, one could eventually deduce the following proposition, proved in Kruse and Meyer (1987).

Proposition 5.3.1 (Identically Distributed Fuzzy Random Variables) *Two fuzzy random variables $\tilde{X}$ and $\tilde{Y}$ are identically distributed on the probability space $(\Omega, \mathcal{A}, \mathbb{P})$ if for $\forall x \in \mathbb{R}$, the fuzzy distribution functions induced by $\tilde{X}$ and $\tilde{Y}$ denoted, respectively, by $\tilde{F}_{\tilde{X}}$ and $\tilde{F}_{\tilde{Y}}$ coincide.*

For the independence of the FRV, we have the following definition.

Definition 5.3.4 (Independent Fuzzy Random Variables) The fuzzy random variables $\tilde{X}_1, \ldots, \tilde{X}_n$ are said to be mutually or completely independent fuzzy random variables if and only if for each $\alpha \in [0; 1]$,

- the random variables $\tilde{X}_{1\alpha}^L, \ldots, \tilde{X}_{n\alpha}^L$ are, respectively, mutually or completely independent random variables and
- the random variables $\tilde{X}_{1\alpha}^R, \ldots, \tilde{X}_{n\alpha}^R$ are, respectively, mutually or completely independent random variables.

5.3.2 The Féron–Puri and Ralescu Approach

For the Puri and Ralescu (1986)—RFV, since the space of fuzzy numbers fails in having an acceptable total ordering procedure, inducing the distribution function of an RFV is less direct. Another method has to be proposed. Indeed, we know that an RFV is measurable in the Borel sense. This latter leads to the following definition of the distribution function as described by Viertl (2011):

Definition 5.3.5 (Fuzzy Probability Distribution of a Random Fuzzy Variable) Consider a random fuzzy variable $\tilde{X}$ defined on the probability space $(\Omega, \mathcal{A}, \mathbb{P})$. Let $\mathcal{B}$ be a Borel σ-algebra. For $B \in \mathcal{B}$, a fuzzy probability distribution $\tilde{F}_{\tilde{X}}$ on the system of the Borel-sets $\mathcal{B}$ is associated with an MF $\mu_{\tilde{F}_{\tilde{X}}}$ such that $\forall x \in \mathbb{R}$

$$\mu_{\tilde{F}_{\tilde{X}}}(x) = \begin{cases} 0 & \text{if } \forall \alpha \in [0; 1], x \notin \left[\tilde{\pi}_\alpha^L(B); \tilde{\pi}_\alpha^R(B)\right], \\ \sup\left\{\alpha \in [0; 1] \mid x \in \left[\tilde{\pi}_\alpha^L(B); \tilde{\pi}_\alpha^R(B)\right]\right\} & \text{otherwise,} \end{cases}$$

$$(5.10)$$

where the sets $\tilde{\pi}_\alpha^L(B)$ and $\tilde{\pi}_\alpha^R(B)$ defined on the Borel σ-field $\mathcal{B}$ are given by

$$\tilde{\pi}_\alpha^L(B) = \mathbb{P}\big(\{\omega \in \Omega \mid \tilde{X}_\alpha \cap B \neq 0\}\big), \tag{5.11}$$

$$\tilde{\pi}_\alpha^R(B) = \mathbb{P}\big(\{\omega \in \Omega \mid \tilde{X}_\alpha \subseteq B\}\big). \tag{5.12}$$

Furthermore, by the Borel assumption on the measurability of the RFV, the definitions of identically distributed and independent random fuzzy variables can analogously be deduced.

Definition 5.3.6 (Identically Distributed Random Fuzzy Variables) The two random fuzzy variables $\tilde{X}$ and $\tilde{Y}$ are identically distributed on the probability space $(\Omega, \mathcal{A}, \mathbb{P})$ if and only if for any Borel set B of the Borel σ-field denoted by $\mathcal{B}$, the following equality is verified:

$$P\big(\omega \in \Omega \mid \tilde{X}_\alpha(\omega) \subseteq B\big) = P\big(\omega \in \Omega \mid \tilde{Y}_\alpha(\omega) \subseteq B\big), \quad \forall \alpha \in [0; 1]. \tag{5.13}$$

Definition 5.3.7 (Independent Random Fuzzy Variables) The random fuzzy variables $\tilde{X}_1, \ldots, \tilde{X}_n$ are said to be mutually or completely independent random fuzzy variables, if $\forall \alpha \in [0; 1]$, they satisfy the following equality:

$$P\big(\tilde{X}_{1\alpha}(\omega) \subseteq B_1, \ldots, \tilde{X}_{n\alpha}(\omega) \subseteq B_n\big) = P\big(\tilde{X}_{1\alpha}(\omega) \subseteq B_1\big) \cdot \ldots \cdot P\big(\tilde{X}_{n\alpha}(\omega) \subseteq B_n\big), \tag{5.14}$$

where $B_1, \ldots, B_n$ are Borel-sets of the Borel σ-algebra $\mathcal{B}$.

From another side, it is expected to be able to write the fuzzy probability distribution function by the fuzzy probability density function, similarly to the classical approach. This transition can be shown in the following proposition, as implicitly seen in Viertl (2011).

Proposition 5.3.2 *On the probability space $(\Omega, \mathcal{A}, \mathbb{P})$, the fuzzy probability distribution function $\tilde{F}_{\tilde{X}}$ of the FRV $\tilde{X}$ can also be expressed in terms of the fuzzy probability density function $\tilde{f}$ as follows:*

$$\big(\tilde{F}_{\tilde{X}}\big)_\alpha = \Big[\int_\Omega \tilde{f}_\alpha^L(x)\mathrm{d}\mathbb{P}(x); \int_\Omega \tilde{f}_\alpha^R(x)\mathrm{d}\mathbb{P}(x)\Big]. \tag{5.15}$$

5.3.3 Synthesis

To extend the concepts of random variables in the fuzzy environment, it is essential to define a fuzzy random vector. Despite the fact that the described approaches are conceptually different, this definition of the random vector is the same as seen in the subsequent definitions. We have to remark that in these definitions, the term fuzzy random variable is used to generalize both the FRV and the RFV.

Definition 5.3.8 (Fuzzy Random Vector) Consider a probability space $(\Omega^n, \mathcal{A}, \mathbb{P})$. A fuzzy random vector is an n-dimensional fuzzy random variable denoted by $\underline{\tilde{X}}$ such that $\underline{\tilde{X}} : \Omega^n \rightarrow \mathbb{F}(\mathbb{R}^n)$.

We can easily see that the properties on the independence of two fuzzy random vectors are exactly the same as the properties related to the random variables following both fuzzy approaches. In addition, the definition of the identical and independently distributed fuzzy random variables induces the definition of a fuzzy random sample.

Definition 5.3.9 (Fuzzy Random Sample) The vector of fuzzy random variables $\underline{\tilde{X}} = (\tilde{X}_1, \ldots, \tilde{X}_n)$ is said to be a fuzzy random sample if and only if the fuzzy random variables $\tilde{X}_i, i = 1, \ldots, n$ are mutually independent and identically distributed.

5.4 Expectation and Variance

Calculating the measures of a given distribution is a prominent step toward analyzing a random variable and describing the associated distributions. The common ones are the location, the dispersion, and the asymmetry measures. We are particularly interested in the expectation and the variance. According to the fuzzy theory, one could suspect that these measures would analogously be of type fuzzy. However, this idea is not always suitable. The procedure depends on the approach chosen for defining the concept of fuzzy random variable. For Kwakernaak (1978) and Kruse and Meyer (1987), since the extension of the measures to the fuzzy environment is based on the extension principle, then these parameters are supposed to be fuzzy, viewed as fuzzy perceptions of the original parameter. From another side, for Puri and Ralescu (1986), the extension is based on the ideas of Fréchet (1948) in the context of random sets as described by Aumann (1965). The parameters are then defined over chosen metric spaces. Therefore, the expectation is conceived to be fuzzy, while the variance is a scalar. Therefore, we could easily see that the nature of the parameters changes between both approaches. These are summarized in Table 5.1.

In this section, we give the definitions of the expectation and variance of the fuzzy random variables of both approaches. The first part of this section is devoted to the fuzzy parameters by the Kwakernaak–Kruse and Meyer approach, while the second one treats of the parameters by the Féron–Puri and Ralescu one.

Table 5.1 Nature of the parameters of the considered distributions by approach

Approach	Expectation	Variance
Kwakernaak–Kruse and Meyer FRV	Fuzzy	Fuzzy
Féron–Puri and Ralescu RFV	Fuzzy	Scalar

5.4.1   The Kwakernaak–Kruse and Meyer Approach

Since the definition of an FRV à la Kwakernaak–Kruse and Meyer approach relies on fuzzy perceptions of the set of originals, it would then be interesting to define the fuzzy perception of a parameter of an FRV. We have to mention that practically calculating this fuzzy entity is perceived as a complicated task. The corresponding definition is given as follows:

Definition 5.4.1 (Fuzzy Perception of a Parameter of a Fuzzy Random Variable) Consider a fuzzy random variable $\tilde{X}$ defined on the probability space $(\Omega, \mathcal{A}, \mathbb{P})$ and seen as a fuzzy perception of the original random variable X of the set of originals $\mathcal{X}$, such that $X \in \mathcal{X}$. We denote by θ a parameter of a given original of $\mathcal{X}$. The fuzzy perception of θ is the fuzzy number $\tilde{\theta}$ such that

$$\tilde{\theta}(t) = \sup_{X \in \mathcal{X} | \theta(X) = t} \inf_{\omega \in \Omega} \tilde{X}\Big(X(\omega) \Big), \quad t \in \mathbb{R}. \tag{5.16}$$

Using this definition, one could define the fuzzy perception of the expectation of an FRV as seen in the following definition.

Definition 5.4.2 (Fuzzy Perception of the Expectation of a Fuzzy Random Variable) Consider now the expectation of a fuzzy random variable to be a fuzzy parameter as described in Definition 5.4.1. Thus, it is seen as a fuzzy perception of the traditional expectation. The fuzzy expectation is then defined by

$$\tilde{\mathbb{E}}(\tilde{X})(t) = \sup_{X \in \mathcal{X} | \mathbb{E}(X) = t} \inf_{\omega \in \Omega} \tilde{X}\Big(X(\omega) \Big), \tag{5.17}$$

such that its MF can be written as follows:

$$\mu_{\tilde{\mathbb{E}}(\tilde{X})}(t) = \sup \Big\{ \mu_{\tilde{X}}(\omega) \mid X \in \mathcal{X}, \mathbb{E}(X) = t \Big\}, \quad \forall \omega \in \Omega \text{ and } t \in \mathbb{R}. \tag{5.18}$$

The fuzzy perception of the expectation can also be written using the α-level sets such that for $\forall \alpha \in [0; 1]$,

$$\Big(\tilde{\mathbb{E}}(\tilde{X}) \Big)_{\alpha} = \Big[\mathbb{E}(\tilde{X}_{\alpha}^{L}); \mathbb{E}(\tilde{X}_{\alpha}^{R}) \Big]. \tag{5.19}$$

Analogously to the fuzzy expectation, Kwakernaak–Kruse and Meyer defended their vision of a fuzzy-valued variance as the perception of the variance of its original. This latter can be expressed in the following manner:

Definition 5.4.3 (Fuzzy Perception of the Variance of a Fuzzy Random Variable) Consider a fuzzy random variable $\tilde{X}$ defined on the probability space $(\Omega, \mathcal{A}, \mathbb{P})$. The fuzzy variance of $\tilde{X}$ denoted by $\widetilde{\text{Var}}(\tilde{X})$ is a fuzzy perception of the variance of the original X which belongs to the set of originals $\mathcal{X}$, such that

$$\widetilde{\mathrm{Var}}(\tilde{X})(\sigma^2) = \sup_{X \in \mathcal{X} | \mathrm{Var}(X) = \sigma^2} \quad \inf_{\omega \in \Omega} \quad \tilde{X}\Big(X(\omega)\Big), \tag{5.20}$$

such that its MF can be given by

$$\mu_{\widetilde{\mathrm{Var}}(\tilde{X})}(\sigma^2) = \sup\Big\{\mu_{\tilde{X}}(\omega) \mid X \in \mathcal{X}, \mathrm{Var}(X) = \sigma^2\Big\}, \quad \forall \omega \in \Omega \text{ and } \sigma \in \mathbb{R}^+. \tag{5.21}$$

This membership could also be written using Eq. (2.12) as follows:

$$\mu_{\widetilde{\mathrm{Var}}(\tilde{X})}(x) = \sup\Big\{\alpha I_{\big[\big(\widetilde{\mathrm{Var}}\tilde{x}\big)^L_\alpha ; \big(\widetilde{\mathrm{Var}}(\tilde{x})\big)^R_\alpha\big]}(x) : \alpha \in [0, 1]\Big\}, \tag{5.22}$$

where $\big(\widetilde{\mathrm{Var}}(\tilde{X})\big)^L_\alpha = \inf_{X \in [\tilde{X}^L_\alpha ; \tilde{X}^R_\alpha]} \big(\mathrm{Var}(X)\big)$ and $\big(\widetilde{\mathrm{Var}}(\tilde{X})\big)^R_\alpha = \sup_{X \in [\tilde{X}^L_\alpha ; \tilde{X}^R_\alpha]} \big(\mathrm{Var}(X)\big)$.

Two important properties of the fuzzy variance can be seen in the following propositions. The detailed proofs can be found in Kruse and Meyer (1987).

Proposition 5.4.1 *Consider a fuzzy random variable $\tilde{X}$, a fuzzy number $\tilde{A}$ of the set $\mathbb{F}^\star_c(\mathbb{R})$, and a real number $b \in \mathbb{R}$. If $\widetilde{\mathrm{Var}}(\tilde{X})$ exists, we have that*

$$\widetilde{\mathrm{Var}}(b \cdot \tilde{X} + \tilde{A}) = b^2 \cdot \widetilde{\mathrm{Var}}(\tilde{X}). \tag{5.23}$$

Proposition 5.4.2 *Consider two fuzzy random variables $\tilde{X}$ and $\tilde{Y}$ associated with the same probability space $(\Omega, \mathcal{A}, \mathbb{P})$. If $\widetilde{\mathrm{Var}}(\tilde{X})$ and $\widetilde{\mathrm{Var}}(\tilde{Y})$ exist, we have that*

$$\widetilde{\mathrm{Var}}(\tilde{X} + \tilde{Y}) = \widetilde{\mathrm{Var}}(\tilde{X}) + \widetilde{\mathrm{Var}}(\tilde{Y}). \tag{5.24}$$

5.4.2 The Féron–Puri and Ralescu Approach

For the Puri and Ralescu (1986) approach, the expectation of a random fuzzy variable, the so-called fuzzy expected value, is based on Fréchet (1948)'s ideas and Aumann's integral defined in a Hausdorff topology seen in Aumann (1965).

Definition 5.4.4 (Expectation of a Random Fuzzy Variable) Let $\tilde{X}$ be an integrable and bounded random fuzzy variable defined on the probability space $(\Omega, \mathcal{A}, \mathbb{P})$. The fuzzy expectation, or fuzzy expected value, also called the Aumann-type expectation $\tilde{\mathbb{E}}(\tilde{X})$ of $\tilde{X}$ is a closed fuzzy set of $\mathbb{F}(\mathbb{R}^n)$ given by the following family set of α-cuts:

$$\big(\tilde{\mathbb{E}}(\tilde{X})\big)_\alpha = \{\mathbb{E}(X) \mid X : \Omega \to \mathbb{R}^n, X \in \mathrm{L}^1(\Omega, \mathcal{A}, \mathbb{P}), X(\omega) = \tilde{X}_\alpha(\omega)\}, \tag{5.25}$$

where $\left(\tilde{\mathbb{E}}(\tilde{X})\right)_\alpha$ is called the Aumann integral of the random set $\tilde{X}_\alpha$, for which $\left(\tilde{\mathbb{E}}(\tilde{X})\right)_\alpha^L$ and $\left(\tilde{\mathbb{E}}(\tilde{X})\right)_\alpha^R$ are its respective left and right continuous α-cuts for $\forall \alpha \in [0; 1]$. Note that if the values of $\tilde{X}$ are defined in $\mathbb{F}_c^\star(\mathbb{R})$, then the expectation in the Aumann sense coincides with the expectation expressed in Eq. (5.19), i.e.

$$\left(\tilde{\mathbb{E}}(\tilde{X})\right)_\alpha = \left[\mathbb{E}(\tilde{X}_\alpha^L), \mathbb{E}(\tilde{X}_\alpha^R)\right] = \left[\int_\Omega \tilde{X}_\alpha^L(\omega)\mathrm{d}\mathbb{P}(\omega); \int_\Omega \tilde{X}_\alpha^R(\omega)\mathrm{d}\mathbb{P}(\omega)\right], \quad \forall \alpha \in [0; 1].$$
(5.26)

Fréchet (1948) gave another definition of the fuzzy expectation. His version, also called the Fréchet expectation, can be written in terms of an L_2 metric ρ such as the ρ_2 distance of Diamond and Kloeden (1990) given in Definition 4.1.1, etc. as follows:

$$\left(\tilde{\mathbb{E}}(\tilde{X})\right) = \arg \min_{\tilde{Y} \in \mathbb{F}(\mathbb{R}^n)} \mathbb{E}\left(\left[\rho(\tilde{X}, \tilde{Y})\right]^2\right),$$
(5.27)

for each $\alpha \in [0; 1]$.

In other words, Puri and Ralescu gave a definition of the expectation of an RFV as a generalization of the expectation of a random variable on the real space, as mentioned in Diaz and Gil (1999).

On the other hand, by the definition given for the Fréchet expectation of a random fuzzy variable, one could similarly construct its variance. In a one-dimensional setting, the Fréchet variance can then be expressed by the following definition.

Definition 5.4.5 (Fréchet Variance of a Random Fuzzy Variable) Consider a probability space $(\Omega, \mathcal{A}, \mathbb{P})$ and an RFV $\tilde{X}$. Let $(\mathbb{F}_c(\mathbb{R}), d)$ be a metric space in the Fréchet sense. If it exists, the variance of $\tilde{X}$ is the real number $\mathrm{Var}(\tilde{X})$ given by

$$\mathrm{Var}(\tilde{X}) = \mathbb{E}\left(\left[d(\tilde{X}, \tilde{\mathbb{E}}(\tilde{X}))\right]^2\right),$$
(5.28)

where $\tilde{\mathbb{E}}(\tilde{X})$ is the Fréchet expectation given in Definition 5.4.4.

Using this variance expressed in Eq. (5.28), it is easy to show that the following properties are fulfilled.

Proposition 5.4.3 *Consider a random fuzzy variable $\tilde{X}$, a fuzzy number $\tilde{A}$ of the set $\mathbb{F}_c^\star(\mathbb{R})$, and a real number $b \in \mathbb{R}$. If $\mathrm{Var}(\tilde{X})$ exists, then*

$$\mathrm{Var}(b \cdot \tilde{X} + \tilde{A}) = b^2 \cdot \mathrm{Var}(\tilde{X}).$$
(5.29)

Proposition 5.4.4 *Consider two random fuzzy variables $\tilde{X}$ and $\tilde{Y}$ defined on the same probability space $(\Omega, \mathcal{A}, \mathbb{P})$. If $\mathrm{Var}(\tilde{X})$ and $\mathrm{Var}(\tilde{Y})$ exist, then*

$$\mathrm{Var}(\tilde{X} + \tilde{Y}) = \mathrm{Var}(\tilde{X}) + \mathrm{Var}(\tilde{Y}).$$
(5.30)

Related to the Puri and Ralescu (1986) approach, several authors have exposed differently the variance of random variables in such environment. Note, for instance, Nather (2006) who gave a scalar multiplication giving a clear expression of the covariance of a random fuzzy variable and deduced its variance.

In a simpler way, Feng et al. (2001) showed other ways of expressing the variance and covariance of random fuzzy variables made in terms of the α-cuts of the FRV.

5.4.3 Synthesis

To sum up, even though the nature of the different variance measures related to both approaches is different, an essential coincidence exists. Thus, both approaches have exactly their expectations seen as a fuzzy parameter. The expectations of the Kwakernaak–Kruse and Meyer and the Féron–Puri and Ralescu approaches satisfy the same usual properties such that:

Proposition 5.4.5 *For a fuzzy random variable $\tilde{X}$ such that its fuzzy expectation exists, a fuzzy number $\tilde{A}$ in $\mathbb{F}_c^\star(\mathbb{R})$, and a real value $a \in \mathbb{R}$, we have that*

$$\tilde{\mathbb{E}}(a \cdot \tilde{X} + \tilde{A}) = a \cdot \tilde{\mathbb{E}}(\tilde{X}) \oplus \tilde{A}. \tag{5.31}$$

Proposition 5.4.6 *For two fuzzy random variables $\tilde{X}$ and $\tilde{Y}$ associated with the same probability space $(\Omega, \mathcal{A}, \mathbb{P})$, if their fuzzy expectations exist, we have that*

$$\tilde{\mathbb{E}}(\tilde{X} + \tilde{Y}) = \tilde{\mathbb{E}}(\tilde{X}) \oplus \tilde{\mathbb{E}}(\tilde{Y}). \tag{5.32}$$

The proofs of these latter can be found in Kruse and Meyer (1987). However, for the variance measure, the case is different. As an instance, following the Kwakernaak–Kruse and Meyer approach, the variance is considered to be of fuzzy nature, while for the Féron–Puri and Ralescu one, it is mainly a scalar. Thus, their properties are evidently expected to be different.

5.5 Estimators of the Fuzzy Distributions Parameters

We often use a variety of measures to describe and summarize a given data set. A location measure is a value that best represents a certain location characteristics of the group of observations, while a dispersion measure quantifies the variability between these observations. A further description of the data includes the asymmetry and kurtosis measures known, respectively, as the skewness and the kurtosis. These measures reflect the symmetry and obviously the asymmetry of the distribution, often compared to the standard normal one. In the case of a fuzzy analysis, particular definitions of these measures should be considered.

A fundamental task is to use estimators of the theoretical parameters which best characterizes a given fuzzy distribution. A condition is then to propose unbiased ones. Several references have discussed this subject. We note, for example, Kwakernaak (1978) who expressed an unbiased estimator of the fuzzy expectation. For the median of vague data, Grzegorzewski (1998) gave the definition of the concept of the median in the fuzzy context. In addition, the notion of an estimator of the fuzzy variance has been very appealing to investigate through decades since its nature is directly related to the chosen approach of defining random variables. Based on this fact, different researches have been constructed. As an instance, Akbari and Rezaei (2007) and Akbari and Khanjari Sadegh (2012) gave a uniformly minimum variance unbiased fuzzy estimator.

We expose below few measures only. We are mainly interested in the sample mean of a fuzzy distribution. This estimator of the expectation is known to be fuzzy in both approaches. Furthermore, since the variance is at the end of a central moment, we will discuss about the scalar central moments of a fuzzy distribution based on Fréchet's ideas. These moments will directly lead us to the definition of the skewness and the kurtosis of a fuzzy distribution in the Fréchet sense. Finally, for the approach of Kwakernaak (1978) where the variance—and evidently the moments—is fuzzy-valued, fuzzy approximations can eventually be proposed. This case will not be treated explicitly in the theoretical part of this thesis.

5.5.1   The Mean

Kwakernaak (1978) proved that the so-called fuzzy sample mean is an unbiased estimator of the fuzzy expectation of a fuzzy random variable. First, let us define an unbiased fuzzy estimator of a fuzzy parameter:

Definition 5.5.1 (Unbiased Fuzzy Estimator) Consider a fuzzy point estimator $\hat{\tilde{\theta}}$ of a given fuzzy parameter $\tilde{\theta}$. The fuzzy estimator $\hat{\tilde{\theta}}$ is said to be unbiased if and only if for a given $\theta_p \in [\hat{\tilde{\theta}}_\alpha^L; \hat{\tilde{\theta}}_\alpha^R]$ and for all $\alpha \in [0; 1]$, there exists a value $\theta_0 \in \tilde{\theta}_\alpha$ such that

$$\mathbb{E}(\theta_p) = \theta_0.$$

We could also define the fuzzy sample mean written as follows:

Definition 5.5.2 (Fuzzy Sample Mean) Consider a fuzzy variable $\tilde{X}_i$, $i = 1, \ldots, n$. The fuzzy sample mean of the fuzzy sample $\{\tilde{X}_i\}_{i=1}^n$ denoted by $\overline{\tilde{X}}$ is given by

$$\overline{\tilde{X}} = \frac{1}{n} \sum_{i=1}^n \tilde{X}_i, \tag{5.33}$$

such that its α-cuts are as follows:

$$\overline{\tilde{X}}_\alpha = \left[\frac{1}{n} \sum_{i=1}^n \tilde{X}_{i\alpha}^L, \quad \frac{1}{n} \sum_{i=1}^n \tilde{X}_{i\alpha}^R \right]. \tag{5.34}$$

Similarly to the classical statistical theory, the sample mean in the fuzzy environment is known to be an important parameter since it can be used in many analyses. An interesting lemma related to the SGD shown in Sect. 4.6 is expressed by the following:

Lemma 5.5.1 *Consider a fuzzy sample $\{\tilde{X}_i\}_{i=1}^n$ with its fuzzy sample mean denoted by $\overline{\tilde{X}}$. The following expression is true:*

$$\overline{d_{SGD}(\tilde{X}, \tilde{0})} = d_{SGD}(\overline{\tilde{X}}, \tilde{0}), \tag{5.35}$$

where $\overline{d_{SGD}(\tilde{X}, \tilde{0})}$ is the mean of the SGD measures of the fuzzy sample measured from the fuzzy origin.

Proof Since the SGD is an additive and distributive measure, the proof of this lemma becomes direct such that

$$d_{SGD}(\overline{\tilde{X}}, \tilde{0}) = d_{SGD}\left(\frac{1}{n} \sum_{i=1}^n \tilde{X}_i, \tilde{0}\right) = \frac{1}{n} d_{SGD}\left(\sum_{i=1}^n \tilde{X}_i, \tilde{0}\right) = \frac{1}{n} \sum_{i=1}^n d_{SGD}\left(\tilde{X}_i, \tilde{0}\right) = \overline{d_{SGD}(\tilde{X}, \tilde{0})}.$$

$\square$

The fuzzy sample mean is seen as an estimator of the expectation $\tilde{\mathbb{E}}(\tilde{X})$ of the fuzzy variable $\tilde{X}$, thanks to the following convergence law seen in Kruse (1982) and Colubi et al. (1999):

Proposition 5.5.1 (Law of Large Numbers) *Consider a probability space $(\Omega, \mathcal{A}, \mathbb{P})$. Let $\{\tilde{X}_i\}_{i=1}^n$ be a random fuzzy sample, composed of n integrable, bounded, and identically distributed and mutually independent random fuzzy variables. The almost surely convergence in probability of the fuzzy sample mean toward the expectation of the fuzzy sample with respect to the Hausdorff metric is verified since*

$$\lim_{n \to \infty} \mathbb{P}\left(\delta_\infty(\overline{\tilde{X}}, \mathbb{E}(\tilde{X})) < \epsilon\right) = 1, \ \text{almost surely}, \tag{5.36}$$

where δ_∞ is the Hausdorff metric given in Eq. (4.2), $\mathbb{E}(\tilde{X})$ is the fuzzy expectation of the distribution of the random sample with a finite variance, and $\overline{\tilde{X}}$ is the fuzzy sample mean. Note that the Hausdorff metric applied to compute the distance of an RFV to the fuzzy origin is also a random fuzzy variable.

Inversely, the subsequent lemma can be easily verified.

Lemma 5.5.2 *If for a sequence of mutually independent and identically distributed random fuzzy variables* $\tilde{X}_i$, $i = 1, \ldots, n$, *defined on the probability space* $(\Omega, \mathcal{A}, \mathbb{P})$, *it exists a fuzzy set* $\tilde{A} \in \mathbb{F}_c(\mathbb{R}^p)$ *such that* $\lim_{n \to \infty} \delta_\infty\left(\overline{\tilde{X}}, \tilde{A}\right) = 0$, *almost surely in probability, then* $\tilde{A}$ *is nothing but the expectation of* $\tilde{X}$, *while this latter is integrable and bounded.*

5.5.2   The Sample Moments

Similarly to the Fréchet variance of a random fuzzy variable, one could eventually propose the following definition of central sample moments in the Fréchet sense.

Definition 5.5.3 (Central Sample Moments of a Random Fuzzy Variable) Consider a probability space $(\Omega, \mathcal{A}, \mathbb{P})$. Let $\tilde{X}$ be a random fuzzy variable and $(\mathbb{F}_c^\star(\mathbb{R}), d)$ a metric space. The k-th Fréchet-type central sample moment of $\tilde{X}$ denoted by $v_k(\tilde{X})$ if the real values given by

$$v_k(\tilde{X}) = \mathbb{E}\left(\left[d(\tilde{X}, \tilde{\mathbb{E}}(\tilde{X}))\right]^k\right). \tag{5.37}$$

Furthermore, v_k can be almost surely in probability expressed by

$$\underline{v}_k(\tilde{X}) = \frac{1}{n} \sum_{i=1}^{n} \left[d\left(\tilde{X}_i, \overline{\tilde{X}}\right)\right]^k, \tag{5.38}$$

where n is the number of observations, $\overline{\tilde{X}}$ is the fuzzy sample mean, and d is a chosen distance. In particular, the first central moment is known to be null, while the second one denoted by v_2 is nothing but the Fréchet-variance as described in Definition 5.4.5. It is given by

$$\underline{v}_2(\tilde{X}) = \frac{1}{n} \sum_{i=1}^{n} \left[d\left(\tilde{X}_i, \overline{\tilde{X}}\right)\right]^2. \tag{5.39}$$

Therefore, the definition of the concept of a variance of an RFV à la Puri-Ralescu differs in terms of the chosen metric d. It is the same for the definition of the expectation.

In a similar way, one could propose the moments denoted by $\underline{v}'_k$ of a random fuzzy variable. They can be written by

$$\underline{v}'_k(\tilde{X}) = \frac{1}{n} \sum_{i=1}^{n} \left[d(\tilde{X}_i, \tilde{0})\right]^k. \tag{5.40}$$

A direct consequence of the previous definition is to see that, for a given metric d, we are now able to construct simultaneously the asymmetry and kurtosis measures using specific central moments, notably the second, third, and fourth ones.

Furthermore, calculating a dispersion measure such as the variance is seen as an essential task. Contrariwise, this latter could eventually face multiple problems in the computation process. A prominent way to calculate a variance is thus by computing its estimator. We usually prefer an unbiased one. Using chosen distances, researchers tended to give more often unbiased estimators of the variance of an RFV à la Féron–Puri and Ralescu. As for us, we proposed in Berkachy and Donzé (2019) an unbiased scalar estimator of the variance with respect to the SGD. It is written in the following manner:

Proposition 5.5.2 (Unbiased Estimator of the Variance with Respect to the SGD) *Consider a random fuzzy sample* $\tilde{X} = (\tilde{X}_1, \ldots, \tilde{X}_n)$ *with* $\tilde{X}_i$, $i = 1, \ldots, n$, *independent and identically distributed and associated with the probability space* $(\Omega, \mathcal{A}, \mathbb{P})$. *An unbiased estimator* S_{SGD}^2 *of the variance* $\mathrm{Var}(\tilde{X})$ *of the distribution of* $\tilde{X}_i$ *with respect to the SGD can be expressed by*

$$S_{SGD}^2 = \frac{1}{n-1} \sum_{i=1}^{n} \left[d_{SGD}(\tilde{X}_i, \overline{\tilde{X}}) \right]^2, \tag{5.41}$$

where $\overline{\tilde{X}}$ *is the fuzzy sample mean shown in Definition 5.5.2.*

Proof Consider $\tilde{X} \in \mathbb{F}^n(\mathbb{R})$ to be the fuzzy random sample. We know that if $\tilde{X}_i$ is a random fuzzy variable, then $d_{SGD}(\tilde{X}_i, \tilde{0})$ is a random variable on $\mathbb{R}$, and $S_{d_{SGD}}^2$ is also a random variable such that

$$S_{d_{SGD}}^2 = \frac{1}{n-1} \sum_{i=1}^{n} \left(d_{SGD}(\tilde{X}_i, \tilde{0}) - \overline{d_{SGD}(\tilde{X}, \tilde{0})} \right)^2. \tag{5.42}$$

Moreover, an estimator is said to be unbiased if its expectation is nothing but the parameter to estimate itself. The entity $S_{d_{SGD}}^2$ can then be regarded as an unbiased estimator of the classical variance of the crisp random variable $d_{SGD}(\tilde{X}_i, \tilde{0})$. Thus, we could write the following equality:

$$\mathbb{E}[S_{d_{SGD}}^2] = \mathrm{Var}(d_{SGD}(\tilde{X}_i, \tilde{0})). \tag{5.43}$$

In the same spirit, let us now consider that the variance of a fuzzy random variable is defined in the Fréchet sense using the SGD. Then, for our case, we have to prove that

$$\mathbb{E}[S_{SGD}^2] = \mathrm{Var}(\tilde{X}).$$

Then, from Proposition 4.6.1 and Lemma 5.5.1, we have the following:

$$\mathbb{E}[S_{SGD}^2] = \mathbb{E}\Big[\frac{1}{n-1}\sum_{i=1}^{n} d_{SGD}^2(\tilde{X}_i, \overline{\tilde{X}})\Big] = \mathbb{E}\Big[\frac{1}{n-1}\sum_{i=1}^{n}\big(d_{SGD}(\tilde{X}_i, \tilde{0}) - d_{SGD}(\overline{\tilde{X}}, \tilde{0})\big)^2\Big],$$

$$= \mathbb{E}\Big[\frac{1}{n-1}\sum_{i=1}^{n}\big(d_{SGD}(\tilde{X}_i, \tilde{0}) - \overline{d_{SGD}(\tilde{X}, \tilde{0})}\big)^2\Big],$$

$$= \mathbb{E}[S_{d_{SGD}}^2] = \mathrm{Var}\big(d_{SGD}(\tilde{X}, \tilde{0})\big) = d_{SGD}\big(\mathrm{Var}(\tilde{X}), \tilde{0}\big) = \mathrm{Var}(\tilde{X}).$$

We remind that the variance denoted by $\mathrm{Var}(\tilde{X})$ in this context is a scalar, and then by Definition 4.6.1, we have that $d_{SGD}\big(\mathrm{Var}(\tilde{X}), \tilde{0}\big) = \mathrm{Var}(\tilde{X})$. □

We note that the abovementioned proposition can eventually be seen as a definition of the sample variance based on the expected value of a fuzzy number.

It is often useful to express the variance of a fuzzy random sample for which we use particular geometric symmetrical shapes as given in the following proposition.

Proposition 5.5.3 (Unbiased Estimators of the Variance with Respect to the SGD—Triangular and Trapezoidal Symmetrical Case) *Let $\tilde{X} = (\tilde{X}_1, \ldots, \tilde{X}_n)$ be a fuzzy random sample such that $\tilde{X}_i$, $i = 1, \ldots, n$, are independent and identically distributed. Suppose that these latter are modeled by triangular or trapezoidal symmetrical fuzzy numbers. According to the chosen modelling shapes, the unbiased estimator of the fuzzy variance denoted by S_{SGD}^2 is a real number expressed as follows:*

- *If the fuzzy random sample $\tilde{X}$ composed of $\tilde{X}_i$, $i = 1, \ldots, n$, is modeled by triangular symmetrical fuzzy numbers, $\tilde{X}_i = (p_i, q_i, r_i)$, with the corresponding fuzzy sample mean $\overline{\tilde{X}}$ given by the tuple $\overline{\tilde{X}} = (p_m, q_m, r_m)$, then the unbiased estimator of the variance is as follows:*

$$S_{SGD}^2 = \frac{1}{n-1}\sum_{i=1}^{n}\big(q_i - q_m\big)^2. \tag{5.44}$$

- *In the case of modelling with trapezoidal symmetrical fuzzy numbers given by $\tilde{X}_i = (p_i, q_i, r_i, s_i)$, the fuzzy sample mean is written by $\overline{\tilde{X}} = (p_m, q_m, r_m, s_m)$. Then, the unbiased estimator of the variance is given by*

$$S_{SGD}^2 = \frac{1}{n-1}\sum_{i=1}^{n}\Big(\frac{q_i + r_i}{2} - \frac{q_m + r_m}{2}\Big)^2. \tag{5.45}$$

Proof From Definition 4.6.4, we have that the SGD of a triangular fuzzy number $\tilde{X}_i$ given by the tuple (p_i, q_i, r_i) is

$$d_{SGD}(\tilde{X}_i, \tilde{0}) = \frac{1}{4}(p_i + 2q_i + r_i).$$

A triangular symmetrical fuzzy number verifies the following property:

$$p_i + r_i = 2q_i.$$

Consequently, the fuzzy sample mean verifies the property $p_m + r_m = 2q_m$, since it is also modeled by a triangular symmetrical fuzzy number. The unbiased estimator $S^2_{d_{SGD}}$ of the variance will then be written by

$$S^2_{SGD} = \frac{1}{n-1} \sum_{i=1}^{n} d^2_{SGD}(\tilde{X}_i, \overline{\tilde{X}}) = \frac{1}{n-1} \sum_{i=1}^{n} \left(d_{SGD}(\tilde{X}_i, \tilde{0}) - d_{SGD}(\overline{\tilde{X}}, \tilde{0})\right)^2,$$

$$= \frac{1}{n-1} \sum_{i=1}^{n} \left(\frac{1}{4}(p_i + 2q_i + r_i) - \frac{1}{4}(p_m + 2q_m + r_m)\right)^2,$$

$$= \frac{1}{n-1} \sum_{i=1}^{n} \frac{1}{16}\left((4q_i)^2 + (4q_m)^2 - 2(4q_i)(4q_m)\right) = \frac{1}{n-1} \sum_{i=1}^{n} \left(q_i - q_m\right)^2.$$

Similarly, if the fuzzy random sample is modeled by trapezoidal symmetrical fuzzy numbers, i.e. $\tilde{X}_i = (p_i, q_i, r_i, s_i)$, $i = 1, \ldots, n$, the following property holds:

$$p_i + s_i = q_i + r_i.$$

Then, we have that

$$S^2_{SGD} = \frac{1}{n-1} \sum_{i=1}^{n} d^2_{SGD}(\tilde{X}_i, \overline{\tilde{X}}) = \frac{1}{n-1} \sum_{i=1}^{n} \left(d_{SGD}(\tilde{X}_i, \tilde{0}) - d_{SGD}(\overline{\tilde{X}}, \tilde{0})\right)^2,$$

$$= \frac{1}{n-1} \sum_{i=1}^{n} \left(\frac{1}{4}(p_i + q_i + r_i + s_i) - \frac{1}{4}(p_m + q_m + r_m + s_m)\right)^2,$$

$$= \frac{1}{n-1} \sum_{i=1}^{n} \left(\frac{q_i + r_i}{2} - \frac{q_m + r_m}{2}\right)^2.$$

$\square$

Analogously to the second central sample moment, i.e. the variance, one could express the unbiased estimators of all other $k - 1$ central sample moments (eventually the sample moments) with respect to the SGD or other convenient metrics.

For the approach by Kwakernaak–Kruse and Meyer, even if we consider the fuzzy k-th central moment of an FRV $\tilde{X}$ as the direct extension of the classical real-based moment to the fuzzy environment, written by

$$\underline{\tilde{v}}^k(\tilde{X}) = \frac{1}{n} \sum_{i=1}^{n} (\tilde{X}_i \ominus \overline{\tilde{X}})^k, \tag{5.46}$$

we can clearly observe two main problems for calculating this fuzzy parameter:

- the fuzzy difference between the occurrence $\tilde{X}_i$ and the fuzzy expectation, eventually the fuzzy sample mean and
- the fuzzy product.

These problems are due to the complexity faced in the arithmetic fuzzy operations. Therefore, an approximation is often needed. One could use different types of operators to approximate these fuzzy arithmetic in a way preserving the fuzzy nature of the parameter. One could eventually use Propositions 4.7.6 and 4.7.7 to accomplish this task. Many of the properties of such operators can be found in Grzegorzewski and Mrowka (2005). On the other hand, unbiased estimators are usually required. These estimators will intuitively be fuzzy in this case. References have already discussed about finding estimators of fuzzy moments. We noted, for instance, in Berkachy and Donzé (2018) the different ways of finding a near fuzzy number in the purpose of facilitating the estimation of the fuzzy variance. However, Akbari and Khanjari Sadegh (2012) gave a method for calculating fuzzy uniformly minimum variance unbiased estimators but based on the SGD and on another L_2 metric. In this script, we limit ourselves to the real-valued case of estimators for the moments and evidently the variance.

5.6 Simulation

We are now interested in showing the use of the Fréchet-type variance and its estimator and in the skewness and excess of kurtosis measures. We remind that these measures can, respectively, be written in terms of the central sample moments by

$$\frac{v_3}{\underline{v}_2^{3/2}} \quad \text{and} \quad \frac{v_4}{\underline{v}_2^2} - 3.$$

For this purpose, we develop a simulation based on the following scenario.

At a museum, a particular painting is getting a lot of attention. For this purpose and in order to understand the actual point of view of the visitors about this painting, we would like to collect the rating of each person. The idea is to ask each visitor to rate the painting on a scale between 1 and 10 (seen as a Likert scale between very bad and very good).

However, the points of view of people for this task are mitigated since a given person can find the painting "good" and "bad" at the same time and from different perspectives. In addition, the rating "good" of a given visitor could eventually be different from a rating "good" of another one regarding their tastes of art. Consequently, in this case, the fuzziness is present in the perception of the people. We intend to model the situation by an FRV denoted by $\tilde{R}$. As discussed in Sect. 5.2, two approaches defining a fuzzy random variable exist, depending on our motivation for each of them. For the sake of illustration of these approaches, we would like to use both of them for our simulation. Thus, we propose a first set of simulation setups where we consider the case as epistemic, i.e. where we model the fuzzy perception of the random rating of the people, and a second case set where we suppose that our randomly chosen data are fuzzy by nature, i.e. the ontic case. Therefore, for a sample size N going from 1000 to 10,000 by step of 100, the two cases of rating are exposed in the following setups:

- **Case 1**: We ask each person to rate the painting by giving a number between 1 and 10, and we will model their perception of the rating by corresponding triangular fuzzy numbers given by the following arbitrary tuples:

$$\tilde{R}_1 = (1, 1, 1.5), \ \tilde{R}_2 = (1, 2, 3.5), \ \tilde{R}_3 = (1, 3, 4), \ \tilde{R}_4 = (3, 4, 6), \ \tilde{R}_5 = (4, 5, 6),$$

$$\tilde{R}_6 = (5, 6, 7), \ \tilde{R}_7 = (5, 7, 8), \ \tilde{R}_8 = (7, 8, 8.5), \ \tilde{R}_9 = (7.5, 9, 10), \ \tilde{R}_{10} = (9.5, 10, 10).$$

For this simulation and considering the sample of size N, we randomly generate a data set of real integers from 1 to 10 representing the rating of a given individual. Note that no requirements are fixed on the distribution of this data set.

- **Case 2**: We ask each visitor to draw himself/herself a triangular fuzzy number and place it on the x–y axis, for which some restrictions were required such that the triangle should be strictly positive where it belongs to the interval $[1; 10]$ on the x-axis.

  In practice, we randomly generated a vector of real data (not necessarily integers) from 1 to 10 constituting the core set of the triangular fuzzy numbers that represent the fuzzy ratings. Correspondingly to each value of the generated vector, we chose two real numbers: one smaller and one bigger. In addition, multiple conditions have been fixed. As an instance, each chosen value should not exceed the value 3 on each of both sides of the core set, such that the amplitude (spread of the fuzzy number) of the support set should not exceed the value 6.

A last crucial condition is that on the y-axis, the triangle should attain the value 1. For example, the following fuzzy numbers could not be accepted for such data set: $(0, 4.2, 4.5)$, $(-1.5, 1.3, 2)$, while the fuzzy numbers $(1.2, 2.65, 4.05)$ and $(4.37, 5.25, 7.19)$ might represent ratings with respect to the pre-defined conditions.

We mention that we used triangular fuzzy numbers in these cases because of their simplicity. By this choice of the shape, we intended to avoid any influence of the complexity of the shapes modelling the fuzziness contained in the data set.

Suppose that we construct synthetic data sets, generated randomly following the previous requirements. The purpose is to calculate the variance, the skewness, and the kurtosis measures for the different constructed distributions. By these calculations, the idea is to investigate two main objectives:

- the influence of the size of the chosen sample on the statistical measures and
- the main differences between using the distances and metrics described in Chap. 4, in terms of these measures.

From the results of our iterative simulations, a first finding related to the first objective is that in our current case, the variance, the skewness, and the kurtosis measures are independent of the size of the sample. This result can be seen as a less time-consuming property. As such, we expose in this example the data sets of size 1000 and 10,000 observations only.

According to the second objective, we will use all of the distances proposed in Chap. 4 in order to calculate the considered measures. The idea is to investigate the difference between the use of each one of them and to highlight clear tendencies. We mention that for the first case, the use of these distances is seen as a crisp approximation of the fuzzy operations, while in the second case, it is based on the Fréchet proposition. The results are exposed in Table 5.2.

We can clearly see that the variance and the kurtosis are slightly different between all distances. However, for the skewness measure, an interesting conclusion arose from these computations: not all the skewness measures of all metrics are the same. Indeed, the skewness measures by the non-directional distances are close, while the ones by the directional distances are very close as well. In addition, the directional distances induce a skewness measure close to the value 0, while the others are not. This result is only applicable for the skewness, obviously not for the variance and the kurtosis for a simple calculation reason: the skewness is calculated by an odd rank of central moments, paving the way to the importance of the directionality of the distances, while the kurtosis and the variance are computed by means of even rank central moments. From another side, since our proposed distances $d_{SGD}^{\theta^\star}$ and d_{GSGD} coincide with the others, we can clearly validate their use and see their relevance compared to the ones already existing. In addition, if we would like to

Table 5.2 Simulation results of Cases 1 and 2 for samples of size 1000 and 10000—Sect. 5.6

Type of distance		Variance		Skewness		Kurtosis	
		Case 1	Case 2	Case 1	Case 2	Case 1	Case 2
ρ_1	1000	7.7667	6.4153	1.3122	1.2653	-1.1733	-1.3042
	10000	7.7901	6.6266	1.2992	1.2464	-1.2127	-1.3597
ρ_2	1000	7.8456	6.5652	1.3101	1.2548	-1.1780	-1.3292
	10000	7.8670	6.7702	1.2973	1.2371	-1.2166	-1.3814
$d_{Bertoluzza}$	1000	7.7987	6.4769	1.3109	1.2620	-1.1765	-1.3128
	10000	7.8214	6.6876	1.2980	1.2434	-1.2154	-1.3674
$d_{mid/spr}$	1000	7.7987	6.4769	1.3109	1.2620	-1.1765	-1.3128
	10000	7.8214	6.6876	1.2980	1.2434	-1.2154	-1.3674
$\rho_3^\star$	1000	9.0055	7.7765	1.3102	1.2392	-1.1686	-1.3583
	10000	9.0076	7.9578	1.2982	1.2224	-1.2038	-1.4084
$\delta_{3,\frac{1}{2}}^\star$	1000	7.9190	6.7047	1.3086	1.2468	-1.1811	-1.3485
	10000	7.9385	6.9046	1.2959	1.2299	-1.2191	-1.3986
$d_{l-wabl/ldev/rdev}$	1000	7.7930	6.4563	1.3115	1.2638	-1.1749	-1.3083
	10000	7.8157	6.6668	1.2985	1.2450	-1.2140	-1.3634
d_{SGD}	1000	7.7667	6.4019	0.1003	-0.0620	-1.1733	-1.2973
	10000	7.7901	6.6150	0.0095	-0.0154	-1.2127	-1.3540
$d_{SGD}^{\theta\star}$	1000	7.7842	6.4350	1.3119	1.2657	-1.1740	-1.3038
	10000	7.8072	6.6460	1.2989	1.2466	-1.2132	-1.3596
d_{GSGD}	1000	7.7842	6.4350	0.0998	-0.0610	-1.1740	-1.3038
	10000	7.8072	6.6460	0.0093	-0.0153	-1.2132	-1.3596

compare Case 1 and Case 2, one could say that no clear interpretations can be done on the differences between both of them. The obtained results based on these cases are quite similar. Finally, note that the calculated skewness for the data sets of both cases is highlighted in Fig. 5.1.

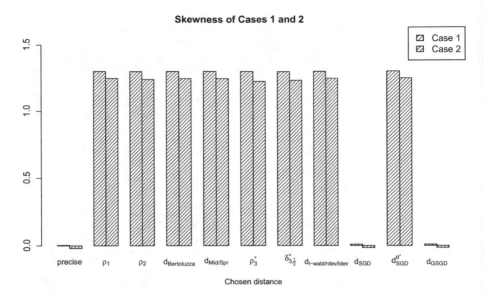

Fig. 5.1 The skewness—**Case 1** and **Case 2** with a 10000 sized-sample—Sect. 5.6

References

Akbari, M. G., & Khanjari Sadegh, M. (2012). Estimators based on fuzzy random variables and their mathematical properties. *Iranian Journal of Fuzzy Systems, 9*(1), 79–95. ISSN: 1735-0654. https://doi.org/10.22111/ijfs.2012.227. http://ijfs.usb.ac.ir/article_227.html

Akbari, M. G., & Rezaei, A. (2007). An uniformly minimum variance unbiased point estimator using fuzzy observations. *Austrian Journal of Statistics, 36*(4), 307–317.

Aumann, R. J. (1965). Integrals of set-valued functions. *Journal of Mathematical Analysis and Applications, 12*(1), 1–12.

Berkachy, R., & Donzé, L. (2018). Estimation of the fuzzy variance by different approximations of the fuzzy product. In *Proceedings of the 23rd International Conference on Computational Statistics COMPSTAT, Iasi, Romania*.

Berkachy, R., & Donzé, L. (2019). Central moments of a fuzzy random variable using the signed distance: A look towards the variance. In S. Destercke et al. (Eds.), *Uncertainty Modelling in Data Science* (pp. 17–24). Cham: Springer International Publishing. ISBN: 978-3-319-97547-4.

Blanco-Fernandez, A., et al. (2014). A distance-based statistical analysis of fuzzy number-valued data. *International Journal of Approximate Reasoning, 55*(7), 1487–1501. ISSN: 0888-613X. https://doi.org/10.1016/j.ijar.2013.09.020. http://www.sciencedirect.com/science/article/pii/S0888613X13002120.

Colubi, A., et al. (1999). A generalized strong law of large numbers. *Probability Theory and Related Fields, 114*(3), 401–417. ISSN: 1432-2064. https://doi.org/10.1007/s004400050229.

Diamond, P., & Kloeden, P. (1990). Metric spaces of fuzzy sets. *Fuzzy Sets and Systems, 35*(2), 241–249. ISSN: 0165-0114. https://doi.org/10.1016/0165-0114(90)90197-E

Diaz, M. L., & Gil, M. A. (1999). An extension of Fubini's theorem for fuzzy random variables. *Information Sciences, 115*, 29–41. https://doi.org/10.1016/S0020-0255(98)10086-5

Feng, Y., Hu, L., & Shu, H. (2001). The variance and covariance of fuzzy random variables and their applications. *Fuzzy Sets and Systems, 120*(3), 487–497. ISSN: 0165-0114. https://doi.org/10.1016/S0165-0114(99)00060-3. http://www.sciencedirect.com/science/article/pii/S0165011499000603

Féron, R. (1976a). Ensembles aléatoires flous. *Comptes Rendus de l'Académie des Sciences de Paris A, 282,* 903–906.

Féron, R. (1976b). Ensembles flous attachés à un ensemble aléatoire flou. *Comptes Rendus de l'Académie des Sciences de Paris A, 9,* 51–66.

Féron, R. (1979). Sur les notions de distance et d'écart dans une structure floue et leurs applications aux ensembles aléatoires flous. *Comptes Rendus de l'Académie des Sciences de Paris A, 289,* 35–38.

Fréchet, M. (1948). Les éléments aléatoires de nature quelconque dans un espace distancié. *Annales de l'institut Henri Poincaré, 10*(4), 215–310.

Gil, M. A. (2018). Fuzzy random variables à la Kruse & Meyer and à la Puri & Ralescu: Key differences and coincidences. In S. Mostaghim, A. Nürnberger, & C. Borgelt (Eds.), *Frontiers in Computational Intelligence* (pp. 21–29). Cham: Springer International Publishing. ISBN: 978-3-319-67789-7. https://doi.org/10.1007/978-3-319-67789-7_2

Gil, M. A., Colubi, A., & Teran, P. (2014). Random fuzzy sets: Why, when, how. *Boletin de Estadistica e Investigacion Operativa, 30,* 5–29.

Gil, M. A., Lopez-Diaz, M., & Ralescu, D. A. (2006). Overview on the development of fuzzy random variables. *Fuzzy Sets and Systems, 157*(19), 2546–2557. ISSN: 0165-0114. https://doi.org/10.1016/j.fss.2006.05.002. http://www.sciencedirect.com/science/article/pii/S016501140600203X

Grzegorzewski, P. (1998). Statistical inference about the median from vague data. *Control and Cybernetics, 27*(3), 447–464.

Grzegorzewski, P., & Mrowka, E. (2005). Trapezoidal approximations of fuzzy numbers. *Fuzzy Sets and Systems, 153*(1), 115–135. ISSN: 0165-0114. https://doi.org/10.1016/j.fss.2004.02.015

Kruse, R. (1982). The strong law of large numbers for fuzzy random variables. *Information Sciences, 28*(3), 233–241. ISSN: 0020-0255. https://doi.org/10.1016/0020-0255(82)90049-4. http://www.sciencedirect.com/science/article/pii/0020025582900494

Kruse, R., & Meyer, K. D. (1987). *Statistics with vague data* (Vol. 6). Springer Netherlands.

Kwakernaak, H. (1978). Fuzzy random variables I. Definitions and theorems. *Information Sciences, 15*(1), 1–29. ISSN: 0020-0255. https://doi.org/10.1016/0020-0255(78)90019-1. http://www.sciencedirect.com/science/article/pii/0020025578900191

Kwakernaak, H. (1979). Fuzzy random variables II. Algorithms and examples for the discrete case. *Information Sciences, 17*(3), 253–278. ISSN: 0020-0255. https://doi.org/10.1016/0020-0255(79)90020-3. http://www.sciencedirect.com/science/article/pii/0020025579900203

Nather, W. (2006). Regression with fuzzy random data. *Computational Statistics & Data Analysis, 51*(1), 235–252. ISSN: 0167-9473. http://dx.doi.org/10.1016/j.csda.2006.02.021

Puri, M. L., & Ralescu, D. A. (1986). Fuzzy random variables. *Journal of Mathematical Analysis and Applications, 114*(2), 409–422. ISSN: 0022-247X. https://doi.org/10.1016/0022-247X(86)90093-4. http://www.sciencedirect.com/science/article/pii/0022247X86900934

Viertl, R. (2011). *Statistical methods for fuzzy data.* John Wiley and Sons, Ltd.

Yager, R. R. (1979). A note on probabilities of fuzzy events. *Information Sciences, 18*(2), 113–129. ISSN: 0020-0255. https://doi.org/10.1016/0020-0255(79)90011-2. http://www.sciencedirect.com/science/article/pii/0020025579900112

Zadeh, L. A. (1968). Probability measures of fuzzy events. *Journal of Mathematical Analysis and Applications, 23*(2), 421–427. ISSN: 0022-247X. https://doi.org/10.1016/0022-247X(68)90078-4

Chapter 6
Fuzzy Statistical Inference

The objective of exposing the previous chapters is to apply fuzzy statistical inference methods. We know that in a given scientific experiment, common tasks of a researcher are often divided into two steps: the first one is to postulate a hypothesis based on a preliminary view of the practitioner, and the second one is to reject or falsify this hypothesis. This procedure is called a statistic test, where in a classical context, a null hypothesis and an alternative one are mandatory. If such hypothesis is doubtedly false, then it will be rejected. Otherwise, it will not be rejected, at a certain significance level.

When fuzziness is taken into account, the traditional methods of testing hypotheses might be in many cases inadequate. These methods are only valid in case of precise data. If this latter is not present, the complexity of such situations increases. Many researches have been interested to study the extension between different classical methods and their homologues in the fuzzy theory. Researchers in the field considered that this uncertainty can mainly have two sources: the data and the hypotheses. Right after the introduction of the concept of fuzzy sets and fuzzy probabilities by Zadeh (1965, 1968), the corresponding fuzzy statistical tests for given parameters were proposed for one or both of these sources.

It is known that to accomplish this task of testing hypotheses, one could eventually construct confidence intervals for the required parameter and use it in a hypothesis testing procedure. Statistical methods based on the confidence intervals are widely used for such problems. In compliance with the fuzzy theory, statistical tests based on fuzzy confidence intervals were displayed. We note for example Grzegorzewski (2000, 2001) who proposed an approach of testing hypotheses by fuzzy confidence intervals where he considered the data as fuzzy. This hypotheses test leads to getting fuzzy decisions. These latter are interpreted by the so-called degree of conviction, i.e. a degree of acceptability of the hypotheses. We highlight that this degree is exclusively related to the statement of "not rejecting" a given hypothesis. We proposed in Berkachy and Donzé (2019c) a hypotheses testing procedure where we considered that the fuzziness not only comes from the data, but also from the hypotheses. The fuzzy decisions were finally defuzzified in order to get interpretable decisions.

Nevertheless, the construction of the confidence intervals themselves presents multiple difficulties. The traditional way of this calculation is easy to use but very limited in terms of parameters and distributions. In the same way, there exist many cases that are not elaborated enough to deal with fuzzy contexts. When uncertainty increases, the "classical" interval has to be determined by means of considering every single value of the support set of the involved fuzzy numbers. Different researchers have already proposed methods estimating confidence intervals in the fuzzy environment: Kruse and Meyer (1987) gave a theoretical definition of fuzzy confidence intervals; Viertl (2011) and Viertl and Yeganeh (2016) proposed a definition of the concept of confidence regions. They were mainly interested in applying them in a Bayesian context; Kahraman et al. (2016) presented different approaches of the construction of fuzzy confidence intervals developed between 1980 and 2015, and they proposed the notion of the so-called hesitant fuzzy confidence intervals; Couso and Sanchez (2011) exposed a method based on inner and outer approximations of confidence intervals for fuzzy data. However, even with the wide spectrum of available approaches, the main limitation of the expressions reflected by a dependence to specific parameters with defined distributions persists. It is then interesting to develop a practical unified approach that generalizes all possible cases of fuzzy confidence intervals. As such, in Berkachy and Donzé (2019b), we proposed a practical procedure of constructing fuzzy confidence intervals by the likelihood ratio method. This is seen as an efficient tool in the construction of confidence intervals. As instance, in the fuzzy environment, Gil and Casals (1988) showed the extension of the likelihood ratio test to the case where the imprecision is contained in the data. In some other contexts, Denoeux (2011) gave a maximum likelihood estimation using the Expectation–Maximization-algorithm from fuzzy data. Note that the contribution shown in Berkachy and Donzé (2019b) enabled us to introduce the obtained confidence interval in the process of fuzzy hypotheses testing.

A decision for a given test can also be made using the concept of the p-value. The idea is to compare it to the significance level of the test. The p-value is defined as the probability of observing a specific sample given that the defended null hypothesis is actually true. Researchers have treated the notion of the p-value in the fuzzy context. As such, Filzmoser and Viertl (2004) proposed a definition of the fuzzy p-value based on the assumption that the fuzziness of the tests is contained in the data. They asserted as well that similarly to the process of decision of Neyman–Pearson, a three-level decision rule can be defined. Thus, they supposed the existence of a region of no rejection of both hypotheses. Afterward, Parchami et al. (2010) considered a similar context, but they asserted that the uncertainty is a matter of hypotheses rather than data. Therefore, in Berkachy and Donzé (2019a), we gave a generalization of both definitions where we provided a method based on the treatment of the fuzziness that might be contained in the data and/or in the hypotheses. The outcome of this method is a fuzzy p-value, which might be in many cases difficult to interpret. Even though the defuzzification operation is considered to be a loss of information, a transformation in the nature of the p-value can be useful. We proposed then to defuzzify the obtained fuzzy p-value in order to get an interpretable crisp decision. In addition, Berkachy and Donzé (2017a, 2017b) showed a method for this operation based on the signed distance operator.

In this chapter, we first show the definition of a given fuzzy hypothesis. We after display the construction of a given fuzzy confidence interval. Then, the process of construction of a fuzzy confidence interval (FCI) by the likelihood ratio is exposed. Moreover, we show in detail the hypotheses testing approaches, based on these intervals, followed by the fuzzy p-values. Both the fuzzy decisions and the fuzzy p-values are afterward defuzzified. We propose to defuzzify them by the SGD operator defined in Sect. 4.6 from one side and from the GSGD given in Sect. 4.7 from another one. The purpose is to sort out the main differences and drawbacks that might occur when using both distances in such contexts. All these procedures are illustrated by multiple detailed examples. Applications on a financial data set are also provided. In addition, a discussion on the comparison between the classical and fuzzy approaches is given. This chapter is closed by some guidelines on the use of each one of both approaches. The choice between the classical and fuzzy hypotheses testing approaches should be well-argued.

Let us first briefly recall the idea behind the classical hypothesis testing approach.

6.1 Testing Hypotheses by the Classical Approach

Consider a population described by a probability distribution P_θ depending on the parameter θ, and belonging to a family of distributions $\mathcal{D} = \{P_\theta : \theta \in \Theta\}$. Let $X_1, \ldots, X_n$ be a random sample. By a classical hypothesis testing approach, and for a given parameter θ defined on the space Θ, we would like to test a null hypothesis denoted by H_0, $H_0 \colon \theta \in \Theta_{H_0}$, against an alternative one denoted by H_1, $H_1 \colon \theta \in \Theta_{H_1}$, on the so-called significance level δ. We remind that this latter is defined as the probability of rejecting the null hypothesis H_0 given it is true (error type I). Note that Θ_{H_0} and Θ_{H_1} are subsets of the parameter space Θ such that $\Theta_{H_0} \cap \Theta_{H_1} = \emptyset$.

A testing problem is nothing but a decision one. A natural decision of such tests made on both hypotheses mainly relies on one of the two following possibilities:

"reject the null hypothesis H_0" or "not reject the null hypothesis H_0".

We note that contrariwise, Neyman and Pearson (1933) highlighted a three-decision procedure where they considered the possibility of emitting one out of three decisions. Their third decision is then

"the null and the alternative hypotheses are neither not rejected nor rejected".

From another side, it is known that issuing a decision relies on the so-called test statistic defined as a function of the considered sample. Let us denote by T a test statistic such that $T\colon \mathbb{R}^n \to \mathbb{R}$. Since the natural decisions of the test are mainly divided into two decisions, i.e. rejecting or not the null hypothesis, the space of possible values of the test statistic T is also divided into two regions: a rejection region called R and its complement R^c. However, it is often usual for us to consider a specific form of the rejection region R based on the defined alternative hypothesis H_1. We recall the following three typical forms of the hypotheses test dilemma. They are named as the "left one-sided," the "right one-sided," and the "two-sided." The three test forms are given, respectively, by

1. $H_0 : \theta \geq \theta_0$ vs. $H_1 : \theta < \theta_0$,   (6.1)

2. $H_0 : \theta \leq \theta_0$ vs. $H_1 : \theta > \theta_0$,   (6.2)

3. $H_0 : \theta = \theta_0$ vs. $H_1 : \theta \neq \theta_0$,   (6.3)

where θ is the parameter to test and θ_0 a particular value of this parameter. The purpose is then to decide whether to reject or not the null hypothesis H_0. Decision rules based on the statistic T are displayed in order to make a decision. Thus, respectively, to the test forms previously given, we reject H_0 if:

1. $T \leq t_l$ (left one-sided test),   (6.4)

2. $T \geq t_r$ (right one-sided test),   (6.5)

3. $T \notin (t_a, t_b)$ (two-sided test),   (6.6)

where t_l, t_r, t_a, and t_b are appropriately chosen quantiles of the distribution of T.

In other words, this testing problem is finally seen as a decision problem based on the statistic T. Equivalently to Eqs. (6.4)–(6.6), DeGroot and Schervish (2012) asserted that at a significance level δ, the conditions on the statistic T of rejecting the null hypothesis are given in terms of the probability P under the null hypothesis, by the following:

1. $P_{H_0}(T \leq t_l) = \delta$,   (6.7)

2. $P_{H_0}(T \geq t_r) = \delta$,   (6.8)

3. $P_{H_0}(T \leq t_a) = P_{H_0}(T \geq t_b) = \dfrac{\delta}{2}$,   (6.9)

where for the third case, a symmetrical probability density function is considered.

It is then clear to see that if the value of the test statistic given by $t = T(X_1, \ldots, X_n)$ falls into the rejection region R, then we tend to reject the null hypothesis H_0.

Another idea of having a decision regarding our hypotheses is to construct confidence intervals around the parameter θ. Suppose we are interested by the mean, we will then construct a confidence interval around it in a way to get for a particular test, the probability $1 - \delta$ that the mean would be contained in this interval. We call $1 - \delta$ the coverage probability or the confidence level.

Let us now give an example to illustrate the construction of such intervals. Consider the random sample $X_1, \ldots, X_n$ where X_i and X_j, $i \neq j$, are independent identically distributed and taken from the well-known normal distribution $N(\mu, \sigma)$ where σ is known. Using the confidence intervals around the mean μ and for a specific value μ_0, we would not reject the null hypothesis "H_0: $\mu = \mu_0$" if:

1. $\mu_0 \geq \overline{X} - u_{1-\delta} \dfrac{\sigma}{\sqrt{n}}$ (left one-sided test), $\qquad\qquad$ (6.10)

2. $\mu_0 \leq \overline{X} + u_{1-\delta} \dfrac{\sigma}{\sqrt{n}}$ (right one-sided test), $\qquad\qquad$ (6.11)

3. $\overline{X} - u_{1-\frac{\delta}{2}} \dfrac{\sigma}{\sqrt{n}} \leq \mu_0 \leq \overline{X} + u_{1-\frac{\delta}{2}} \dfrac{\sigma}{\sqrt{n}}$ (two-sided test), $\qquad$ (6.12)

where $\overline{X}$ is the sample mean, $u_{1-\delta}$ (respectively, $u_{1-\frac{\delta}{2}}$) is the $1 - \delta$ (respectively, $1 - \frac{\delta}{2}$) quantile of the normal distribution at the significance level δ, n is the sample size, and σ the standard deviation.

One could calculate the so-called p-value, indicating whether to reject or not the hypothesis H_0. This concept denoted by p is defined by Everitt (2006) as "The probability of the observed data (or data showing a more extreme departure from the null hypothesis) when the null hypothesis is true."

One can express the respective p-values p of the tests (6.4)–(6.6) by means of the probability distribution P depending on the null hypothesis H_0 as follows:

1. $p = P_{H_0}(T \leq t_0)$, $\qquad\qquad\qquad\qquad\qquad\qquad\qquad$ (6.13)

2. $p = P_{H_0}(T \geq t_0)$, $\qquad\qquad\qquad\qquad\qquad\qquad\qquad$ (6.14)

3. $p = 2 \min[P_{H_0}(T \leq t_0), P_{H_0}(T \geq t_0)]$, $\qquad\qquad\qquad$ (6.15)

where t_0 is the observed value of the test statistic T.

Once we calculate the p-value related to a given null hypothesis, we would be able to decide correspondingly to the following decision rule.

Decision Rule The decision is based on comparing the p-value to the pre-defined significance level δ in the following sense:

- We reject the null hypothesis H_0, if the p-value is smaller than δ.
- We do not reject it, otherwise.

6.2   Fuzzy Hypotheses

Despite the fact that the previous procedures are powerful in testing hypotheses, they could be in many cases limited. Thus, if uncertainty is taken into account in the data sets or in the hypotheses we define, the classical methods become less convenient. Methods taking into account the fuzziness in both components should be shown. We displayed in the previous chapters the definitions and properties of modelling vague data by fuzzy methods. Let us now define a hypothesis modelled by such methods.

Definition 6.2.1 (Fuzzy Hypothesis) Consider a parameter θ defined on the space Θ. A fuzzy hypothesis denoted by $\tilde{H}$ on the parameter θ, also written as "$\tilde{H}$: θ is H," is a fuzzy subset of Θ and associated to the membership function $\mu_{\tilde{H}}$.

We note that a fuzzy hypothesis can be seen as a generalization of the precise hypothesis simply denoted by H, for which $\mu_{\tilde{H}} = I_{\{\theta \in \Theta\}}$. For the fuzzy context, the hypotheses are often modelled by common fuzzy numbers similarly to the ones given in Sect. 2.3. For example, if we would like to model them by simple triangles, consider three real values $p < q < r \in \mathbb{R}$. By respect to the tests (6.1)–(6.3), the fuzzy hypotheses seen in Fig. 6.1 can be written by

1. $\tilde{H}^{OL} = (p, q, q)$ (fuzzy left one-sided hypothesis),   (6.16)

2. $\tilde{H}^{OR} = (p, p, q)$ (fuzzy right one-sided hypothesis),   (6.17)

3. $\tilde{H}^{T} = (p, q, r)$ (fuzzy two-sided hypothesis).   (6.18)

Furthermore, it is interesting to define the so-called fuzzy boundary of a given fuzzy hypotheses as shown in Parchami et al. (2010). It is given by:

Definition 6.2.2 (Boundary of a Fuzzy Hypothesis) Consider a fuzzy hypothesis $\tilde{H}$ for a parameter θ defined on the space Θ such that $\tilde{H} : \theta$ is H. The boundary $\tilde{H}^*$ of $\tilde{H}$ is a fuzzy subset of Θ, with its membership function denoted by $\mu_{\tilde{H}^*}$. The

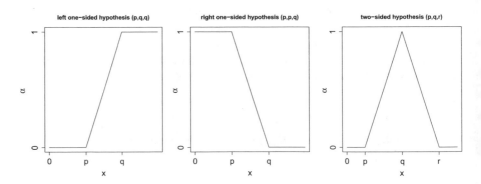

Fig. 6.1 The fuzzy hypotheses—Sect. 6.2

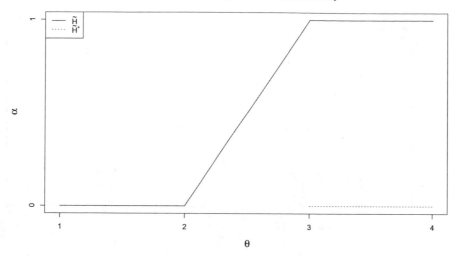

Fig. 6.2 Fuzzy hypothesis $\tilde{H}$ and its boundary $\tilde{H}^*$—Example 6.2.1

fuzzy boundaries related to the tests (6.1)–(6.3) are, respectively, expressed by

1. $\tilde{H}^* = H$ if $\theta \leq \theta_0$, 0 otherwise, ($\tilde{H}$ is left one-sided),   (6.19)

2. $\tilde{H}^* = H$ if $\theta \geq \theta_0$, 0 otherwise, ($\tilde{H}$ is right one-sided),   (6.20)

3. $\tilde{H}^* = H$, ($\tilde{H}$ is two-sided).   (6.21)

We propose the following example to illustrate this definition.

Example 6.2.1 Let the hypothesis $\tilde{H} : \theta$ be $H(\theta)$ to be a fuzzy one-sided hypothesis such that $\tilde{H}$ is given by $\tilde{H} = (2, 3, 3)$. Then, by Eq. (6.19) of Definition 6.2.2, the fuzzy hypothesis $\tilde{H}$ and its boundary can be seen in Fig. 6.2.

For further purposes, it is also important to show the definition of the median of the distribution of a test statistic in the fuzzy context.

Definition 6.2.3 (Median of the Distribution of the Test Statistic) A fuzzy set $\tilde{M}$ is said to be the median of the distribution of the test statistic T under the boundary of the fuzzy null hypothesis $\tilde{H}_0$, such that its membership function is nothing but the boundary of the null hypothesis itself $\mu_{\tilde{M}} = \tilde{H}_0^*$ and its α-cuts are given by

$$\tilde{M}_\alpha^L = \inf\{m \mid m \in \text{supp } \tilde{M}\}, \tag{6.22}$$

$$\tilde{M}_\alpha^R = \sup\{m \mid m \in \text{supp } \tilde{M}\}, \tag{6.23}$$

where m is the median of the distribution of T related to the parameter to test θ.

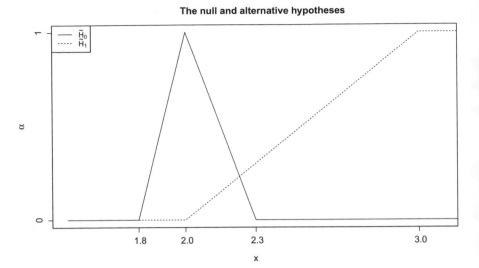

Fig. 6.3 The fuzzy null and alternative hypotheses—Example 6.2.2

An example of a fuzzy hypothesis clarifying the concept of fuzziness in the hypotheses and how to model it can be seen in the following example:

Example 6.2.2 We are interested in collecting the approximative real feel of the weather's temperature of a number of persons in a given city. Suppose we get different answers like: "near 5 degrees," "approximatively 2," "between 0 and 2," "maybe 0," etc. These answers are seen to be uncertain since these persons are not expressing their opinion precisely.

We would like to test the null hypothesis for the mean of the temperature denoted by μ. Regarding the alternative hypothesis, suppose that the test is left one-sided and given by the following triangular fuzzy numbers as given in Eqs. (6.16)–(6.18):

$$\tilde{H}_0^T = (1.8, 2, 2.3) \quad \text{against} \quad \tilde{H}_1^{OL} = (2, 3, 3).$$

These latter are shown in Fig. 6.3.

6.3   Fuzzy Confidence Intervals

It is natural to express a confidence interval for a specific parameter θ in the fuzzy environment. Indeed, we know that the parameter in this case is fuzzy type as defined in Definition 5.4.1. Therefore, getting a FCI is a direct consequence of the fuzziness of the parameter. Fuzzy confidence intervals can be defined by different approaches. Since it is obviously an epistemic situation, Kruse and Meyer

(1987) gave a definition of this interval when fuzziness occurs. Many computation procedures can then be derived from the Kruse and Meyer (1987) definition. A known one is the approach based on a pre-defined distribution around a specific parameter. Thus, by this method, one has to always fix the parameter and the distribution to be used. This leads to the fact that this method appears to be somehow limited in some contexts. We propose a method of constructing a FCI by the so-called likelihood ratio. Our proposition presents a generalization of the previous construction and gives a practical tool to estimate a FCI for any type of parameter with any type of distribution.

Let us first define the known fuzzy confidence interval, given as follows.

6.3.1 Fuzzy Confidence Intervals for Pre-defined Parameter

Consider a random sample $X_1, \ldots, X_n$ of size n with the probability distribution P_θ. Its fuzzy perception is denoted by $\tilde{X}_1, \ldots, \tilde{X}_n$. We would like to test the following hypotheses for a given parameter θ:

$$H_0 : \theta = \theta_0, \quad \text{against} \quad H_1 : \theta \neq \theta_0.$$

This task can be accomplished by constructing a FCI for θ. At a given significance level δ and according to the considered fuzzy sample, a two-sided fuzzy confidence interval denoted by $\tilde{\Pi}$ can be defined as seen in Kruse and Meyer (1987) by:

Definition 6.3.1 (Fuzzy Confidence Interval) At the significance level δ, consider a parameter θ and a symmetrical confidence interval for it denoted by $[\pi_1, \pi_2]$. A fuzzy confidence interval denoted by $\tilde{\Pi}$ is a convex and normal fuzzy set, for which its respective left and right α-cuts designated by $\tilde{\Pi}_\alpha = [\tilde{\Pi}_\alpha^L, \tilde{\Pi}_\alpha^R]$ are defined as

$$\tilde{\Pi}_\alpha^L = \inf\{a \in \mathbb{R} : \exists x_i \in (\tilde{X}_i)_\alpha, \forall i = 1, \ldots, n, \text{ such that } \pi_1(x_1, \ldots, x_n) \leq a\}, \quad (6.24)$$

$$\tilde{\Pi}_\alpha^R = \sup\{a \in \mathbb{R} : \exists x_i \in (\tilde{X}_i)_\alpha, \forall i = 1, \ldots, n, \text{ such that } \pi_2(x_1, \ldots, x_n) \geq a\}. \quad (6.25)$$

By Eq. (2.1.6), its membership function $\mu_{\tilde{\Pi}}(x)$ can be written as

$$\mu_{\tilde{\Pi}}(x) = \sup\{\alpha I_{[\tilde{\Pi}_\alpha^L, \tilde{\Pi}_\alpha^R]} : \alpha \in [0, 1]\}. \quad (6.26)$$

This fuzzy confidence interval is said to belong to $1 - \delta$ confidence region such that for all parameter θ, we have that

$$P(\tilde{\Pi}_\alpha^L \leq \theta \leq \tilde{\Pi}_\alpha^R) \geq 1 - \delta, \quad \forall \alpha \in [0; 1]. \quad (6.27)$$

From the previous definition, one could directly derive the definitions of the one-sided fuzzy confidence intervals, expressed in the following remark:

Remark 6.3.1 At a significance level δ, the left one-sided fuzzy confidence interval $\tilde{\Pi}_\alpha$ is given by

$$\tilde{\Pi}_\alpha = [\tilde{\Pi}_\alpha^L, \infty],$$

and the right one-sided one is written as

$$\tilde{\Pi}_\alpha = [-\infty, \tilde{\Pi}_\alpha^R].$$

We often draw a fuzzy sample from a known theoretical distribution. It is therefore always convenient to expose the confidence interval for a particular parameter related to this known distribution. To illustrate, we show the two-sided fuzzy confidence interval for the mean in the context of a normal distribution. It is given in the following remark:

Remark 6.3.2 Consider a fuzzy sample $\tilde{X}_1, \ldots, \tilde{X}_n$ of size n and drawn from the normal distribution with known variance. We denote by $\overline{\tilde{X}}$ its fuzzy sample mean, for which the left and right α-cuts are, respectively, written as $(\overline{\tilde{X}})_\alpha^L$ and $(\overline{\tilde{X}})_\alpha^R$. The two-sided FCI for the mean of the considered fuzzy sample is given by its α-cuts as follows:

$$\tilde{\Pi}_\alpha = \left[\tilde{\Pi}_\alpha^L, \tilde{\Pi}_\alpha^R\right] = \left[(\overline{\tilde{X}})_\alpha^L - u_{1-\frac{\delta}{2}}\frac{\sigma}{\sqrt{n}}, \ (\overline{\tilde{X}})_\alpha^R + u_{1-\frac{\delta}{2}}\frac{\sigma}{\sqrt{n}}\right], \tag{6.28}$$

where σ is the standard deviation and $u_{1-\frac{\delta}{2}}$ is the $1 - \frac{\delta}{2}$ ordered quantile from the standard normal distribution.[1]

Accordingly, it is interesting to show in Example 6.3.1, an example of computation based on such methods where we treat data as fuzzy and hypotheses as well. Despite the fact that the hypotheses of the normal distribution might not be fully fulfilled, this very known distribution will be used to only simplify the understanding of the different steps.

Example 6.3.1 We are interested in the scenario of Example 6.2.2. Consider a random sample composed by 10 observations $X_1, \ldots, X_{10}$ describing the real feel of weather temperature expressed by people in a given city. Suppose this sample to be taken from a normal distribution with its mean μ and a known variance $\sigma^2 = 1.29$, i.e. $N(\mu, \sigma^2 = 1.29)$. The estimation of the FCI for the mean μ at

---

[1]By extending the classical statistical theory to the fuzzy one, the use of such expressions for calculating the fuzzy confidence interval is actually a convention as seen in Kruse and Meyer (1987).

Table 6.1 The data set and the corresponding fuzzy number of each observation—Example 6.3.1

Index	X_i	Triangular fuzzy number
1	4	$(3, 4, 5)$
2	1	$(0, 1, 2)$
3	3	$(2, 3, 4)$
4	2	$(1, 2, 3)$
5	3	$(2, 3, 4)$
6	2	$(1, 2, 3)$
7	5	$(4, 5, 6)$
8	2	$(1, 2, 3)$
9	3	$(2, 3, 4)$
10	3	$(2, 3, 4)$
	$\overline{X} = 2.8$	$\tilde{X} = (1.8, 2.8, 3.8)$

the confidence level $1 - \delta$ can be done by the following steps:

- **Model the data:** The considered sample is supposed to be uncertain. We would like to model the fuzzy perception of it by fuzzy numbers. The simplest shape that we could use is the triangles. Thus, if a given answer is "approximately 2," then we could model it by the triangular fuzzy number written by the arbitrary tuple $(1, 2, 3)$. Suppose that we want to model the value "1" by the triangular fuzzy number given by the tuple $\tilde{L}_1 = (0, 1, 2)$, the value "2" by $\tilde{L}_2 = (1, 2, 3)$, "3" by $\tilde{L}_3 = (2, 3, 4)$, "4" by $\tilde{L}_4 = (3, 4, 5)$, and "5" by $\tilde{L}_5 = (4, 5, 6)$. This description leads to modelling our sample observations by fuzzy numbers. The original data and the fuzzified ones are provided in Table 6.1.
- **Define the required test:** At the significance level δ, suppose we want to test the following hypotheses for the mean μ where we consider a particular value $\mu_0 = 2$ such that

$$\tilde{H}_0^T = (1.8, 2, 2.3) \quad \text{against} \quad \tilde{H}_1^T = (1.25, 2, 5).$$

The purpose is then to test on a $\delta = 0.05$ significance level the fuzzy null hypothesis $\tilde{H}_0$ against the fuzzy alternative one $\tilde{H}_1$ associated, respectively, to the following α-cuts:

$$(\tilde{H}_0^T)_\alpha = \begin{cases} (\tilde{H}_0^T)_\alpha^L = 1.8 + 0.2\alpha; \\ (\tilde{H}_0^T)_\alpha^R = 2.3 - 0.3\alpha. \end{cases} \quad \text{and} \quad (\tilde{H}_1^T)_\alpha = \begin{cases} (\tilde{H}_1^T)_\alpha^L = 1.25 + 0.75\alpha; \\ (\tilde{H}_1^T)_\alpha^R = 5 - 3\alpha. \end{cases}$$

$$(6.29)$$

- **Calculate the fuzzy sample average:** We now have to calculate the fuzzy sample average $\tilde{X}$ as seen in Definition 5.4.1, where n is the number of observations $n = 10$ and $\tilde{X}_i, i = 1, \ldots, n$, are the fuzzy perceptions of the observations. The

α-cuts of the fuzzy sample average $\overline{\tilde{X}} = (1.8, 2.8, 3.8)$ are given by

$$(\overline{\tilde{X}})_\alpha = \begin{cases} (\overline{\tilde{X}})_\alpha^L = 1.8 + \alpha; \\ (\overline{\tilde{X}})_\alpha^R = 3.8 - \alpha. \end{cases} \tag{6.30}$$

- **Calculate the confidence intervals for the mean:** We finally have to calculate the fuzzy confidence intervals described in Definition 6.3.1. Using the calculated α-cuts of the fuzzy sample average $\overline{\tilde{X}}$ given in Eq. (6.30), we get the following lower FCI on the confidence level 1–0.05:

$$\tilde{\Pi}_\alpha = [\tilde{\Pi}_\alpha^L; \tilde{\Pi}_\alpha^R] = \left[(\overline{\tilde{X}})_\alpha^L - u_{1-\frac{\delta}{2}} \frac{\sigma}{\sqrt{n}}; (\overline{\tilde{X}})_\alpha^R + u_{1-\frac{\delta}{2}} \frac{\sigma}{\sqrt{n}} \right] = [1.0965 + \alpha; 4.5035 - \alpha],$$

$$\tag{6.31}$$

where $n = 10$ is the number of observations, $\sigma = 1.135$ is the standard deviation, and $u_{1-\frac{\delta}{2}} = u_{0.975} = 1.96$ is the 0.975-quantile of the normal distribution. This interval is shown in Fig. 6.4.

Similarly to the FCI for the mean of a normal distribution, it is then always necessary to consider the parameter from one side and the distribution from another one. In the fuzzy context, it is an asset to have a clear methodology that can be adapted to any type of parameter and any distribution. A practical method generalizing the construction of fuzzy confidence intervals for any parameter could be advantageous.

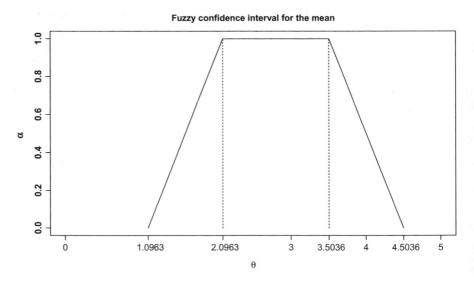

Fig. 6.4 Fuzzy confidence interval for the mean $\tilde{\Pi}$—Example 6.3.1

6.3.2 Fuzzy Confidence Intervals with the Likelihood Function

Fuzzy confidence intervals have been seen as a powerful tool in testing the null hypothesis against the alternative one for a given parameter. In the fuzzy context, following Definition 6.3.1, constructing a FCI for a particular parameter depends on a specific distribution. However, constructing a fuzzy confidence interval by a practical method where all possible cases of parameters can be applied could be welcome. In classical statistical theory, this task can be done using the so-called likelihood ratio method. For our fuzzy context, we proposed in Berkachy and Donzé (2019b) a new approach of constructing fuzzy confidence intervals based on the concept of likelihood ratio, but with considering conveniently the fuzziness contained in the variables. We mention that the likelihood ratio has been used several times in the fuzzy environment. As such, we note Gil and Casals (1988) that discussed on extending the tests by this ratio to the fuzzy context. From another side, Denoeux (2011) proposed a maximum likelihood estimation using the Expectation–Maximization method when imprecision intervenes.

First of all, let us recall the likelihood function in the classical theory, given by:

Definition 6.3.2 (Likelihood Function) Let X_i, $i = 1, \ldots, n$, be a sequence of random variables, independent identically distributed (i.i.d) and x_i, $i = 1, \ldots, n$, be the corresponding realizations. The variable X_i is drawn from a pre-defined distribution with its probability density function (pdf) denoted by $f(x_i; \underline{\theta})$. Consider $\underline{\theta}$ to be a vector of unknown parameters in the parameter space Θ. The likelihood function $L(\underline{\theta}; x_i)$ is defined by

$$L(\underline{\theta}; x_i) = f(x_i; \underline{\theta}). \tag{6.32}$$

We mention that $f(x_i; \underline{\theta})$ is called the likelihood function in this case since it is actually a function of the vector of parameters $\underline{\theta}$ rather than x_i.

When fuzziness in the variable X_i occurs, we consider its fuzzy perception i.e. the FRV $\tilde{X}_i$ with its corresponding fuzzy realization $\tilde{x}_i$. This latter is associated to a measurable MF denoted by $\mu_{\tilde{x}_i}$ in the sense of Borel, such that

$$\mu_{\tilde{x}_i} : x \to [0; 1].$$

Based on the probability concepts defined by Zadeh (1968) and seen in Sect. 5.1, we are now able to propose the likelihood function extended to the fuzzy environment. It is defined as follows:

Definition 6.3.3 (Likelihood Function of a Fuzzy Observation) Let $\tilde{\underline{\theta}}$ be a vector of fuzzy parameters in the parameter space Θ. For a single fuzzy observation $\tilde{x}_i$, the likelihood function can be written by

$$L(\tilde{\underline{\theta}}; \tilde{x}_i) = P(\tilde{x}_i; \tilde{\underline{\theta}}) = \int_{\mathbb{R}} \mu_{\tilde{x}_i}(x) f(x; \tilde{\underline{\theta}}) dx. \tag{6.33}$$

Note that one can analogously express this previous probability by means of the α-cuts of the involved fuzzy numbers.

Consider now the fuzzy sample $\underline{\tilde{x}}$ of all fuzzy realizations $\tilde{x}_i$ of the FRV $\tilde{X}_i$. Using the definition of the joint membership function of independent and identically distributed fuzzy numbers given in Definition 5.1.5, the likelihood function denoted by $L(\underline{\tilde{\theta}}; \underline{\tilde{x}})$ can be expressed by

$$L(\underline{\tilde{\theta}}; \underline{\tilde{x}}) = P(\underline{\tilde{x}}; \underline{\tilde{\theta}}) = \int_{\mathbb{R}} \mu_{\tilde{x}_1}(x) f(x; \underline{\tilde{\theta}}) dx \cdot \ldots \cdot \int_{\mathbb{R}} \mu_{\tilde{x}_n}(x) f(x; \underline{\tilde{\theta}}) dx$$

$$= \prod_{i=1}^{n} \int_{\mathbb{R}} \mu_{\tilde{x}_i}(x) f(x; \underline{\tilde{\theta}}) dx. \tag{6.34}$$

We can then deduce the log-likelihood function denoted by $l(\underline{\tilde{\theta}}; \underline{\tilde{x}})$ defined as

$$l(\underline{\tilde{\theta}}; \underline{\tilde{x}}) = \log L(\underline{\tilde{\theta}}; \underline{\tilde{x}}) \tag{6.35}$$

$$= \log \int_{\mathbb{R}} \mu_{\tilde{x}_1}(x) f(x; \underline{\tilde{\theta}}) dx + \ldots + \log \int_{\mathbb{R}} \mu_{\tilde{x}_n}(x) f(x; \underline{\tilde{\theta}}) dx. \tag{6.36}$$

Let $\hat{\underline{\tilde{\theta}}}$ be a maximum likelihood estimator of the fuzzy parameter $\underline{\tilde{\theta}}$. The likelihood ratio is the ratio

$$\frac{L(\underline{\tilde{\theta}}; \underline{\tilde{x}})}{L(\hat{\underline{\tilde{\theta}}}; \underline{\tilde{x}})},$$

where $L(\underline{\tilde{\theta}}; \underline{\tilde{x}})$ is the likelihood function depending on the fuzzy parameter $\underline{\tilde{\theta}}$, and $L(\hat{\underline{\tilde{\theta}}}; \underline{\tilde{x}})$ is the likelihood function evaluated at the estimator $\hat{\underline{\tilde{\theta}}}$ such that $L(\hat{\underline{\tilde{\theta}}}; \underline{\tilde{x}}) \neq 0$ and finite.

We express after the logarithm of the likelihood ratio. In this context, it is defined as the difference between the log-likelihood functions evaluated, respectively, at $\hat{\underline{\tilde{\theta}}}$ and $\underline{\tilde{\theta}}$, also called the deviance. The LR can be defined by

$$LR = -2 \log \frac{L(\underline{\tilde{\theta}}; \underline{\tilde{x}})}{L(\hat{\underline{\tilde{\theta}}}; \underline{\tilde{x}})} = 2\left[l(\hat{\underline{\tilde{\theta}}}; \underline{\tilde{x}}) - l(\underline{\tilde{\theta}}; \underline{\tilde{x}})\right], \tag{6.37}$$

where $L(\hat{\underline{\tilde{\theta}}}; \underline{\tilde{x}}) \neq 0$, $L(\underline{\tilde{\theta}}; \underline{\tilde{x}}) \neq 0$ and are both finite.

In fact, in the classical statistical approach, this ratio is proved to be asymptotically Chi-squared distributed with k degrees of freedom. Note that the parameter k is chosen in accordance with the number of constraints expressed by the null hypothesis. However, in a fuzzy context, it is not clear if this asymptotical property holds. If this question has to be elaborated, multiple methods can be used to solve such problems and can be used to consistently estimate the distribution of a given statistic, such as asymptotic or bootstrap techniques. This latter will be further investigated in the next subsection (Sect. 6.3.3).

Constructing a $100(1 - \delta)\%$ fuzzy confidence interval by the likelihood ratio consists of finding every value $\tilde{\theta}$ at which the null hypothesis is at the point of not being rejected (accepted). Let us consider η to be the $1 - \delta$-quantile of the distribution of LR. Once η is estimated, we could express the confidence interval in terms of the following:

$$2\left[l(\hat{\tilde{\theta}}; \underline{\tilde{x}}) - l(\tilde{\theta}; \underline{\tilde{x}})\right] \leq \eta. \tag{6.38}$$

This latter can also be written as

$$l(\tilde{\theta}; \underline{\tilde{x}}) \geq l(\hat{\tilde{\theta}}; \underline{\tilde{x}}) - \frac{\eta}{2}, \tag{6.39}$$

which means practically that the constructed confidence interval includes all possible values of $\tilde{\theta}$ where the variation of the log-likelihood maximum is about $\frac{\eta}{2}$ no more.

Consequently, one could clearly see the importance of the principle of the likelihood ratio in the field of hypothesis testing, namely in the likelihood ratio test statistic. Based on the likelihood ratio, our main purpose is then to construct a fuzzy confidence interval $\tilde{\Pi}_{LR}$ given by its left and right α-cuts $[(\tilde{\Pi}_{LR})_\alpha^L; (\tilde{\Pi}_{LR})_\alpha^R]$ such that for all parameters θ the following equation

$$P\left((\tilde{\Pi}_{LR})_\alpha^L \leq \theta \leq (\tilde{\Pi}_{LR})_\alpha^R\right) \geq 1 - \delta, \quad \forall \alpha \in [0; 1], \tag{6.40}$$

holds. Accordingly, we propose the following procedure of construction of fuzzy confidence intervals.

Procedure

Since the data set is assumed to be imprecise, it is a clear consequence to see that the log-likelihood is a function of fuzzy entities. Indeed, the maximum likelihood estimator is modelled by a fuzzy number as well. Thus, this estimator is no more reduced to a single element only, as in the classical approach. One should rather consider every element of the support set of the fuzzy parameter. Our idea is then to revisit the likelihood procedure of constructing fuzzy confidence intervals by introducing the fact of modelling the fuzziness contained in the distribution. However, this task is seen as a tedious one, since the computational burden induced in the calculation of the log-likelihood function by considering every value of the set defining the support of the fuzzy parameter is serious. To solve this problem, we will choose particular values for which we will calculate the so-called threshold points. We will be interested in the intersection points between these latter and the log-likelihood function. By combining all these information, we will be able to propose fuzzy confidence intervals for any parameter with any type of distribution.

To detail, let us first introduce the so-called standardizing function:

Definition 6.3.4 (Standardizing Function) Consider a fuzzy number $\tilde{\theta}$ with its MF $\mu_{\tilde{\theta}}$. Let $\theta \in$ supp $(\tilde{\theta})$. The standardizing function denoted by I_{stand} is given by

$$I_{stand} : \quad \mathbb{R} \to \mathbb{R}$$

$$l(\theta, \underline{\tilde{x}}) \mapsto I_{stand}\big(l(\theta, \underline{\tilde{x}})\big) = \frac{l(\theta, \underline{\tilde{x}}) - I_a}{I_b - I_a},$$

where I_a, I_b are arbitrary real values such that $I_a < I_b$, $I_a \leq l(\theta, \underline{\tilde{x}}) \leq I_b$, and $I_{stand}\big(l(\theta, \underline{\tilde{x}})\big)$ is bounded. We mention that $0 \leq I_{stand}(l(\theta, \underline{\tilde{x}})) \leq 1$. Thus, this function is intentionally introduced to preserve the $[0; 1]$-interval property of the α-cuts of a given fuzzy number.

The calculation procedure can now on be given by the subsequent steps.

For a given fuzzy parameter $\tilde{\theta}$, the very first task is to calculate the log-likelihood function $l(\tilde{\theta}; \underline{\tilde{x}})$ given in Eq. (6.36).

Second, we have to carefully choose the needed values for the calculations. The support set of any fuzzy parameter is composed of an infinity of elements. From the set of values, we propose to take into account the lower and upper bounds of the support and the core sets only. Thus, the number of considered elements is reduced to four values. Let supp$(\hat{\tilde{\theta}})$ and core$(\hat{\tilde{\theta}})$ be, respectively, the support and the core sets of $\hat{\tilde{\theta}}$. The four values denoted by $p \leq q \leq r \leq s$ are written by

$$p = \min(\text{supp}(\hat{\tilde{\theta}})); \tag{6.41}$$

$$q = \min(\text{core}(\hat{\tilde{\theta}})); \tag{6.42}$$

$$r = \max(\text{core}(\hat{\tilde{\theta}})); \tag{6.43}$$

$$s = \max(\text{supp}(\hat{\tilde{\theta}})). \tag{6.44}$$

Since the fuzzy parameter is bounded and the sets supp$(\hat{\tilde{\theta}})$ and core$(\hat{\tilde{\theta}})$ are not empty, then the four values p, q, r, and s always exist.

We highlight that in our situation we consider the case of a symmetrical probability function where each part of the log-likelihood function (left or right hand sides) is monotonic and continuous. As such, the left hand side is always increasing and the right one is decreasing.

We afterward estimate η. This is done by the bootstrap technique as detailed in the next subsection. For the considered values p, q, r, and s, we after have to calculate their corresponding threshold values denoted by I_1, I_2, I_3, and I_4. It is done by affecting the value θ by each of the four values in the right hand side of

Eq. (6.39) as follows:

$$I_1 = l(p; \tilde{x}) - \frac{\eta}{2};$$

(6.45)

$$I_2 = l(q; \tilde{x}) - \frac{\eta}{2};$$

(6.46)

$$I_3 = l(r; \tilde{x}) - \frac{\eta}{2};$$

(6.47)

$$I_4 = l(s; \tilde{x}) - \frac{\eta}{2}.$$

(6.48)

In addition, it is important to find the minimum and maximum thresholds denoted, respectively, by $I_{\min}$ and $I_{\max}$. They are given by

$$I_{\min} = \min(I_1, I_2, I_3, I_4),$$

(6.49)

$$\text{and } I_{\max} = \max(I_1, I_2, I_3, I_4).$$

(6.50)

We actually consider $I_{\min}$ and $I_{\max}$ in the purpose of covering the interval of all the possible values that could verify Eq. (6.39).

Next, an intersection exists between the log-likelihood function and the threshold values I_1, I_2, I_3, and I_4. We denote by $\theta_1^{\star L}, \theta_2^{\star L}, \theta_3^{\star L}, \theta_4^{\star L}$ and $\theta_1^{\star R}, \theta_2^{\star R}, \theta_3^{\star R}, \theta_4^{\star R}$ the obtained intersection abscisses. These values can be found by solving the equations written by

$$l^L(\theta_1^{\star L}; \tilde{x}) = I_1 \quad \text{and} \quad l^R(\theta_1^{\star R}; \tilde{x}) = I_1,$$

(6.51)

$$l^L(\theta_2^{\star L}; \tilde{x}) = I_2 \quad \text{and} \quad l^R(\theta_2^{\star R}; \tilde{x}) = I_2,$$

(6.52)

$$l^L(\theta_3^{\star L}; \tilde{x}) = I_3 \quad \text{and} \quad l^R(\theta_3^{\star R}; \tilde{x}) = I_3,$$

(6.53)

$$l^L(\theta_4^{\star L}; \tilde{x}) = I_4 \quad \text{and} \quad l^R(\theta_4^{\star R}; \tilde{x}) = I_4.$$

(6.54)

Note that the symbols "L" and "R" refer to the left and right sides of a given entity. We have to introduce also the minimum and maximum left intersection abscisses given by

$$\theta_{\inf}^{\star L} = \inf(\theta_1^{\star L}, \theta_2^{\star L}, \theta_3^{\star L}, \theta_4^{\star L}),$$

(6.55)

$$\text{and } \theta_{\sup}^{\star L} = \sup(\theta_1^{\star L}, \theta_2^{\star L}, \theta_3^{\star L}, \theta_4^{\star L}).$$

(6.56)

For the right side, the minimum and maximum intersection abscisses are written as

$$\theta_{\inf}^{\star R} = \inf(\theta_1^{\star R}, \theta_2^{\star R}, \theta_3^{\star R}, \theta_4^{\star R}),$$

(6.57)

$$\text{and } \theta_{\sup}^{\star R} = \sup(\theta_1^{\star R}, \theta_2^{\star R}, \theta_3^{\star R}, \theta_4^{\star R}).$$

(6.58)

Additionally, let us mention that the calculated abscisses are single real values. They are used in the construction of the α-cuts of the FCI denoted by $\tilde{\Pi}_{LR}$. Combining all the aforecalculated components, this fuzzy confidence interval using the likelihood ratio can now on be expressed. It is given by its left and right α-cuts $(\tilde{\Pi}_{LR})_\alpha = \left[(\tilde{\Pi}_{LR})_\alpha^L; (\tilde{\Pi}_{LR})_\alpha^R\right]$ defined, respectively, as follows:

$$(\tilde{\Pi}_{LR})_\alpha^L = \left\{\theta \in \mathbb{R} \mid \theta_{\inf}^{\star L} \leq \theta \leq \theta_{\sup}^{\star L} \text{ and } \alpha = I_{stand}\left(l(\theta, \underline{\tilde{x}})\right) = \frac{l(\theta, \underline{\tilde{x}}) - I_{\min}}{I_{\max} - I_{\min}}\right\}, \quad (6.59)$$

$$(\tilde{\Pi}_{LR})_\alpha^R = \left\{\theta \in \mathbb{R} \mid \theta_{\inf}^{\star R} \leq \theta \leq \theta_{\sup}^{\star R} \text{ and } \alpha = I_{stand}\left(l(\theta, \underline{\tilde{x}})\right) = \frac{l(\theta, \underline{\tilde{x}}) - I_{\min}}{I_{\max} - I_{\min}}\right\}. \quad (6.60)$$

It is important now to prove that the fuzzy confidence interval $\tilde{\Pi}_{LR}$ verifies Definition 6.3.1.

Proof First of all, it is important to see that Eqs. (6.59) and (6.60) correspond, respectively, to Eqs. (6.24) and (6.25).

Consider the interval $[\pi_1^{LR}; \pi_2^{LR}]$ to be a crisp confidence interval calculated using the classical likelihood ratio method. The values π_1^{LR} and π_2^{LR} correspond to the values of any interval given for instance by the respective values $\theta_1^{\star L}$ and $\theta_1^{\star R}$. Therefore, it is evident to see that the couples of values taken two by two, respectively, from the vectors $(\theta_1^{\star L}, \ldots, \theta_4^{\star L})$ and $(\theta_1^{\star R}, \ldots, \theta_4^{\star R})$—in addition to any other couple of crisp real values constructed in the same way—constitute a family of crisp confidence intervals for every particular value of θ.

In addition, we could write the FCI given in Eqs. (6.59) and (6.60) by the fundamental expressions of its α-cuts $(\tilde{\Pi}_{LR})_\alpha^L$ and $(\tilde{\Pi}_{LR})_\alpha^R$ as shown in Eq. (2.8). They are given by

$$(\tilde{\Pi}_{LR})_\alpha^L = \inf\left\{x \in \mathbb{R} \mid \mu_{\tilde{\Pi}_{LR}}(x) \geq \alpha\right\} \text{ and } (\tilde{\Pi}_{LR})_\alpha^R = \sup\left\{x \in \mathbb{R} \mid \mu_{\tilde{\Pi}_{LR}}(x) \geq \alpha\right\}.$$

Following the construction of $\theta_{\inf}^{\star L}$, $\theta_{\inf}^{\star R}$, $\theta_{\sup}^{\star L}$, and $\theta_{\sup}^{\star R}$ given in Eqs. (6.55)–(6.58) and by consequence of generating the family of crisp confidence intervals, one could directly deduce that Eqs. (6.59) and (6.60) can equivalently be written by

$$(\tilde{\Pi}_{LR})_\alpha^L = \inf\left\{b \in \mathbb{R} : \exists x_i \in (\tilde{X}_i)_\alpha, \forall i = 1, \ldots, n, \text{ such that } \pi_1^{LR}(x_1, \ldots, x_n) \leq b\right\}, \quad (6.61)$$

$$(\tilde{\Pi}_{LR})_\alpha^R = \sup\left\{b \in \mathbb{R} : \exists x_i \in (\tilde{X}_i)_\alpha, \forall i = 1, \ldots, n, \text{ such that } \pi_2^{LR}(x_1, \ldots, x_n) \geq b\right\}. \quad (6.62)$$

We have now to prove that $\tilde{\Pi}_{LR}$ is nothing but a convex and normal fuzzy set. Thus, by effect of the introduction of the standardizing function I_{stand} and the upper semi-continuity of the log-likelihood function on its corresponding definition domain, it is clear to assert that $\tilde{\Pi}_{LR}$ is a normal fuzzy set. In other terms, by

Definition 6.3.4, the following equalities are verified $\forall \theta \in \mathbb{R}$,

$$\sup(\mu_{\tilde{\Pi}_{LR}}(\theta)) = \sup\left(I_{stand}(l(\theta, \underline{\tilde{x}}))\right) = 1, \tag{6.63}$$

$$\inf(\mu_{\tilde{\Pi}_{LR}}(\theta)) = \inf\left(I_{stand}(l(\theta, \underline{\tilde{x}}))\right) = 0. \tag{6.64}$$

For the convexity of $\tilde{\Pi}_{LR}$ as proposed in Definition 2.1.1, we have to prove that

$$\mu_{\tilde{\Pi}_{LR}}(t\theta_1 + (1-t)\theta_2) \geq \min\left(\mu_{\tilde{\Pi}_{LR}}(\theta_1), \mu_{\tilde{\Pi}_{LR}}(\theta_2)\right), \quad \forall \theta_1, \theta_2 \in \mathbb{R} \text{ and } t \in [0; 1]. \tag{6.65}$$

This equation can be proved as follows:
Consider two values θ_1 and θ_2 in $\mathbb{R}$. By Definition 6.3.4, we know that the membership function of $\tilde{\Pi}_{LR}$ can be written by

$$\mu_{\tilde{\Pi}_{LR}}(t\theta_1 + (1-t)\theta_2) = I_{stand}\left(l(t\theta_1 + (1-t)\theta_2, \underline{\tilde{x}})\right).$$

We know that the likelihood function is concave, i.e.

$$l(t\theta_1 + (1-t)\theta_2, \underline{\tilde{x}}) \geq t\, l(\theta_1, \underline{\tilde{x}}) + (1-t)\, l(\theta_2, \underline{\tilde{x}}).$$

Then, using Definition 6.3.4 and since the application I_{stand} is monotone, we get that

$$\mu_{\tilde{\Pi}_{LR}}(t\theta_1 + (1-t)\theta_2) \geq I_{stand}\left(t\, l(\theta_1, \underline{\tilde{x}}) + (1-t)\, l(\theta_2, \underline{\tilde{x}})\right)$$

$$\geq \frac{t\, l(\theta_1, \underline{\tilde{x}}) + (1-t)\, l(\theta_2, \underline{\tilde{x}}) - I_{min}}{I_{max} - I_{min}}$$

$$\geq \frac{t\, l(\theta_1, \underline{\tilde{x}}) - t\, I_{min}}{I_{max} - I_{min}} + \frac{(1-t)\, l(\theta_2, \underline{\tilde{x}}) - (1-t)\, I_{min}}{I_{max} - I_{min}}$$

$$\geq t\, I_{stand}(l(\theta_1, \underline{\tilde{x}})) + (1-t)\, I_{stand}(l(\theta_2, \underline{\tilde{x}}))$$

$$\geq t\, \mu_{\tilde{\Pi}_{LR}}(\theta_1) + (1-t)\, \mu_{\tilde{\Pi}_{LR}}(\theta_2)$$

$$\geq \min(\mu_{\tilde{\Pi}_{LR}}(\theta_1), \mu_{\tilde{\Pi}_{LR}}(\theta_2)),$$

with $\mu_{\tilde{\Pi}_{LR}}(\theta_1)$ and $\mu_{\tilde{\Pi}_{LR}}(\theta_2) \in [0; 1]$, $\mu_{\tilde{\Pi}_{LR}}(\theta_1) = I_{stand}\left(l(\theta_1, \underline{\tilde{x}})\right)$ and $\mu_{\tilde{\Pi}_{LR}}(\theta_2) = I_{stand}(l(\theta_2, \underline{\tilde{x}}))$. $\square$

The previous procedure of construction of the fuzzy confidence interval by the likelihood ratio method is summarized in the following algorithm:

Algorithm:

1. Calculate the log-likelihood function $l(\tilde{\theta}; \underline{\tilde{x}})$—Eq. (6.36).
2. Find the needed values for computations p, q, r, and s—Eqs. (6.41)–(6.44).
3. Estimate the quantile η by the bootstrap technique—Eq. (6.37).
4. Calculate the threshold points I_1, I_2, I_3, and I_4—Eqs. (6.45)–(6.48).
5. Calculate the minimum and maximum thresholds $I_{\min}$ and $I_{\max}$—Eqs. (6.49) and (6.50).
6. Find the intersection points $\theta_1^{\star L}$, $\theta_2^{\star L}$, $\theta_3^{\star L}$, $\theta_4^{\star L}$ and $\theta_1^{\star R}$, $\theta_2^{\star R}$, $\theta_3^{\star R}$, $\theta_4^{\star R}$ obtained from Eqs. (6.51)–(6.54).
7. Find the minimum and maximum intersection abscisses $\theta_{\inf}^{\star L}$, $\theta_{\sup}^{\star L}$, $\theta_{\inf}^{\star R}$, and $\theta_{\sup}^{\star R}$—Eqs. (6.55)–(6.58).
8. Define the fuzzy confidence interval by writing its left and right α-cuts—Eqs. (6.59) and (6.60).

Concerning the coverage level of our fuzzy confidence interval, from a theoretical point of view, we would like to prove that Eq. (6.40) is verified. It can be done in the following manner:

Proof Suppose that $\tilde{\Pi}_{LR}$ is a two-sided FCI. By definition, this FCI is bounded. We would like to prove that for every value θ

$$P_H\left((\tilde{\Pi}_{LR})_\alpha^L \leq \theta \leq (\tilde{\Pi}_{LR})_\alpha^R\right) \geq 1 - \delta, \quad \forall \delta \in [0; 1]. \tag{6.66}$$

We know that

$$P_H\left(\theta \in (\tilde{\Pi}_{LR})_\alpha\right) = 1 - P_H\left(\theta \notin (\tilde{\Pi}_{LR})_\alpha\right).$$

If every value of θ does not belong to the α-level set of $\tilde{\Pi}_{LR}$, it is equivalent to saying that this parameter does not belong to its support set nor to its closure denoted by $(\tilde{\Pi}_{LR})_{\alpha=0}$. Therefore, we could write this equivalence as follows:

$$P_H\left(\theta \in (\tilde{\Pi}_{LR})_\alpha\right) = 1 - P_H\left(\theta \notin (\text{supp}(\tilde{\Pi}_{LR})) \cup \theta \notin (\tilde{\Pi}_{LR})_{\alpha=0}\right).$$

Then, let π_{min} and π_{max} be, respectively, the lower and upper bounds of the support set of the fuzzy confidence interval $\tilde{\Pi}_{LR}$ written by $[\pi_{min}; \pi_{max}]$. We denote also by $[\pi_{min}^0; \pi_{max}^0]$ the lower and upper bounds of the closure set, i.e. $(\tilde{\Pi}_{LR})_{\alpha=0}$. It is natural to see that $[\pi_{min}; \pi_{max}] \subset [\pi_{min}^0; \pi_{max}^0]$. Since our construction is generated as a family of classical confidence intervals by the likelihood ratio method, it is evident to see that the interval $[\pi_{min}^0; \pi_{max}^0]$ is nothing but the largest crisp confidence interval related to a particular value $\theta \in \text{supp}(\tilde{\theta})$ at the significance level $1 - \delta$. Therefore, we have that $P_H(\theta \notin (\tilde{\Pi}_{LR})_\alpha) \leq \delta$, which is equivalent to $P_H(\theta \notin [\pi_{min}^0; \pi_{max}^0]) \leq \delta$, for every value of θ. Accordingly, verifying Eq. (6.66) is given by

$$P_H(\theta \in (\tilde{\Pi}_{LR})_\alpha) = P_H(\theta \in [\pi_{min}^0; \pi_{max}^0]) \geq 1 - \delta, \quad \forall \delta \in [0; 1].$$

For the cases of one-sided FCI, the proofs can be similarly conceived.   □

6.3.3 Bootstrap Technique for the Approximation of the Likelihood Ratio

The bootstrap technique has been mainly proposed by Efron (1979). Considering a random sample taken from an unknown distribution, Efron (1979) described a complete procedure of estimation of a particular sampling distribution based on some observed data. This method consists of drawing a large number of samples based on a particular primary data set. Accordingly, a bootstrap distribution is constructed and further studied. The aim of such techniques is then to grant the approximation of a sampling distribution using random sampling tools. Such methods have been considered as satisfactory in various situations. The bootstrap has been afterward re-considered in the fuzzy environment. As such, Gonzalez-Rodriguez et al. (2006) studied the hypotheses testing procedure related to the mean of fuzzy random variables using bootstrap. The fuzzy bootstrap is also seen as easier and computationally lighter than other techniques, i.e. the asymptotic one, as stated in Montenegro et al. (2004).

We propose to use the bootstrap technique defined in the fuzzy environment in order to empirically estimate the distribution of the required deviance of the log-likelihood function evaluated at $\hat{\tilde{\theta}}$ and the one evaluated at $\tilde{\theta}$ denoted by LR and given in Eq. (6.37).

In a general manner, let us consider η to be an appropriate quantile of the distribution of LR. Two bootstrap approaches for imprecisely valued samples are exposed:

- The first one is to simply generate with replacement a large number D of bootstrap samples and consequently calculate the needed deviance.
- The second one is to generate D bootstrap samples by preserving the characteristics of location and dispersion s_0 and ϵ of the nearest symmetrical trapezoidal fuzzy numbers related to the initial data set. These characteristics have been closely described in Proposition 4.7.6.

For such approaches, two algorithms can be proposed. The first one, i.e. Algorithm 1, consists of simply generating with replacement D random bootstrap samples as follows:

Algorithm 1:

1. Considering a particular estimator $\hat{\tilde{\theta}}$, compute the value of the deviance $2\left[l(\hat{\tilde{\theta}}; \tilde{x}) - l(\tilde{\theta}; \tilde{x})\right]$ related to the considered original fuzzy sample.
2. From the primary data set, draw randomly with replacement of a set of fuzzy observations to construct the bootstrap data set.
3. Calculate the bootstrap deviance $2\left[l(\hat{\tilde{\theta}}; \tilde{x}) - l(\tilde{\theta}; \tilde{x})\right]^{\text{boot}}$.
4. Recursively repeat Steps 2 and 3 a large number D of times to get a set of D values constructing the distribution of bootstrap deviance.
5. Find η the $1 - \delta$-quantile of the bootstrap distribution of LR.

The algorithm using the couple of characteristics (s_0, ϵ) can also be given as follows:

Algorithm 2:

1. Consider a fuzzy sample. Calculate for each observation the couple of characteristics (s_0, ϵ).
2. From the calculated set of couples (s_0, ϵ), draw randomly with replacement and with equal probabilities a sample of couples (s_0, ϵ). Construct the bootstrap sample.
3. Calculate the bootstrap deviance $2\left[l(\hat{\tilde{\theta}}; \tilde{x}) - l(\tilde{\theta}; \tilde{x})\right]^{\text{boot}}$ for each generated sample.
4. Recursively repeat Steps 2 and 3 a large number D of times to get a set of D values constructing the distribution of bootstrap deviance.
5. Find η the $1 - \delta$-quantile of the bootstrap distribution of LR.

It is clear that a maximum likelihood estimator has to be calculated. As such, we propose to use the EM-algorithm applied in the fuzzy environment as proposed by Denoeux (2011). We highlight that this estimation procedure gives a crisp estimator. It is interesting to model the fuzzy perception of such estimators. For this purpose, we propose to model the obtained crisp estimator by a triangular fuzzy number for which this estimator constitutes the core of the chosen fuzzy number. In order to reduce the complexification of the procedure related to a possible effect caused by the shape of the fuzzy number, we consider symmetrical shapes only. We then get a triangular symmetrical fuzzy number modelling the maximum likelihood estimator (MLE) estimator. For such calculations, the R package EM.Fuzzy described in Parchami (2018) can be used.

6.3.4 Simulation Study

To illustrate our calculation procedures, we present a simulation study where we consider several randomly generated data sets from different sample sizes. To simplify the case, we generated data sets taken from a normal distribution $N(5, 1)$ and composed by 100–1000 observations. The corresponding observations are modelled by triangular symmetrical fuzzy numbers with a support set of spread 2.

We are interested in estimating the bootstrap distribution of the likelihood ratio for the theoretical mean of the considered data set. The corresponding bootstrap quantiles η have been calculated for each data set by both Algorithms 1 and 2 and using different numbers of iterations going from 100 to 1000. For our specific case, we reduce ourselves to the case of $D = 1000$ iterations only. In addition, we propose two cases of modelling the MLE estimator under the normality assumption:

- The first case is by a triangular symmetrical fuzzy number of spread 2.
- The second one is by a triangular symmetrical fuzzy number of spread 1.

We would be interested to investigate the influence of the degree of fuzziness in the modelling procedure of the estimator. As a matter of vulgarization, we use also the fuzzy sample mean as a fuzzy estimator in the process of calculation.

At the significance level 0.05, the quantiles of the distribution of the likelihood ratio LR estimated by the bootstrap technique can be found in Table 6.2. From this latter, it is clear to see that the distributions calculated using both bootstrap algorithms give close quantiles. From another side, the quantiles related to each size of the generated samples are somehow close. A remark could arise also is that less fuzziness (i.e. fuzzy number modelling the MLE estimator with spread 1) yields to a smaller quantile compared to the case where the fuzziness is higher (i.e. fuzzy number modelling the MLE estimator with spread 2). We remark also that in the conventional approach, the deviance in the same setups is majorated by the quantile 3.84 taken from the χ^2 distribution at the significance level 0.05 with 1 constraint.

Table 6.2 The 95%-quantiles of the bootstrap distribution of LR—Case of a data set taken from a normal distribution $N(5, 1)$ modelled by triangular symmetrical fuzzy numbers at 1000 iterations

Algorithm 1			
Sample size	$N = 50$	$N = 100$	$N = 500$
Bootstrap quantile using the sample mean	1.990	2.038	2.342
Bootstrap quantile using the MLE estimator (Spread 2)	1.809	1.927	2.181
Bootstrap quantile using the MLE estimator (Spread 1)	1.523	1.626	1.825
Algorithm 2			
Sample size	$N = 50$	$N = 100$	$N = 500$
Bootstrap quantile using the sample mean	1.802	1.845	2.118
Bootstrap quantile using the MLE estimator (Spread 2)	1.854	1.971	2.201
Bootstrap quantile using the MLE estimator (Spread 1)	1.563	1.671	1.864

Table 6.3 The fuzzy confidence interval by the likelihood ratio at the 95% significance level—Case of 500 observations taken from a normal distribution $N(5, 1)$ modelled by triangular symmetrical fuzzy numbers

Algorithm 1	Support set		Core set	
	Lower	Upper	Lower	Upper
FCI using the sample mean	3.991	5.996	4.940	5.047
FCI using the MLE estimator (Spread 2)	3.804	6.184	4.795	5.193
FCI using the MLE estimator (Spread 1)	4.303	5.685	4.797	5.191
Algorithm 2				
	Lower	Upper	Lower	Upper
FCI using the sample mean	3.991	5.996	4.945	5.042
FCI using the MLE estimator (Spread 2)	3.803	6.184	4.795	5.193
FCI using the MLE estimator (Spread 1)	4.303	5.685	4.797	5.191

Using Table 6.2 giving the estimated bootstrap quantiles, we propose now to calculate the fuzzy confidence intervals by the likelihood ratio as shown in the procedure of Sect. 6.3.2. Note that for sake of simplification, we will only discuss the case with 500 observations. The obtained fuzzy confidence intervals are exposed in Table 6.3. This latter gives the lower and upper bounds of the support and the core sets of the calculated FCI.

From Table 6.3, we could easily remark that Algorithms 1 and 2 give somehow the same support and core sets of the FCI. In other terms, we could say that no effect of choosing any of the two defended bootstrap algorithms clearly arose. Consequently, adopting Algorithm 2 where the location and dispersion characteristics are preserved gives similar results using Algorithm 1. Therefore, we highlight that the fluctuations in the estimated bootstrap quantiles between both algorithms have no big differences on the obtained FCI.

In addition, we could evidently see that the fuzziness of the estimator has a direct influence on the calculated FCI. Thus, the FCI related to the MLE estimator with lower fuzziness (Spread 1) has the smallest support set between all. Accordingly, we could conclude that the degree of fuzziness of the calculated FCI is directly affected by the degree of fuzziness present in the modelling process of the estimator using fuzzy numbers. Conveniently modelling the MLE estimator plays then an important role.

Furthermore, let us consider $\tilde{\Pi}$ the traditional fuzzy confidence interval as described in Sect. 6.3.1. We obtain the following FCI:

$$\tilde{\Pi} = (3.907, 4.907, 5.080, 6.080).$$

If we would be interested to compare the bootstrap fuzzy confidence intervals given in Table 6.3 to $\tilde{\Pi}$, we remark that the FCI using the MLE estimators has slightly larger core sets. This could be due to the method of calculation of the MLE estimators. However, the support sets vary between the cases using the lowest spread and the highest spread of the fuzzy estimator.

Considering the fuzzy sample mean as an estimator used in the calculations of the LR-FCI, we observe that the core and the support sets of this latter are smaller than the ones of $\tilde{\Pi}$. It is then clear that having a broad idea about FCI calculated in different modelling scenarios with some other setups, estimators, and distributions is important to elaborate in further studies.

Concerning the coverage rates in the context previously described, we performed a simulation study by generating data sets composed by 100–1000 observations. The main purpose was to estimate a fuzzy confidence interval for the mean of such distributions at the significance level 0.05 and consequently to investigate the corresponding coverage rates.

Since the distribution of the likelihood ratio is unknown, we have performed the bootstrap techniques using Algorithms 1 and 2. In addition, similarly to the cases already provided, we have considered calculations using the fuzzy sample mean as an estimator, as well as the MLE estimator with spread 2, and the one with spread 1. We add that we calculated the fuzzy confidence intervals by the traditional fuzzy method as shown in Eq. (6.28) in the purpose of comparison between all.

Briefly, the obtained coverage rate of all types of confidence intervals seemed to be very interesting to explore. The coverage rates of the fuzzy confidence intervals by Algorithms 1 and 2 are almost the same for all chosen setups. For this purpose, we will only interpret the intervals based on Algorithm 1.

We found that the coverage levels of our confidence intervals are very close to the ones calculated by the traditional method. In comparing the traditional FCI to ours, we got an overall difference in coverage rates not exceeding 1.4%. We expose here the case using the same characteristics described in the abovementioned analysis.

The data sets are composed of 100–1000 observations. These data sets are supposed to be imprecise and modelled by triangular symmetrical fuzzy numbers such that the spread of each fuzzy number is equal to 2. For instance, for the data set composed by 500 fuzzy observations, the coverage rate of the traditional fuzzy confidence interval is about 94.4% in the core set and about 100% in the support set, while the one by the likelihood ratio method for the fuzzy sample mean is exactly the same. However, for the fuzzy confidence intervals using the small and large spreads of the MLE estimator, i.e. spread 1 and spread 2, respectively, we obtained a coverage rate of 95.6% in the core for both intervals and about 100% in the support set. It is clear that such rates are approximatively acceptable since a 95% confidence level has to be insured. A direct interpretation of this latter is that the fuzziness of the MLE estimator does not affect the coverage rate of the expected fuzzy confidence interval but affects strongly its spread.

From our simulation studies, a general conclusion could be that using the fuzzy sample mean as an estimator, our interval seems to be slightly more restrictive than the traditional fuzzy one. Contrariwise, once the MLE estimator is considered, the LR-intervals appeared to be moderately less restrictive. However, additional studies have to be performed to confirm or not our coverage rates in various setups with some other calculations of the MLE estimators.

6.3.5   *Numerical Example*

Let us now illustrate our methodology. We re-consider the same data set used for Example 6.3.1 and calculate in Example 6.3.2 the FCI by the likelihood ratio.

Example 6.3.2 Consider again the fuzzified sample of 10 observations given in Table 6.1 such that its fuzzy sample mean is given by $\overline{\tilde{X}} = (1.8, 2.8, 3.8)$. We remind that this primary sample is derived from a normal distribution with a standard deviation of 1.135.

Similarly to Example 6.3.1, we would like to calculate the FCI for the mean by the likelihood ratio method. By the aforementioned procedure, the fuzzy confidence interval $\tilde{\Pi}_{LR}$ at the confidence level $1 - \delta = 1 - 0.05$ can be found as follows.

First of all, we have to calculate the log-likelihood function. For the fuzzy sample $\tilde{X}_i, i = 1, \ldots, 10$, this function is given by

$$
l(\tilde{\theta}; \underline{\tilde{x}}) = \log \int_{\mathbb{R}} \mu_{\tilde{X}_1}(x) f(x; \tilde{\theta}) dx + \ldots + \log \int_{\mathbb{R}} \mu_{\tilde{X}_{10}}(x) f(x; \tilde{\theta}) dx
$$

$$
= \log \int_3^4 (3 + x) f(x; \tilde{\theta}) dx + \log \int_4^5 (5 - x) f(x; \tilde{\theta}) dx + \ldots
$$

$$
+ \log \int_2^3 (2 + x) f(x; \tilde{\theta}) dx + \log \int_3^4 (4 - x) f(x; \tilde{\theta}) dx.
$$

Mention that the function denoted by $f(x; \tilde{\theta})$ is the probability density function of the standard normal distribution where $\sigma = 1.135$.

Since the sample is assumed to be fuzzy, we can deduce that the considered parameter is fuzzy as well. We calculate the crisp maximum likelihood estimator for the mean by the EM-algorithm applied in the fuzzy environment, and we get $\hat{\tilde{\theta}} = 3.6568$. Assume now that this crisp estimator contains vagueness and is modelled by the triangular symmetrical fuzzy number $(3.1568, 3.6568, 4.1568)$. Therefore, its support set is the interval $[p; s]$, where $p = 3.1568$ and $s = 4.1568$. In addition, its core set is composed of the value $q = 3.6568$ only.

Next, by Algorithm 1 of the bootstrap technique, we estimate the quantile η of the bootstrap distribution of the deviance LR at the significance level $\delta = 0.05$. We get that $\eta = 1.4778$, and $\frac{\eta}{2} = \frac{1.4778}{2} = 0.7389$. We have to calculate after the threshold points I_1, I_2, I_3, and I_4 as given in Eqs. (6.45)–(6.48). Then, we get the following points:

$$
I_1 = l(3.1568; \underline{\tilde{x}}) - 0.7389 = -16.2258,
$$

$$
I_2 = l(3.6568; \underline{\tilde{x}}) - 0.7389 = -18.3079,
$$

$$
I_3 = l(3.6568; \underline{\tilde{x}}) - 0.7389 = -18.3079,
$$

$$
I_4 = l(4.1568; \underline{\tilde{x}}) - 0.7389 = -22.1101,
$$

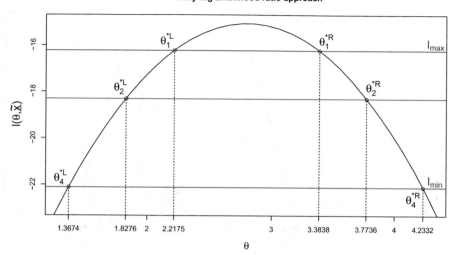

Fig. 6.5 Fuzzy log-likelihood function for the mean and the intersection with the upper and lower bounds of the fuzzy parameter $\overline{\overline{X}}$—Example 6.3.2

where the minimum and maximum thresholds seen in Fig. 6.5 are given by

$$I_{\min} = \min(I_1, I_2, I_3, I_4) = -22.1101, \quad \text{and} \quad I_{\max} = \max(I_1, I_2, I_3, I_4) = -16.2258.$$

We have to find now the intersection points $\theta_1^{\star L}, \theta_2^{\star L}, \theta_3^{\star L}, \theta_4^{\star L}$ and $\theta_1^{\star R}, \theta_2^{\star R}, \theta_3^{\star R}, \theta_4^{\star R}$ as shown in Eqs. (6.51)–(6.54). They are given by

$$\theta_1^{\star L} = 2.2175, \qquad \theta_1^{\star R} = 3.3838,$$
$$\theta_2^{\star L} = \theta_3^{\star L} = 1.8276, \qquad \theta_2^{\star R} = \theta_3^{\star R} = 3.7736,$$
$$\theta_4^{\star L} = 1.3674, \qquad \theta_4^{\star R} = 4.2332,$$

with the minimum and maximum intersection abscisses $\theta_{\inf}^{\star L}, \theta_{\sup}^{\star L}, \theta_{\inf}^{\star R},$ and $\theta_{\sup}^{\star R}$ expressed by

$$\theta_{\inf}^{\star L} = \inf(\theta_1^{\star L}, \theta_2^{\star L}, \theta_3^{\star L}, \theta_4^{\star L}) = 1.3674, \qquad \theta_{\inf}^{\star R} = \inf(\theta_1^{\star R}, \theta_2^{\star R}, \theta_3^{\star R}, \theta_4^{\star R}) = 3.3838,$$
$$\theta_{\sup}^{\star L} = \sup(\theta_1^{\star L}, \theta_2^{\star L}, \theta_3^{\star L}, \theta_4^{\star L}) = 2.2175, \qquad \theta_{\sup}^{\star R} = \sup(\theta_1^{\star R}, \theta_2^{\star R}, \theta_3^{\star R}, \theta_4^{\star R}) = 4.2332.$$

Finally, by standardizing the obtained function to the y-interval $[0; 1]$, we are able to write the FCI $\tilde{\Pi}_{LR}$ given by its left and right α-cuts $(\tilde{\Pi}_{LR})_\alpha = \left[(\tilde{\Pi}_{LR})_\alpha^L, (\tilde{\Pi}_{LR})_\alpha^R\right]$ as seen in Eqs. (6.59) and (6.60). The obtained FCI shown in

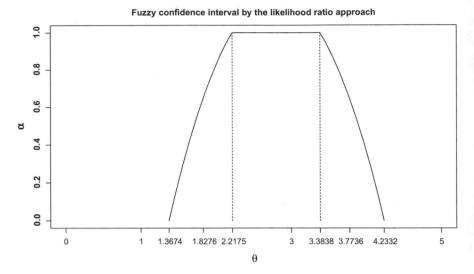

Fig. 6.6 Fuzzy confidence interval by likelihood ratio method $\tilde{\Pi}_{LR}$—Example 6.3.2

Fig. 6.6 is written by the following:

$$(\tilde{\Pi}_{LR})^L_\alpha = \left\{\theta \in \mathbb{R} \mid 1.3674 \leq \theta \leq 2.2175 \text{ and } \alpha = \frac{l(\theta, \tilde{x}) + 22.1101}{5.8843}\right\},$$

$$(\tilde{\Pi}_{LR})^R_\alpha = \left\{\theta \in \mathbb{R} \mid 3.3838 \leq \theta \leq 4.2332 \text{ and } \alpha = \frac{l(\theta, \tilde{x}) + 22.1101}{5.8843}\right\}.$$

It is important to briefly compare the two approaches of computation of the fuzzy confidence interval (the traditional one versus the one using the likelihood method). In terms of the shape of the "traditional" fuzzy confidence interval (its membership function), we would easily see that it depends firmly on the shape of the fuzzy numbers modelling the data set. It is the same for the shape of the fuzzy sample mean since this fact is a direct consequence. However, the FCI by LR is directly related to the chosen probability density function. In addition, it is clear to see that regarding the shapes of both intervals, the LR one is more elaborated, this latter taking into account multiple knots. This fact is seen as an increase of the accuracy of such methods. Furthermore, in the particular proposed setups, if we superpose Figs. 6.4 and 6.6, we can directly notice that $\tilde{\Pi}$ has a larger support set than $\tilde{\Pi}_{LR}$, such that $\tilde{\Pi}_{LR} \subseteq \tilde{\Pi}$. In our case, our approach is said to be more restrictive since a given hypothesis tends to be more often rejected. However, once we assume that the fuzzy number modelling the maximum likelihood estimator is fuzzier, we could expect to get a FCI with a larger support set.

Finally, for such elaborated methods, the computations are known to be tedious. Yet, by the defended approach, we proposed an affordable tool to reduce this complexity. Another advantage is that this approach can be applied to a widespread of estimators and for any pre-defined distribution. No restrictions on the choice of the parameter exist. One can construct a confidence interval by adapting only the type of estimator with the chosen distribution. Thus, our procedure of calculation of the FCI by the likelihood ratio is seen in some sense general. Hence, the influence of the choice of the fuzzy number modelling the crisp maximum likelihood estimator would be interesting to investigate, in addition to the method and adequacy of the calculation of the crisp estimator.

6.4 Fuzzy Hypotheses Testing

As asserted before, we used to perform the statistical inference by considering fuzzy data. Using a test statistic, we test the null hypothesis H_0 against the alternative one H_1, based on the data sets containing fuzziness and given an unknown parameter θ. We remind that the classical outcome of such tests is a binary decision reflecting whether H_0 is rejected or not against H_1. We note for example Grzegorzewski (2000) who gave in a series of papers a general method for constructing hypotheses tests with vague data, based on fuzzy confidence intervals. Accordingly, the decisions made are vague as well. In other terms, the fuzzy test leads to a non-binary fuzzy decision giving a level of conviction of rejecting each of the two hypotheses. However, since human thoughts are imprecise, one could often face the case where the hypotheses are fuzzy as well as the data. For this reason, we proposed in Berkachy and Donzé (2017a, 2019c) an extension of the Grzegorzewski (2000) approach to the case considering fuzzy hypotheses instead of precise ones. Thus, we supposed that the fuzziness is not only contained in the data set, but also in the hypotheses. In addition, since the fuzzy decision is not always interpretable, we showed a defuzzification procedure of this decision based on the SGD operator described in Sect. 4.6. In this section, we recall the studies described above, and we illustrate them by empirical examples. Moreover, we will not only use the SGD, but also the GSGD for the defuzzification step in order to help in decision making.

6.4.1 Fuzzy Hypotheses Testing Using Fuzzy Confidence Intervals

The aim is to extend the classical approach of hypotheses testing to the fuzzy approach when fuzziness occurs also in the hypotheses. The purpose is then to test the parameter θ and get a fuzzy decision. An appropriate method consists of constructing a two-sided confidence interval $\tilde{\Pi}$ built on the considered fuzzy sample with the confidence level $1 - \delta$. Both approaches of estimating a FCI given in Definition 6.3.1 and in Sect. 6.3.2 can be efficiently used. Indeed, we have to stress that the FCI by the likelihood ratio method was intentionally proposed in order to be able to integrate it in the present procedure. From another side, note that we will describe theoretically a two-sided test, but a one-sided one can be easily written.

Let $X_1, \ldots, X_n$ be a random sample from a population described by a probability distribution P_θ depending on the parameter θ and belonging to a family of distributions $\mathcal{D} = \{P_\theta : \theta \cdot \in \Theta\}$. We consider $X_1, \ldots, X_n$ to be mutually independent and identically distributed. We assume that this sample is fuzzy, which will be modelled by fuzzy numbers. Therefore, we define the fuzzy random sample denoted by $\underline{\tilde{X}} = (\tilde{X}_1, \ldots, \tilde{X}_n)$, as the fuzzy perception of the involved random sample. The fuzzy random sample is characterized by its membership function $\mu_{\underline{\tilde{X}}}$ such that $\mu_{\underline{\tilde{X}}} : \mathbb{R}^n \to [0, 1]^n$. The objective is to test the two-sided hypothesis "θ is near θ_0," against the alternative hypothesis "θ is away from θ_0."

In this case, we consider that the null and alternative hypotheses are fuzzy. This setup permits to treat not only the data as fuzzy but also the hypotheses. We could model the fuzzy null hypothesis by a triangular fuzzy number for example denoted by $\tilde{H}_0 = (p, \theta_0, r)$. In this case, its left and right α-cuts $(\tilde{H}_0)_\alpha^L$ and $(\tilde{H}_0)_\alpha^R$ are as follows:

$$\begin{cases} (\tilde{H}_0)_\alpha^L = p + (\theta_0 - p)\alpha, \\ (\tilde{H}_0)_\alpha^R = r - (r - \theta_0)\alpha. \end{cases} \tag{6.67}$$

Analogously to the classical approach and after considering the hypotheses, the next step is to define the test statistic. Indeed, when the sample is assumed to be fuzzy, then we could evidently expect the test statistic to be fuzzy as well. Viertl (2011) provided a definition of a fuzzy test statistic denoted by $\tilde{Z}$ given in the following sense.

Consider ϕ to be a real-valued function, $\phi : \mathbb{R}^n \to \mathbb{R}$. The fuzzy test statistic $\tilde{Z}$ is a function of the fuzzy random sample $\tilde{X}_1, \ldots, \tilde{X}_n$ and given by $\tilde{Z} = \phi(\tilde{X}_1, \ldots, \tilde{X}_n)$. Its membership function $\mu_{\tilde{Z}}$ is written by

$$\mu_{\tilde{Z}}(z) = \begin{cases} \sup \{\mu_{\underline{\tilde{X}}}(\underline{x}) : \phi(\underline{x}) = z\} & \text{if } \exists \underline{x} : \phi(\underline{x}) = z, \\ 0 & \text{if } \nexists \underline{x} : \phi(\underline{x}) = z, \end{cases} \tag{6.68}$$

where $\underline{x}$ is, if it exists, a n-dimensional vector such that it exists a fuzzy element of this vector for which its MF attains the value 1 and its α-cuts are compact, convex, and bounded subsets of $\mathbb{R}^n$, $\forall z \in \mathbb{R}$. For all $\alpha \in (0; 1]$, the fuzzy number $\tilde{Z}$ can also be written using its α-cuts as

$$\tilde{Z}_\alpha = \left[\min_{\underline{\tilde{X}} \in \underline{\tilde{X}}_\alpha} \phi(\underline{\tilde{X}}), \; \max_{\underline{\tilde{X}} \in \underline{\tilde{X}}_\alpha} \phi(\underline{\tilde{X}}) \right]. \tag{6.69}$$

From another side, inspired by Kruse and Meyer (1987) a fuzzy test can be defined in the following way.

Definition 6.4.1 (Fuzzy Test) Consider a fuzzy random vector $\tilde{X}_1, \ldots, \tilde{X}_n$ related to the original $X_1, \ldots, X_n$ in the space Ω. Let δ be a particular value of the interval $[0; 1]$. A function $\tilde{\phi}$ is called a fuzzy test for the null and the alternative hypotheses at the significance level δ if

$$\sup_{\alpha \in [0;1]} P_H\{\omega \in \Omega : \tilde{\phi}_\alpha(\tilde{X}_1(\omega), \ldots, \tilde{X}_n(\omega)) \subseteq \{1\}\} \leq \delta, \tag{6.70}$$

where $\tilde{\phi}_\alpha$ is the α-cut of $\tilde{\phi}$.

For a given parameter θ and at the significance level δ, let $\tilde{\Pi}$ be a FCI as described in Definition 6.3.1. The null and alternative hypotheses related to the test are assumed to be fuzzy. They are written, respectively, by $\tilde{H}_0$ and $\tilde{H}_1$. We define the geometric intersection points between $\tilde{\Pi}$, and these fuzzy hypotheses designated by $a_L, a_R, r_L,$ and r_R such that

$$a_L = \inf(\tilde{H}_0 \cap \tilde{\Pi}_\alpha), \tag{6.71}$$

$$a_R = \sup(\tilde{H}_0 \cap \tilde{\Pi}_\alpha), \tag{6.72}$$

$$r_L = \inf(\tilde{H}_0 \cap \neg \tilde{\Pi}_\alpha), \tag{6.73}$$

$$r_R = \sup(\tilde{H}_0 \cap \neg \tilde{\Pi}_\alpha), \tag{6.74}$$

where the operator "$\neg$" refers to the negation. In addition, we denote by $\alpha_1, \alpha_2, \alpha_3,$ and $\alpha_4 \in [0; 1]$, the α-levels corresponding to the values $a_L, a_R, r_L,$ and r_R, with $\alpha_1 \leq \alpha_2$ and $\alpha_3 \leq \alpha_4$.

We are now able to propose the fuzzy test statistic $\tilde{\phi}$ defined on the n-dimensional space of fuzzy numbers by

$$\tilde{\phi} : \left(\mathbb{F}_c^\star(\mathbb{R})\right)^n \mapsto \mathbb{F}([0; 1]),$$

such that its α-cuts are given by

$$\tilde{\phi}_\alpha(\tilde{X}_1, \ldots, \tilde{X}_n) = \begin{cases} \{0\} & \text{if } \left(\tilde{\Pi}_\alpha \setminus (\neg \tilde{\Pi})_\alpha\right) \cap \tilde{H}_0 = \theta_0; \\ \{1\} & \text{if } \left((\neg \tilde{\Pi})_\alpha \setminus \tilde{\Pi}_\alpha\right) \cap \tilde{H}_0 = \theta_0; \\ ([\alpha_1; \alpha_2]) \cup ([\alpha_3; \alpha_4]) & \text{if } \left(\tilde{\Pi}_\alpha \cap (\neg \tilde{\Pi})_\alpha\right) \cap \tilde{H}_0 = [a_L; a_R] \cup [r_L; r_R]; \\ \emptyset & \text{if } \left(\tilde{\Pi}_\alpha \cup (\neg \tilde{\Pi})_\alpha\right) \cap \tilde{H}_0 = \emptyset, \end{cases}$$

$$\tag{6.75}$$

where $\tilde{\Pi}_\alpha \setminus (\neg \tilde{\Pi})_\alpha$ is the relative complement of $(\neg \tilde{\Pi})_\alpha$ with respect to the set $\tilde{\Pi}_\alpha$.

Proposition 6.4.1 $\tilde{\phi}$ *is a fuzzy test.*

Proof In order to prove that $\tilde{\phi}$ given in Eq. (6.75) is a fuzzy test, it should verify Definition 6.4.1 of a fuzzy test. It can be done in the following manner.

Consider the two-sided symmetrical confidence interval $\tilde{\Pi}$ given by its α-cuts $[\tilde{\Pi}_\alpha^L, \tilde{\Pi}_\alpha^R]$ on the significance level δ. For this FCI, Eq. (6.27) holds for every value of θ and can be expressed by

$$P_H\left(\tilde{\Pi}_\alpha^L \leq \theta \leq \tilde{\Pi}_\alpha^R\right) \geq 1 - \delta, \quad \forall \alpha \in [0; 1]. \tag{6.76}$$

From another side, we could write based on the conditions expressed in Eq. (6.75) the following:

$$P_H\left\{\omega \in \Omega : \tilde{\phi}_\alpha(\tilde{X}_1(\omega), \ldots, \tilde{X}_n(\omega)) \subseteq \{1\}\right\} = 1 - P_H\left\{\omega \in \Omega : \tilde{\phi}_\alpha(\tilde{X}_1(\omega), \ldots, \tilde{X}_n(\omega))\right.$$

$$\left. \in \{\{0\}, \{([\alpha_1; \alpha_2]) \cup ([\alpha_3; \alpha_4])\}\}\right\}. \tag{6.77}$$

Then, by construction of $\tilde{\phi}$ as shown in Eq. (6.75), we could easily see that the following expression is true:

$$P_H\left\{\omega \in \Omega : \tilde{\phi}_\alpha(\tilde{X}_1(\omega), \ldots, \tilde{X}_n(\omega)) \in \{\{0\}, \{([\alpha_1; \alpha_2]) \cup ([\alpha_3; \alpha_4])\}\}\right\} \tag{6.78}$$

$$= P_H\left\{\omega \in \Omega : \theta_0 \in \tilde{\Pi}_\alpha(\omega)\right\}$$

$$= P_H\left\{\omega \in \Omega : \tilde{\Pi}_\alpha^L(\omega) \leq \theta_0 \leq \tilde{\Pi}_\alpha^R(\omega)\right\},$$

where the interpretation of such cases is that the parameter θ_0 belongs to the α-level set of the fuzzy confidence interval $\tilde{\Pi}$ at the significance level $1 - \delta$.

Therefore, we could write that

$$P_H\left\{\omega \in \Omega : \tilde{\phi}_\alpha(\tilde{X}_1(\omega), \ldots, \tilde{X}_n(\omega)) \subseteq \{1\}\right\} = 1 - P_H\left\{\omega \in \Omega : \tilde{\Pi}_\alpha^L(\omega) \leq \theta_0 \leq \tilde{\Pi}_\alpha^R(\omega)\right\}$$

$$\leq 1 - (1 - \delta) = \delta. \tag{6.79}$$

Hence, by Eq. (6.76), we could directly deduce that

$$\sup_{\alpha \in [0; 1]} P_H\left\{\omega \in \Omega : \tilde{\phi}_\alpha(X_1(\omega), \ldots, X_n(\omega)) \subseteq \{1\}\right\} \leq \delta. \tag{6.80}$$

□

Our idea for the fuzzy test is to construct new fuzzy decisions based on the intersection between the fuzzy null hypothesis and the fuzzy confidence interval, and eventually the negation of it. Therefore, we denote by $\tilde{D}_0$ the triangular fuzzy number derived from the intersection of $\tilde{H}_0$ and $\tilde{\Pi}_\alpha$ written as $\tilde{D}_0 = (p_0, q_0, r_0)$, such that

$$p_0 = \min\left(\mu_{\tilde{\Pi}}(a_L), \mu_{\tilde{\Pi}}(a_R)\right), \tag{6.81}$$

$$q_0 = \mu_{\tilde{\Pi}}(\theta_0), \tag{6.82}$$

$$r_0 = \max\left(\mu_{\tilde{\Pi}}(a_L), \mu_{\tilde{\Pi}}(a_R)\right). \tag{6.83}$$

Similarly, the triangular fuzzy number $\tilde{D}_1 = (p_1, q_1, r_1)$ is derived from the intersection of $\tilde{H}_0$ and $\neg\tilde{\Pi}_\alpha$ such that

$$p_1 = \min\left(\mu_{\neg\tilde{\Pi}}(r_L), \mu_{\neg\tilde{\Pi}}(r_R)\right), \tag{6.84}$$

$$q_1 = \mu_{\neg\tilde{\Pi}}(\theta_0), \tag{6.85}$$

$$r_1 = \max\left(\mu_{\neg\tilde{\Pi}}(r_L), \mu_{\neg\tilde{\Pi}}(r_R)\right). \tag{6.86}$$

It is important to see that $\tilde{D}_0$ and $\tilde{D}_1$ are convex fuzzy numbers on $\mathbb{R}$ with $\mu_{\tilde{D}_0}$: $\mathbb{R} \to [0; 1]$ and $\mu_{\tilde{D}_1} : \mathbb{R} \to [0; 1]$. The proof can be written as follows:

Proof By construction, it is easy to prove that

$$\forall x \in \mathbb{R}, \mu_{\tilde{D}_0}(x) \in [0; 1],$$

where $p_0 \leq q_0 \leq r_0$.

For the convexity property of $\tilde{D}_0$, by choosing $x_1 = a_L$ and $x_2 = a_R$, the proof of Definition 2.1.1 becomes straightforward. It is the same for $\tilde{D}_1$. □

These fuzzy numbers are constructed with the intention of assuring that their membership functions are continuous and their supports are bounded. In addition, we mention that the letter "D" is used to refer to the term "decision," the number "0" to the "null" hypothesis, and the number "1" to the "alternative" one. A complete list of figures illustrating different situations of intersection between fuzzy confidence intervals and the fuzzy null hypothesis can be found in Tables 9.3, 9.4, and 9.5.

Consequently, the membership function of the fuzzy test statistic $\tilde{\phi}$ is seen as a union of the membership functions $\mu_{\tilde{D}_0}$ and $\mu_{\tilde{D}_1}$, characterizing, respectively, the fuzzy decisions of the null and alternative hypotheses. The membership function of $\tilde{\phi}$ can then be written as

$$\mu_{\tilde{\phi}}(t) = \mu_{\tilde{D}_0} I_{\{\text{do not reject } H_0\}}(t) + \mu_{\tilde{D}_1} I_{\{\text{do not reject } H_1\}}(t), \ t \in \{0, 1\}. \tag{6.87}$$

It is clear to see that this latter does not lead to a binary decision. In other terms, we would not necessarily strictly reject or not a given null hypothesis. Contrariwise, the obtained decision will be also fuzzy. Thus, we may get a decision interpreted as a degree of conviction of accepting the null hypothesis given by $\mu_{\tilde{D}_0}$, or rejecting it by a degree of $\mu_{\tilde{D}_1}$.

The proposed fuzzy test is seen as a generalization of the case where the hypotheses are not supposed fuzzy. Thus, the subsequent remark shows that our fuzzy test, in specific conditions, could return the test given in Grzegorzewski (2000) and Berkachy and Donzé (2017a).

Remark 6.4.1 If the null hypothesis is crisp, the intersections are reduced to θ_0 and we get that $\mu_{\tilde{D}_1}(\theta_0) = 1 - \mu_{\tilde{D}_0}(\theta_0)$ as seen in Berkachy and Donzé (2017a). Then, $\mu_{\tilde{\phi}}(t)$ is written by

$$\mu_{\tilde{\phi}}(t) = \mu_{\tilde{D}_0}(\theta_0) \ I_{\{\text{do not reject } H_0\}}(t) + (1 - \mu_{\tilde{D}_0}(\theta_0)) \ I_{\{\text{do not reject } H_1\}}(t), \ t \in \{0, 1\}. \tag{6.88}$$

Finally, and most importantly, by combining all the abovementioned information, we display the decision rule, seen to be fuzzy as well. Based on the support and the kernel of the fuzzy confidence interval $\tilde{\Pi}$, denoted, respectively, by supp $\tilde{\Pi}$ and core $\tilde{\Pi}$, the fuzzy decision rule $\tilde{\phi}$ on the parameter θ is written by

$$\tilde{\phi}(\tilde{X}_1, \ldots, \tilde{X}_n) = \begin{cases} \{1|H_0, \quad 0|H_1\} & \text{if } \theta_0 \in \text{core } \tilde{\Pi}, \\ \{0|H_0, \quad 1|H_1\} & \text{if } \theta_0 \notin \text{supp } \tilde{\Pi}, \\ \{\mu_{\tilde{D}_0}(t)|H_0, \ \mu_{\tilde{D}_1}(t)|H_1\} & \text{if else,} \end{cases} \tag{6.89}$$

where "$|H_0$" and "$|H_1$" mean, respectively, "decision related to the null hypothesis" and "decision related to the alternative hypothesis." We add that in the first two cases of Eq. (6.89), the left side of each term, i.e. 0 or 1, has a specific significance as well. Thus, the value 0 indicates the rejection, and the value 1 indicates the acceptance of a given hypothesis. For the last case, the decision rule is seen as a degree of conviction of accepting H_0 with $\mu_{\tilde{D}_0}(t)$ or rejecting it with $\mu_{\tilde{D}_1}(t)$.

We display the subsequent simple numerical example to clarify the proposed methodology:

Example 6.4.1 Consider the data set given in Table 6.1 and the modelling procedure presented for calculating the fuzzy confidence interval for the mean given in Sect. 6.3.1. Suppose now that for a particular value $\mu_0 = 2$ and a significance $\delta = 0.05$, we want to test the following hypotheses:

$$\tilde{H}_0^T = (1.8, 2, 2.3) \quad \text{against} \quad \tilde{H}_1^{OR} = (2, 2, 5),$$

with their respective α-cuts:

$$(\tilde{H}_0^T)_\alpha = \begin{cases} (\tilde{H}_0^T)_\alpha^L = 1.8 + 0.2\alpha; \\ (\tilde{H}_0^T)_\alpha^R = 2.3 - 0.3\alpha. \end{cases} \quad \text{and} \quad (\tilde{H}_1^{OR})_\alpha = \begin{cases} (\tilde{H}_1^{OR})_\alpha^L = 2; \\ (\tilde{H}_1^{OR})_\alpha^R = 5 - 3\alpha. \end{cases} \tag{6.90}$$

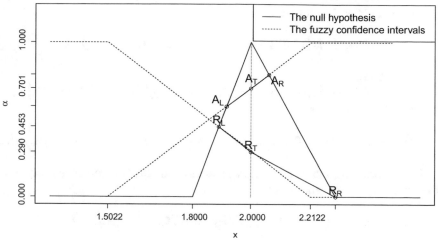

Fig. 6.7 The membership functions of the fuzzy null hypothesis and the fuzzy confidence intervals—Example 6.4.1

By adopting the same procedure of computing the FCI for the mean given in Sect. 6.3.1, we get that the FCI for the mean μ is

$$\tilde{\Pi}_\alpha = [1.5022 + 0.71\alpha; +\infty).$$

In order to obtain the fuzzy decisions related to this test, we have to construct the intersection fuzzy numbers denoted by $\tilde{D}_0$ and $\tilde{D}_1$. First, we can see that the fuzzy confidence interval and its complement overlap with the fuzzy null hypothesis. Therefore, we have to find the values A_L, A_R, R_L, and R_R, i.e. the abscissas of the intersection points between $\tilde{\Pi}$ and $\neg\tilde{\Pi}$ from one side, and $\tilde{H}_0$ from another side as seen in Fig. 6.7. We after have to calculate their α-levels, $\mu_{\tilde{\Pi}}(A_L)$, $\mu_{\tilde{\Pi}}(\mu_0)$, $\mu_{\tilde{\Pi}}(A_R)$, $\mu_{\neg\tilde{\Pi}}(R_L)$, $\mu_{\neg\tilde{\Pi}}(\mu_0)$ and $\mu_{\neg\tilde{\Pi}}(R_R)$, where $\neg\tilde{\Pi}$ is the complement of the fuzzy confidence interval. To clarify the steps followed for this purpose, we will express in detail the calculations related to the membership functions of the fuzzy confidence intervals affected to the values a_L, μ_0, and a_R:

- For the intersection between $\tilde{\Pi}_\alpha^L$ and $(\tilde{H}_0^T)_\alpha^L$, we have to solve the equation $\tilde{\Pi}_\alpha^L = (\tilde{H}_0^T)_\alpha^L$. We then get the following:

$$\tilde{\Pi}_\alpha^L = (\tilde{H}_0)_\alpha^L \Leftrightarrow 1.5022 + 0.71\alpha = 1.8 + 0.2\alpha \Leftrightarrow \alpha = 0.584 = \mu_{\tilde{\Pi}}(A_L).$$

- The intersection between $\tilde{\Pi}_\alpha^L$ and the line $x = \mu_0$ can be found by solving the equation $\tilde{\Pi}_\alpha^L = 2$ as follows:

$$\tilde{\Pi}_\alpha^L = 2 \Leftrightarrow 1.5022 + 0.71\alpha = 2 \Leftrightarrow \alpha = 0.701 = \mu_{\tilde{\Pi}}(A_T),$$

where $A_T = \mu_0$.

- Finally, for the intersection between $\tilde{\Pi}_\alpha^L$ and $(\tilde{H}_0^T)_\alpha^R$, we solve $\tilde{\Pi}_\alpha^L = (\tilde{H}_0^T)_\alpha^R$ in the following manner:

$$\tilde{\Pi}_\alpha^L = (\tilde{H}_0)_\alpha^R \Leftrightarrow 1.5022 + 0.71\alpha = 2.3 - 0.3\alpha \Leftrightarrow \alpha = 0.789 = \mu_{\tilde{\Pi}}(A_R).$$

In the same way, the membership functions of the fuzzy confidence intervals at the abscissas R_L, R_T, and R_R are

$$\mu_{\neg\tilde{\Pi}}(R_L) = 0.453 \; ; \; \mu_{\neg\tilde{\Pi}}(R_T) = 0.299 \; ; \; \mu_{\neg\tilde{\Pi}}(R_R) = 0.$$

Consequently, the fuzzy intersection numbers $\tilde{D}_0$ and $\tilde{D}_1$ are given by the following tuples:

$$\tilde{D}_0 = \left(\mu_{\tilde{\Pi}}(A_L), \mu_{\tilde{\Pi}}(\mu_0), \mu_{\tilde{\Pi}}(A_R)\right) = (0.584, 0.701, 0.789); \qquad (6.91)$$

$$\tilde{D}_1 = \left(\mu_{\neg\tilde{\Pi}}(R_R), \mu_{\neg\tilde{\Pi}}(\mu_0), \mu_{\neg\tilde{\Pi}}(R_L)\right) = (0, 0.299, 0.453). \qquad (6.92)$$

Thus, we can say that the fuzzy number $\tilde{D}_0$ reflects the decision of not rejecting the fuzzy null hypothesis $\tilde{H}_0$. We would not reject $\tilde{H}_0$ at a degree given by the fuzzy number $(0.584, 0.701, 0.789)$. The fuzzy number $\tilde{D}_1$ is related to the alternative hypothesis. In other terms, we would not reject the alternative hypothesis at a degree given by the fuzzy number $(0, 0.299, 0.453)$.

One last step is then to write the fuzzy test statistic using the membership functions of these fuzzy numbers. We get the following fuzzy test statistic:

$$\tilde{\phi}(\tilde{X}_1, \ldots, \tilde{X}_{10}) = \left\{\mu_{\tilde{D}_0}(t)|H_0, \; \mu_{\tilde{D}_1}(t)|H_1\right\}, \quad t \in \{0, 1\}.$$

Accordingly, even if fuzzy decisions are seen as more flexible and contain more information than precise ones, one can clearly see that these fuzzy decisions might in many cases be difficult to be interpreted. For this reason, a defuzzification operation can be proposed.

6.4.2   Defuzzification of the Fuzzy Decisions

If a fuzzy decision does not satisfy a given practitioner because of his intention to precisely reject or not a null hypotheses, he should proceed to a defuzzification by an adequate operator as described in Sect. 3.3. For such tasks, Grzegorzewski (2001) has used two operators called the maximum value operator and the randomized operator. Since the SGD and the GSGD presented in Sects. 4.6 and 4.7 are directional, we believe that these operators might be interesting in this process. In like manner, we showed in Berkachy and Donzé (2017a, 2019c) that defuzzifying the fuzzy decisions by the original SGD could eventually be interesting to investigate. We will thus recall the previous studies described as follows.

The idea is to treat the testing problem as a linguistic variable. Thus, we first decompose the fuzzy test statistic into two linguistic terms, i.e. rejection or acceptance of a given hypothesis. These terms can be modelled by fuzzy numbers. It is obvious to see that for the fuzzy hypothesis test $\tilde{\phi}$, the required fuzzy numbers are nothing but the fuzzy decisions $\tilde{D}_0$ and $\tilde{D}_1$ obtained previously. Now consider a distance operator d given by either the SGD, or the GSGD, since their properties are quite similar. The distances of the fuzzy terms $\tilde{D}_0$ and $\tilde{D}_1$ to the fuzzy origin are then $d(\tilde{D}_0, \tilde{0})$ and $d(\tilde{D}_1, \tilde{0})$. We denote by δ_0 and δ_1, respectively, the indicators of rejection and acceptance of a given hypothesis as follows:

$$\delta_0 = \begin{cases} 1 & \text{if the null hypothesis is accepted;} \\ 0 & \text{otherwise.} \end{cases} \quad \text{and}$$

$$\delta_1 = \begin{cases} 1 & \text{if the null hypothesis is rejected;} \\ 0 & \text{otherwise.} \end{cases} \tag{6.93}$$

The distance defuzzification operator O of $\tilde{D}$ is then defined by

$$O(\tilde{D}) = \delta_0 \cdot d(\tilde{D}_0, \tilde{0}) + \delta_1 \cdot d(\tilde{D}_1, \tilde{0}), \tag{6.94}$$

where the distance d is either the SGD or its GSGD.

If we consider the operator d to be the original SGD given in Definition 4.6.3, the expressions of the signed distances of $\tilde{D}_0$ and $\tilde{D}_1$ are given in the following remark:

Remark 6.4.2 Consider d to be the signed distance presented in Definition 4.6.3, the signed distances of $\tilde{D}_0$ and $\tilde{D}_1$ are written as $d(\tilde{D}_0, \tilde{0}) = d_{SGD}(\tilde{D}_0, \tilde{0})$ and $d(\tilde{D}_1, \tilde{0}) = d_{SGD}(\tilde{D}_1, \tilde{0})$. The fuzzy decisions $\tilde{D}_0$ and $\tilde{D}_1$ are defined as triangular fuzzy number such that their elements are given in Eqs. (6.81)–(6.86). From Eq. (4.22), we get

$$d(\tilde{D}_0, \tilde{0}) = d_{SGD}(\tilde{D}_0, \tilde{0}) = \frac{1}{4}\big(\mu_{\tilde{\Pi}}(A_L) + 2\mu_{\tilde{\Pi}}(\theta_0) + \mu_{\tilde{\Pi}}(A_R)\big), \tag{6.95}$$

$$d(\tilde{D}_1, \tilde{0}) = d_{SGD}(\tilde{D}_1, \tilde{0}) = \frac{1}{4}\big(\mu_{\neg\tilde{\Pi}}(R_L) + 2\mu_{\neg\tilde{\Pi}}(\theta_0) + \mu_{\neg\tilde{\Pi}}(R_R)\big). \tag{6.96}$$

Consequently, by Eqs. (6.89), (6.95), and (6.96) to decompose the fuzzy test statistic $\tilde{\phi}$, we denote by $O \circ \tilde{\phi}$ the precise decision of the considered fuzzy test.

This decision is defined by

$$\left(O \circ \tilde{\phi}\right)(\tilde{D}) = \begin{cases} 1 & \text{if } \mu_{\tilde{\Pi}}(\theta_0) = 1 (\text{No rejection of } H_0, \text{rejection of } H_1); \\ 0 & \text{if } \mu_{\tilde{\Pi}}(\theta_0) = 0 (\text{Rejection of } H_0, \text{no rejection of } H_1); \\ \delta_0 \cdot d(\tilde{D}_0, \tilde{0}) + \delta_1 \cdot d(\tilde{D}_1, \tilde{0}) & \text{if else.} \end{cases}$$

(6.97)

This decision can be interpreted as follows: If $\mu_{\tilde{\Pi}}(\theta_0) = 1$, then we strictly do not reject the null hypothesis. If $\mu_{\tilde{\Pi}}(\theta_0) = 0$, the decision will be to reject strictly H_0. Finally, if $0 < \mu_{\tilde{\Pi}}(\theta_0) < 1$, based on the defuzzification of the fuzzy decisions, the null hypothesis will not be rejected at a degree of conviction of $d(\tilde{D}_0, \tilde{0})$ and rejected at a degree of conviction $d(\tilde{D}_1, \tilde{0})$.

Remark that our method can be perfectly adopted in the case of precise hypotheses. Then, we will get $\mu_{\tilde{\Pi}}(a_L) = \mu_{\tilde{\Pi}}(a_R) = \mu_{\tilde{\Pi}}(\theta_0)$ and $\mu_{\neg\tilde{\Pi}}(r_L) = \mu_{\neg\tilde{\Pi}}(r_R) = \mu_{\neg\tilde{\Pi}}(\theta_0)$, all taking values on $\mathbb{R}$. Thus, in this case, we can write:

$$d\big(\mu_{\tilde{\Pi}}(\theta_0), 0\big) = \mu_{\tilde{\Pi}}(\theta_0) \quad \text{and} \quad d\big((1 - \mu_{\tilde{\Pi}}(\theta_0), 0\big) = 1 - \mu_{\tilde{\Pi}}(\theta_0).$$

(6.98)

Moreover, by substituting Eq. (6.98) in Eq. (6.97), we get the crisp hypotheses-related decision of the fuzzy test expressed as follows:

$$\left(O \circ \tilde{\phi}\right)(D) = \begin{cases} 1 & \text{if } \mu_{\tilde{\Pi}}(\theta_0) = 1 (\text{No rejection of } H_0, \text{rejection of } H_1); \\ \delta_0 \cdot \mu_{\tilde{\Pi}}(\theta_0) + \delta_1 \cdot (1 - \mu_{\tilde{\Pi}}(\theta_0)) & \text{if } \mu_{\tilde{\Pi}}(\theta_0) > 0 \text{ and } \mu_{\tilde{\Pi}}(\theta_0) \neq 1; \\ 0 & \text{otherwise (Rejection of } H_0, \text{no rejection of } H_1). \end{cases}$$

(6.99)

The case of crisp hypotheses gives back a similar result of Grzegorzewski (2001). Furthermore, one can conclude that when fuzziness occurs, the case of inference tests with fuzzy hypotheses is seen as a generalization of the case using precise ones. Our proposed procedure is then considered to be advantageous because of its flexibility and capacity of treatment of the problem of uncertainty in data and/or hypotheses.

The very last step of a hypotheses testing procedure is to emit a decision based on a specific rule, regarding rejecting or not a given null hypothesis.

Decision Rule At the significance level δ, we propose the following decision rule:

- We reject the null hypothesis if $d(\tilde{D}_0, \tilde{0}) < d(\tilde{D}_1, \tilde{0})$.
- We do not reject the null hypothesis if $d(\tilde{D}_0, \tilde{0}) > d(\tilde{D}_1, \tilde{0})$.
- Both null and alternative hypotheses are neither rejected nor not rejected if $d(\tilde{D}_0, \tilde{0}) = d(\tilde{D}_1, \tilde{0})$. In this case, we are not able to decide whether to reject or not the null hypothesis. Note that this case is seen to be rare.

Finally, the hypothesis testing procedure by the fuzzy confidence interval is summed up in the following algorithm:

Algorithm:

1. Consider the null and alternative hypotheses for a given parameter θ and model them by fuzzy numbers.
2. Estimate the fuzzy confidence interval for the concerned parameter. Two approaches are shown—Eq. (6.24) and Procedure of Sect. 6.3.2.
3. Find the intersection values a_L, a_R, r_L, and r_R given in Eqs. (6.71)–(6.74), respectively, and their corresponding α-levels α_1, α_2, α_3, and α_4.
4. Write the fuzzy test statistic $\tilde{\phi}_\alpha$—Eq. (6.75).
5. Construct the triangular fuzzy numbers $\tilde{D}_0 = (p_0, q_0, r_0)$ and $\tilde{D}_1 = (p_1, q_1, r_1)$—Eqs. (6.81)–(6.86).
6. Write the fuzzy decision rule $\tilde{\phi}$—Eq. (6.89).
7. Issue fuzzy decisions related to the null and alternative hypotheses and conclude.
8. Defuzzify the fuzzy decisions by an adequate defuzzification operator.
9. Issue a crisp decision—Eq. (6.97) and conclude.

Example 6.4.2 Since the decisions obtained in Example 6.4.1 are fuzzy, they are difficult to read. We propose to defuzzify them by the SGD and the GSGD.

Consider again the fuzzy decisions $\tilde{D}_0$ and $\tilde{D}_1$ related, respectively, to not rejecting $\tilde{H}_0$ or not rejecting $\tilde{H}_1$. They are written by

$$\tilde{D}_0 = (0.584, 0.701, 0.789) \quad \text{and} \quad \tilde{D}_1 = (0, 0.299, 0.453).$$

We can draw the membership functions of $\tilde{D}_0$ and $\tilde{D}_1$ as shown in Fig. 6.8. Note that these fuzzy numbers are triangular by construction. Then, from Eq. (4.19), their signed distances are as follows:

$$d_{SGD}(\tilde{D}_0, \tilde{0}) = \frac{1}{4}(0.587 + 2 \times 0.701 + 0.789) = 0.694;$$

$$d_{SGD}(\tilde{D}_1, \tilde{0}) = \frac{1}{4}(0 + 2 \times 0.29 + 0.453) = 0.258,$$

and for $\theta^\star = 1$, their generalized signed distances as seen in Eq. (4.28) are given by

$$d_{GSGD}(\tilde{D}_0, \tilde{0}) = 0.696 \quad \text{and} \quad d_{GSGD}(\tilde{D}_1, \tilde{0}) = 0.282.$$

Therefore, we write the fuzzy test statistic using Eq. (6.89) for the SGD by

$$\tilde{\phi}_{SGD}(\tilde{X}_1, \ldots, \tilde{X}_{10}) = \{0.694 | H_0, \quad 0.258 | H_1\},$$

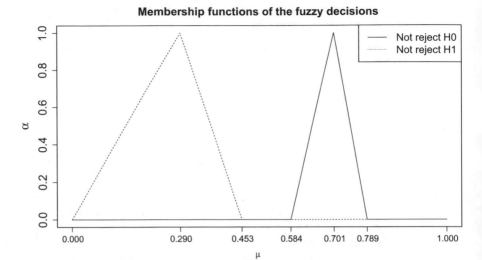

Fig. 6.8 The membership functions of the fuzzy decisions—Example 6.4.2

and for the GSGD by

$$\tilde{\phi}_{GSGD}(\tilde{X}_1, \ldots, \tilde{X}_{10}) = \{0.696|H_0, \quad 0.282|H_1\}.$$

A crisp decision can be obtained by defuzzifying the fuzzy test statistic as given in Eq. (6.104), and we get the following crisp decision:

$$\left(O^{SGD} \circ \tilde{\phi}_{SGD}\right)(\tilde{X}_1, \ldots, \tilde{X}_{10}) = \delta_0 \cdot d(\tilde{D}_0, 0) + \delta_1 \cdot d(\tilde{D}_1, 0) = \delta_0 \cdot 0.694 + \delta_1 \cdot 0.258,$$
(6.100)

and for the GSGD, it is

$$\left(O^{SGD} \circ \tilde{\phi}_{GSGD}\right)(\tilde{X}_1, \ldots, \tilde{X}_{10}) = \delta_0 \cdot 0.696 + \delta_1 \cdot 0.282.$$
(6.101)

The last step is to decide according to the rule given above. The decision is interpreted as follows: since $\mu_{\tilde{\Pi}}(\mu_0) > 0$ and $\mu_{\tilde{\Pi}}(\mu_0) \neq 1$, then we cannot reject $\tilde{H}_0$ with a degree of $\delta_0 \cdot d(\tilde{D}_0, \tilde{0}) + \delta_1 \cdot d(\tilde{D}_1, \tilde{0}) = 1 \cdot d(\tilde{D}_0, \tilde{0}) + 0 \cdot d(\tilde{D}_1, \tilde{0}) = 0.694$ (respectively, 0.696 for the d_{GSGD}). We could add that we cannot reject $\tilde{H}_1$ with a degree of $0 \cdot d(\tilde{D}_0, \tilde{0}) + 1 \cdot d(\tilde{D}_1, \tilde{0}) = 0.258$ (respectively, 0.282 for the d_{GSGD}).

Decision By the SGD, the degree 0.694 of not rejecting $\tilde{H}_0$ is bigger than the degree 0.258 of rejecting it. Then, we will tend to not reject the null hypothesis for this example.

We observe that the decision is the same with the generalized signed distance. It is because of the slight difference in the obtained defuzzified values. One could interpret this fact as a similarity in the use of both distances. However, it would be interesting to treat a case where this difference could eventually change the decision.

6.4.3 Fuzzy p-Values

Once we assume that the fuzziness occurs in a hypothesis test, one has to expect that the p-value associated to the test to be fuzzy also. Furthermore, we could suspect that the uncertainty could not only be contained in the data set, but also in the defined hypotheses. Hryniewicz (2018) gave an overview of several known approaches of calculating a p-value in the fuzzy context. For instance, Filzmoser and Viertl (2004) and Viertl (2011) built a fuzzy p-value by considering the fuzziness of the data. In addition, their procedure is based on a three-case decision rule where the third possibility of Neyman and Pearson (1933) was considered. From another side, Parchami et al. (2017) asserted that the fuzziness of their version of the p-value is a matter of hypotheses, rather than of data. In Berkachy and Donzé (2017b, 2019a, 2019c), we re-considered both approaches and proposed a procedure based simultaneously on fuzzy data and fuzzy hypotheses. Related to the tests (6.1)–(6.3), and to their rejection regions (6.4)–(6.6), we recall the developed approach, and we give the corresponding fuzzy p-values with their α-cuts, these latter being mandatory in the process of assessment of the test statistics results. The p-values written in terms of their α-cuts are given in the subsequent proposition proposed by Berkachy and Donzé (2019c).

Proposition 6.4.2 (Fuzzy p-Value) *Consider a test procedure where the fuzziness is a matter of data and of hypotheses. The null hypothesis H_0 is then supposed to be fuzzy and modelled by the fuzzy number $\tilde{H}_0$. We define by θ_L and θ_R the α-cuts of the boundary of $\tilde{H}_0$. Let $\tilde{t} = \phi(\tilde{X}_1, \ldots, \tilde{X}_n)$ be a fuzzy test statistic that is a function of the fuzzy random sample $\tilde{X}_1, \ldots, \tilde{X}_n$ associated with a membership function $\mu_{\tilde{t}}$. We denote by $\tilde{t}_\alpha^L$ and $\tilde{t}_\alpha^R$ its left and right borders. According to the three rejection regions (6.4)–(6.6), we denote by $\tilde{p}_\alpha$ the α-cuts of the fuzzy p-value $\tilde{p}$. They are written as follows:*

1. $\tilde{p}_\alpha = \left[P_{\theta_R}(T \leq \tilde{t}_\alpha^L), P_{\theta_L}(T \leq \tilde{t}_\alpha^R) \right],$ (6.102)

2. $\tilde{p}_\alpha = \left[P_{\theta_L}(T \geq \tilde{t}_\alpha^R), P_{\theta_R}(T \geq \tilde{t}_\alpha^L) \right],$ (6.103)

3. $\tilde{p}_\alpha = \begin{cases} \left[2P_{\theta_R}(T \leq \tilde{t}_\alpha^L), 2P_{\theta_L}(T \leq \tilde{t}_\alpha^R) \right] & \text{if} \quad A_l > A_r, \\ \left[2P_{\theta_L}(T \geq \tilde{t}_\alpha^R), 2P_{\theta_R}(T \geq \tilde{t}_\alpha^L) \right] & \text{if} \quad A_l \leq A_r, \end{cases}$ (6.104)

for all $\alpha \in (0, 1]$, where A_l and A_r are the areas under the MF $\mu_{\tilde{t}}$ of the fuzzy number $\tilde{t}$ on the left and, respectively, right sides of the median of the distribution of ϕ. Note that the exact location of the median (left or right side) depends on the biggest amount of fuzziness.

Proof We suppose that the fuzziness is coming from the data and the hypotheses. We would like to find the α-cuts of the fuzzy p-value $\tilde{p}$, denoted by $\tilde{p}_\alpha$. The proof will be composed of three main points:

1. Consider a fuzzy test statistic $\tilde{t} = \phi(\tilde{X}_1, \ldots, \tilde{X}_n)$, where $\tilde{t}$ is the fuzzy value obtained from applying the test statistic ϕ on the fuzzy random sample $\tilde{X}_1, \ldots, \tilde{X}_n$. The fuzzy test $\tilde{t}$ is associated to a MF $\mu_{\tilde{t}}$ such that its support $\mathrm{supp}(\mu_{\tilde{t}})$ is given by

$$\mathrm{supp}(\mu_{\tilde{t}}) = \{x \in \mathbb{R} : \mu_{\tilde{t}}(x) > 0\}.$$

Filzmoser and Viertl (2004) asserted that when the fuzziness comes only from the data, the problem can be solved using the definition of the classical precise p-value p. This latter can be extended to the fuzzy context by using the extension principle. For a one-sided test, the first step is to define the uncertain p-values corresponding to the tests (6.1) and (6.2) as a precise value written as follows:

$$\text{1.} \quad p = P\big(T \le t = \max \mathrm{supp}(\mu_{\tilde{t}})\big); \tag{6.105}$$

$$\text{2.} \quad p = P\big(T \ge t = \min \mathrm{supp}(\mu_{\tilde{t}})\big). \tag{6.106}$$

Since fuzziness is assumed to be in the data, as defended by Filzmoser and Viertl (2004), then it would be logical to see the p-value as a fuzzy entity and not a precise one. Denote by $\tilde{p}_{FV}$ the fuzzy p-value based on the approach of Filzmoser and Viertl (2004). It is as well important to write the α-cuts $\tilde{p}_\alpha$. We know that for all $\alpha \in [0; 1]$, the α-cuts of $\tilde{t}$, i.e. $\tilde{t}_\alpha = [\tilde{t}_\alpha^L; \tilde{t}_\alpha^R]$, are compact and closed on the space $\mathbb{R}$. These α-cuts will be used to define the respective α-cuts of the fuzzy p-value $\tilde{p}_{FV}$. With respect to Eqs. (6.105) and (6.106), the α-cuts $\tilde{p}_{FV_\alpha}$ are expressed by

$$\text{1.} \quad \tilde{p}_{FV_\alpha} = \big[P(T \le \tilde{t}_\alpha^L), P(T \le \tilde{t}_\alpha^R)\big]; \tag{6.107}$$

$$\text{2.} \quad \tilde{p}_{FV_\alpha} = \big[P(T \ge \tilde{t}_\alpha^R), P(T \ge \tilde{t}_\alpha^L)\big]. \tag{6.108}$$

The procedure for the two-sided case can be similarly conceivable.

2. The second step is dedicated to the approach of Parchami et al. (2010) that related the fuzziness to the hypothesis and accordingly gave their fuzzy p-value denoted by $\tilde{p}_{PA}$. Its α-cuts $\tilde{p}_{PA_\alpha}$ are given by

$$\text{1.} \quad \tilde{p}_{PA_\alpha} = \big[P_{\theta_R}(T \le t), P_{\theta_L}(T \le t)\big], \tag{6.109}$$

$$\text{2.} \quad \tilde{p}_{PA_\alpha} = \big[P_{\theta_L}(T \ge t), P_{\theta_R}(T \ge t)\big], \tag{6.110}$$

$$\text{3.} \quad \tilde{p}_{PA_\alpha} = \begin{cases} \big[2P_{\theta_R}(T \le t), 2P_{\theta_L}(T \le t)\big] & \text{if } A_l > A_r, \\ \big[2P_{\theta_L}(T \ge t), 2P_{\theta_R}(T \ge t)\big] & \text{if } A_l \le A_r, \end{cases} \tag{6.111}$$

where θ_L and θ_R are the boundaries of the fuzzy null hypothesis, and t is an observed crisp test statistic.

3. By combining the approaches by Filzmoser and Viertl (2004) and the one by Parchami et al. (2010) with the classical definition of the p-value— Eqs. ((6.107), (6.108)), ((6.109), (6.110)), and ((6.13), (6.14))—we get the left and right α-cuts of $\tilde{p}$ as seen in Proposition 6.4.2.

The last step is then to verify that the properties of the membership function are fulfilled. We know that $\mu_{\tilde{t}}$ and $\mu_{\tilde{H}_0}$ are the respective membership functions of $\tilde{t}$ and $\tilde{H}_0$. In addition, the used probabilities are defined on $[0; 1]$. Then, we can deduce that the membership function of $\tilde{p}$ is defined on $[0; 1]$ as well, such that for a particular value, it attains the value 1. Furthermore, we can clearly see that the α-cuts of each of the cases are closed and finite intervals. Thus, for all $\alpha \in (0; 1]$, they are convex and compact subsets of the space $\mathbb{R}$.

□

This fuzzy p-value is then interpreted as the fuzzy number written by its corresponding α-cuts such that each side is calculated by considering a particular α-cut of the fuzzy test statistic, on a given α-cut of the boundary of the null hypothesis. We remind that this fuzzy p-value refers to the case where the test statistic is assumed to be fuzzy and given by its α-level sets $[\tilde{t}_\alpha^L; \tilde{t}_\alpha^R]$, corresponding to a particular fuzzy null hypothesis represented by the α-cuts of its boundary $[\tilde{\theta}_\alpha^L; \tilde{\theta}_\alpha^R]$.

For the ordering of the evoked areas A_l and A_r under the MF of $\tilde{t}$ on the left and right sides of the median $\tilde{M}$ of the distribution of the test statistic as given in Definition 6.2.3, two cases are possible: $A_l > A_r$ and $A_l \leq A_r$. As such, one could depict five main possible positions of the fuzzy numbers $\tilde{t}$ and $\tilde{M}$. The possible positions are given as follows:

- For the cases of Fig. 9.1a and b, we have that $A_l > A_r$.
- For the cases of Fig. 9.2a and b, we have that $A_l \leq A_r$.
- For cases similar to Fig. 9.2c, note that the fuzzy numbers can vary regarding the preference (obviously the need), and the required amount of fuzziness in the fuzzy modelling procedure. For the situation of Fig. 9.2c, we have that $A_l \leq A_r$.

We are now able to define a decision rule based on the developed fuzzy p-value. We mention that similarly to the Neyman and Pearson (1933) approach, Filzmoser and Viertl (2004) adopted a three-decision rule with respect to the left and right α-cuts of the fuzzy p-value. As for us, we consider the same principle of the three-decision problem, and we get the following decision rule:

Decision Rule At a significance level δ, the decision rule for a given test where the data and the hypotheses contain fuzziness is given by:

- If $\tilde{p}_\alpha^R < \delta$, the null hypothesis is rejected.
- If $\tilde{p}_\alpha^L > \delta$, the null hypothesis is not rejected.
- If $\delta \in [\tilde{p}_\alpha^L; \tilde{p}_\alpha^R]$, both null and alternative hypothesis are neither rejected or not.

Table 6.4 The fuzzy data set—Example 6.4.3

Index	Triangular fuzzy number $\bar{\bar{X}}$
1	(1.82,4.23,5.97)
2	(3.38,5.88,6.03)
3	(1.00,2.47,3.74)
4	(5.06,5.88,7.43)
5	(1.00,1.81,2.77)
6	(9.54,9.91,10.00)
7	(4.91,6.14,8.51)
8	(2.98,5.01,6.37)
9	(1.00,3.89,4.53)
10	(5.81,6.37,8.56)
	$\bar{\bar{X}} = (3.65, 5.159, 6.391)$

As an illustration of this procedure, we display an example based on a random fuzzy set, taken from a normal distribution to simplify the understanding of the proposed ideas. We mention that in our setups, we consider the case of known variance only. The case of an unknown variance is not treated.

Example 6.4.3 Consider a sample of 10 observations $\tilde{X}_i$, $i = 1, \ldots, 10$, of the population derived from a normal distribution $N(\mu, 2.2848)$, with known variance. The sample is simulated in the same way as described in Example 5.6. The population is then a group of people expressing their rating for a given painting in a museum. The fuzzy observations are randomly chosen. The data set is given in Table 6.4.

A given expert in arts asserted that a painting having a rate of around 8 is considered to be overall impressive. At a significance level $\delta = 0.05$, the idea is then to know whether the statement "the rating of the chosen painting is approximately 8" would be rejected or not in order to be able to classify it as impressive (or not). Therefore, we assume the hypotheses and the data set to be fuzzy. A fuzzy testing procedure should then be used. We would like in this case to calculate the fuzzy p-value of this test. Let us first consider the following fuzzy hypotheses given by fuzzy numbers—triangular ones to simplify the case—as follows:

$$\tilde{H}_0^T = (7, 8, 9) \quad \text{against} \quad \tilde{H}_1^T = (8, 9, 10).$$

The α-cuts of $\tilde{H}_0^T$ are given by

$$(\tilde{H}_0^T)_\alpha = \begin{cases} (\tilde{H}_0^T)_\alpha^L = 7 + \alpha; \\ (\tilde{H}_0^T)_\alpha^R = 9 - \alpha. \end{cases} \tag{6.112}$$

We have to calculate now the fuzzy sample mean $\overline{\widetilde{X}}$. As shown in Definition 5.4.1, we get $\overline{\widetilde{X}} = (3.65, 5.159, 6.391)$, with its α-cuts written as

$$\overline{\widetilde{X}}_\alpha = \begin{cases} \overline{\widetilde{X}}_\alpha^L = 3.65 + 1.509\alpha; \\ \overline{\widetilde{X}}_\alpha^R = 6.391 - 1.232\alpha. \end{cases} \tag{6.113}$$

Furthermore, the rejection region for this test is defined by Eq. (6.3). Using Proposition 6.4.2, since $\max(\text{supp } \overline{\widetilde{X}}) \leq \min(\text{supp } \tilde{H}_0)$, we can easily deduce that the areas A_l and A_r given in Eq. (6.104) are such that $A_l \leq A_r$. The fuzzy p-value $\tilde{p}$ is then given by its α-cuts $\tilde{p}_\alpha$ as follows:

$$\tilde{p}_\alpha = \left[2P_{\theta_L}(T \geq \tilde{t}_\alpha^R), 2P_{\theta_R}(T \geq \tilde{t}_\alpha^L) \right]. \tag{6.114}$$

In addition, since the sample is assumed to be normally distributed and $T \sim N(0, 1)$, the α-cuts $\tilde{p}_\alpha$ of the corresponding fuzzy p-value $\tilde{p}$ are then written as:

$$\tilde{p}_\alpha = \left[2\int_{\theta_L(\alpha)}^\infty (2\pi)^{-\frac{1}{2}} \exp(\frac{-u^2}{2}) du, \ 2\int_{\theta_R(\alpha)}^\infty (2\pi)^{-\frac{1}{2}} \exp(\frac{-u^2}{2}) du \right], \tag{6.115}$$

where the α-cuts of the boundary of the fuzzy null hypothesis are given by $\left[\theta_L(\alpha); \theta_R(\alpha) \right]$ for which the functions $\theta_L(\alpha)$ et $\theta_R(\alpha)$ depend on α. Therefore, using Eqs. (6.112) and (6.113), we get the following expressions:

$$\theta_L(\alpha) = \frac{\overline{\widetilde{X}}_\alpha^L - (\tilde{H}_0^T)_\alpha^R}{\sigma / \sqrt{n}} = -7.4079 + 3.4741\alpha \quad \text{and}$$

$$\theta_R(\alpha) = \frac{\overline{\widetilde{X}}_\alpha^R - (\tilde{H}_0^T)_\alpha^L}{\sigma / \sqrt{n}} = -0.8432 - 3.0905\alpha.$$

By replacing $\theta_L(\alpha)$ et $\theta_R(\alpha)$ in Eq. (6.115), we get the fuzzy p-value $\tilde{p}$ shown in Fig. 6.9. The decision rule of such p-values is given by comparing the threshold δ with the α-cuts of $\tilde{p}$.

Decision In the present situation, the line $p = \delta$ overlaps with the fuzzy p-value such that $\delta \in \left[\tilde{p}_\alpha^L; \tilde{p}_\alpha^R \right]$; thus no decision can be made. Both null and alternative hypotheses are neither rejected nor not rejected.

In this case, no information on the required decision is made. This decision is difficult to interpret in terms of rejecting or not the statement related to the rating of the painting. One could proceed by defuzzifying the fuzzy p-value.

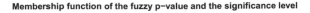

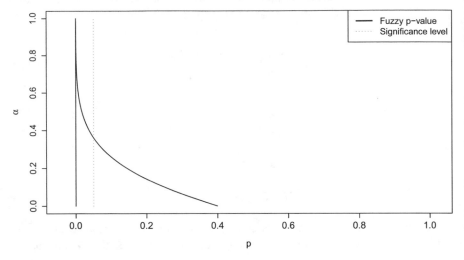

Fig. 6.9 Fuzzy p-value $\tilde{p}$ for the mean—Example 6.4.3

6.4.4 Defuzzification of the Fuzzy p-Value

A crisp p-value resulting from any hypotheses test (classical or fuzzy) could eventually be useful for practitioners. Analogously to the defuzzification of the fuzzy decisions given in Sect. 6.4.2, we propose to defuzzify the fuzzy p-value by a distance d, notably the SGD or the GSGD, as seen for example in Berkachy and Donzé (2017b) for the original SGD. An important fact to identify is the possible similarities in the interpretations of both the fuzzy and defuzzified p-values. For this task, we will then consider Eqs. (6.102)–(6.104) and defuzzify them by the operator d by writing the distance of $\tilde{p}$ to the fuzzy origin $\tilde{0}$, denoted by $d(\tilde{p}, \tilde{0})$.

If d is the original SGD given in Eq. (4.19), the defuzzified p-values are as follows:

$$1.\ d_{SGD}(\tilde{p}, \tilde{0}) = \frac{1}{2} \int_0^1 \left(P_{\theta_R}(T \leq \tilde{t}_\alpha^L) + P_{\theta_L}(T \leq \tilde{t}_\alpha^R) \right) d\alpha, \tag{6.116}$$

$$2.\ d_{SGD}(\tilde{p}, \tilde{0}) = \frac{1}{2} \int_0^1 \left(P_{\theta_L}(T \geq \tilde{t}_\alpha^R) + P_{\theta_R}(T \geq \tilde{t}_\alpha^L) \right) d\alpha, \tag{6.117}$$

$$3.\ d_{SGD}(\tilde{p}, \tilde{0}) = \begin{cases} \frac{1}{2} \int_0^1 \left(2P_{\theta_R}(T \leq \tilde{t}_\alpha^L) + 2P_{\theta_L}(T \leq \tilde{t}_\alpha^R) \right) d\alpha, & \text{if} \quad A_l > A_r, \\ \frac{1}{2} \int_0^1 \left(2P_{\theta_L}(T \geq \tilde{t}_\alpha^R) + 2P_{\theta_R}(T \geq \tilde{t}_\alpha^L) \right) d\alpha, & \text{if} \quad A_l \leq A_r. \end{cases} \tag{6.118}$$

Correspondingly, the decision rule associated to the defuzzified p-values is needed. If d is the SGD or the GSGD, one can expect to get a decision rule similar to the one by the classical p-value.

Decision Rule   The decision rule related to the defuzzified p-values is written by:

- If $d(\tilde{p}, \tilde{0}) < \delta$, the null hypothesis is rejected.
- If $d(\tilde{p}, \tilde{0}) > \delta$, the null hypothesis is not rejected.
- If $d(\tilde{p}, \tilde{0}) = \delta$ (a rare case), the null hypothesis is not rejected nor accepted. One should decide whether or not to reject it.

Although the third case is present in the previous decision rule, it is evident to see that this situation is extremely rare when dealing with crisp p-values. Thus, a main difference between the decision rule of the fuzzy p-value and the one of the defuzzified p-value is that a no-decision region can only be detected in the case of fuzzy p-values. Whenever the fuzzy p-value is defuzzified, this case will no longer be possible. Therefore, since by defuzzifying we get back to the crisp environment for which a no-rejection region can no longer be detected, this operation is seen as disadvantageous.

From another perspective, one could say that with the defuzzified p-value, the overlapping possibility is avoided. The closest decision regarding the distance of the fuzzy p-value to the origin is made. Therefore, even though the no-rejection region is lost, by the crisp p-value, we are almost surely getting a decision.

For Example 6.4.3, since no decision has been made, a defuzzification is needed. The sequel of this example is then given in the following example:

Example 6.4.4   Consider the fuzzy p-value calculated in Example 6.4.3 at the level $\delta = 0.05$ and shown in Fig. 6.9. We would like to defuzzify it in order to make a decision regarding whether to reject or not the null hypothesis. By the SGD, the defuzzified p-value given in Eq. (6.118) is calculated as

$$d_{SGD}(\tilde{p}, \tilde{0}) = \frac{1}{2} \int_0^1 (2P_{\theta_L}(T \geq \tilde{t}_\alpha^R) + 2P_{\theta_R}(T \geq \tilde{t}_\alpha^L)) d\alpha$$

$$= \frac{1}{2} \int_0^1 \left(\int_{-7.4079+3.4741\alpha}^{\infty} (2\pi)^{-\frac{1}{2}} \exp(\frac{-u^2}{2}) du \right.$$

$$\left. + \int_{-0.8432-3.0905\alpha}^{\infty} (2\pi)^{-\frac{1}{2}} \exp(\frac{-u^2}{2}) du \right) d\alpha$$

$$= 0.0360,$$

and by the GSGD given in Eq. (4.28), if we suppose that the weight $\theta^\star$ is equal to $\frac{1}{3}$ for example, the defuzzified p-value is

$$d_{GSGD}(\tilde{p}, \tilde{0}) = 0.0416.$$

Finally, the following decision rule is applied:

Decision Since the defuzzified p-value by the SGD 0.0360 is smaller than the significance level 0.05, then the decision is to reject the fuzzy null hypothesis.

For the defuzzified p-value by the GSGD, the decision is obviously the same. However, this latter is closer to the significance level. Thus, one could expect that the particular case where the null hypothesis is rejected by the SGD and not rejected by the generalized one exists. A further investigation is needed in this purpose.

The computation of the fuzzy p-value can then be summarized in the following algorithm:

Algorithm:

1. Consider a fuzzy sample. Introduce a test by defining the null and alternative hypotheses for a given parameter θ. Model them by fuzzy numbers—Sect. 6.2.
2. Calculate an estimate of the parameter to be tested.
3. Define the rejection region related to the test—Eqs. (6.7)–(6.9).
4. Calculate the α-cuts of the boundary of the fuzzy null hypothesis $\tilde{H}_0$ denoted by $\left[\theta_L(\alpha); \theta_R(\alpha)\right]$.
5. Write the fuzzy p-value by calculating its α-cuts—Eqs. (6.102)–(6.104).
6. Decide, conclude, and eventually go to 7.
7. Defuzzify the fuzzy p-value by an appropriate defuzzification operator:

   • For the signed distance, use Eqs. (6.116)–(6.118).
   • For the generalized signed distance, defuzzify using the expression of the distance given in Eq. (4.28).

8. Decide. Conclude.

6.5   Applications: Finanzplatz

The aim now is to show how practically one can use the previously discussed theories. It is obvious that real life data should be used. For this purpose, we apply our inference methods on a data set coming from the "Finanzplatz: Umfrage 2010" survey described in Appendix B, and composed of $n = 234$ observations, i.e. firms. We are interested in having information regarding the state of the businesses in Zurich. The existing variables in this data set are supposed to be uncertain. Fuzziness is then induced in such variables, and our procedures of fuzzy hypothesis testing can be of good use.

We expose in this section two main applications where we model the data set differently. For each modelled data set, we choose a different method of testing. By that, we would like to show that our procedures are applicable in all modelling contexts. We will mainly use triangular fuzzy numbers for the sake of simplicity, but we have to mention that other shapes modelling the fuzzy numbers can analogously be used.

6.5.1 Application 1

For the first application, we are interested in the present state of business of the considered firms. In order to simplify the situation, let us suppose that this data set is normally distributed with a mean μ and a standard deviation $\sigma = 0.972$. The involved variable is categorical, having five possible answers going from bad to good. Thus, we assume that fuzziness occurs in the collected data set.

Suppose we want to model the fuzzy perception of the data set. This latter is then fuzzified using the corresponding fuzzy numbers given in Table 6.5, and we get the fuzzy random sample written as $\tilde{X}_1, \ldots, \tilde{X}_{n=234}$. We would like to test the hypotheses that "the average of the present state of business" is near the value $\mu_0 = 4$ at the significance level $\delta = 0.05$. These hypotheses are supposed to be imprecise and can be modelled by the following triangular fuzzy numbers:

$$\tilde{H}_0^T = (3.8, 4, 4.2) \quad \text{vs.} \quad \tilde{H}_1^T = (4, 4.5, 5),$$

with the α-cuts of $\tilde{H}_0^T$ given by

$$(\tilde{H}_0^T)_\alpha = \begin{cases} (\tilde{H}_0^T)_\alpha^L = 3.8 + 0.2\alpha; \\ (\tilde{H}_0^T)_\alpha^R = 4.2 - 0.2\alpha. \end{cases} \tag{6.119}$$

The fuzzy sample mean $\overline{\tilde{X}}$ is written by the tuple $\overline{\tilde{X}} = (3.24, 3.765, 4.22)$ for which the α-cuts are expressed by

$$(\overline{\tilde{X}})_\alpha = \begin{cases} (\overline{\tilde{X}})_\alpha^L = 3.24 + 0.525\alpha; \\ (\overline{\tilde{X}})_\alpha^R = 4.22 - 0.455\alpha. \end{cases} \tag{6.120}$$

Table 6.5 Modelling fuzzy numbers—Sect. 6.5.1

Linguistic	Modality	Triangular fuzzy number
X_1	Bad	$\tilde{X}_1 = (1, 1, 1.7)$
X_2	Fairly bad	$\tilde{X}_2 = (1.5, 2, 2.6)$
X_3	Fair	$\tilde{X}_3 = (2.5, 3, 3.5)$
X_4	Fairly good	$\tilde{X}_4 = (3.4, 4, 4.8)$
X_5	Good	$\tilde{X}_5 = (4.5, 5, 5)$

The idea is to construct the fuzzy confidence interval for the mean μ. Two methods can be used: a construction of the FCI using the traditional expression for the mean, and another one based on the likelihood ratio method. By the traditional expression for the mean, the α-cuts of a fuzzy two-sided confidence interval $\tilde{\Pi}$ can be written by

$$\tilde{\Pi}_\alpha = \left[\tilde{\Pi}_\alpha^L, \tilde{\Pi}_\alpha^R\right] = \left[(\tilde{\overline{X}})_\alpha^L - u_{1-\frac{\delta}{2}}\frac{\sigma}{\sqrt{n}}, (\tilde{\overline{X}})_\alpha^R + u_{1-\frac{\delta}{2}}\frac{\sigma}{\sqrt{n}}\right]$$

$$= \left[3.24 + 0.525\alpha - 1.96\frac{0.972}{\sqrt{234}}, 4.22 - 0.455\alpha + 1.96\frac{0.972}{\sqrt{234}}\right]$$

$$= \left[3.115 + 0.525\alpha, 4.344 - 0.455\alpha\right], \qquad (6.121)$$

where $n = 234$, $\sigma = 0.972$, $u_{1-\frac{\delta}{2}} = u_{0.975} = 1.96$ is the $1 - \frac{\delta}{2}$ ordered quantile derived from the standard normal distribution, and $(\tilde{\overline{X}})_\alpha^L$ and $(\tilde{\overline{X}})_\alpha^R$ are, respectively, the left and right α-cuts of the fuzzy sample mean $\tilde{\overline{X}}$.

Note that the FCI $\tilde{\Pi}_{LR}$ by the likelihood ratio method can analogously be computed. For sake of comparison between both intervals, we assumed that the fuzzy sample mean is used as a fuzzy estimator. The case using the fuzzy-modelled MLE estimator is similarly conceivable and gives very close results in terms of interpretations. In addition, we have estimated by the bootstrap technique the quantile η using Algorithm 2, and we got a value 2.1225 for the 95% ordered quantile η.

Figure 6.10 shows the intersection between the "traditional" FCI and the fuzzy null hypothesis. In the same way, Fig. 6.11 shows the intersection between the LR FCI from one side and the fuzzy null hypothesis from another one.

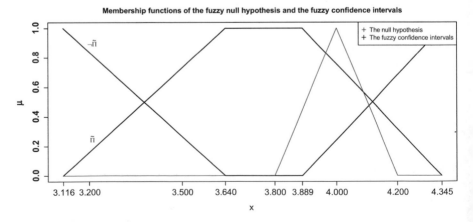

Fig. 6.10 Membership functions of the fuzzy null hypothesis and the fuzzy confidence intervals—Sect. 6.5.1

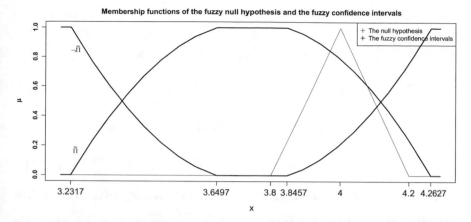

Fig. 6.11 Membership functions of the fuzzy null hypothesis and the fuzzy confidence intervals by LR—Sect. 6.5.1

We now have to find, if it exists, the intersection between the fuzzy null hypothesis and the FCI (and eventually its complement) in order to construct the fuzzy decisions $\tilde{D}_0$ and $\tilde{D}_1$. We get the following:

- For $\tilde{\Pi}$ (see Fig. 6.10), the fuzzy decisions are $\tilde{D}_0 = (0.567, 0.757, 0.831)$ and $\tilde{D}_1 = (0, 0.243, 0.474)$, with their respective distances $d_{SGD}(\tilde{D}_0, \tilde{0}) = 0.7281$ and $d_{SGD}(\tilde{D}_1, \tilde{0}) = 0.2399$ by the SGD. For the GSGD with $\theta^\star = 1$, the distances are as follows:

$$d_{GSGD}(\tilde{D}_0, \tilde{0}) = 0.7311 \text{ and } d_{GSGD}(\tilde{D}_1, \tilde{0}) = 0.2676.$$

According to this latter, the fuzzy test statistic can be written by

$$\tilde{\phi}_{GSGD}(\tilde{X}_1, \ldots, \tilde{X}_{234}) = \{0.7311|H_0, \quad 0.2676|H_1\}.$$

- For $\tilde{\Pi}_{LR}$ (see Fig. 6.11), the fuzzy decisions related to the null and alternative hypotheses are given by $\tilde{D}_0 = (0.5040, 0.7654, 0.8464)$ and $\tilde{D}_1 = (0, 0.2346, 0.4988)$, for which their signed distances are $d_{SGD}(\tilde{D}_0, \tilde{0}) = 0.7203$, and $d_{SGD}(\tilde{D}_1, \tilde{0}) = 0.2420$. Their generalized signed distances are

$$d_{GSGD}(\tilde{D}_0, \tilde{0}) = 0.7254 \text{ and } d_{GSGD}(\tilde{D}_1, \tilde{0}) = 0.2722.$$

Its corresponding fuzzy test statistic is consequently expressed by

$$\tilde{\phi}_{GSGD}(\tilde{X}_1, \ldots, \tilde{X}_{234}) = \{0.7254|H_0, \quad 0.2722|H_1\}.$$

We would like now to defuzzify both fuzzy test statistics. The SGD can be used in this purpose, as seen in Eq. (6.99). Thus, we can write the following:

- For $\tilde{\Pi}$, since $\mu_{\tilde{\Pi}}(\mu_0) > 0$ and $\mu_{\tilde{\Pi}}(\mu_0) \neq 1$, we have that

$$\left(o^{SGD} \circ \tilde{\phi}_{GSGD} \right)(\tilde{X}_1, \ldots, \tilde{X}_{234}) = \delta_0 \cdot d(\tilde{D}_0, \tilde{0}) + \delta_1 \cdot d(\tilde{D}_1, \tilde{0}) = \delta_0 \cdot 0.7311 + \delta_1 \cdot 0.2676.$$

- For $\tilde{\Pi}_{LR}$, since $\mu_{\tilde{\Pi}_{LR}}(\mu_0) > 0$ and $\mu_{\tilde{\Pi}_{LR}}(\mu_0) \neq 1$, then we can get that:

$$\left(o^{SGD} \circ \tilde{\phi}_{GSGD} \right)(\tilde{X}_1, \ldots, \tilde{X}_{234}) = \delta_0 \cdot d(\tilde{D}_0, \tilde{0}) + \delta_1 \cdot d(\tilde{D}_1, \tilde{0}) = \delta_0 \cdot 0.7254 + \delta_1 \cdot 0.2722.$$

Therefore, by this operation, one could deduce the following decision for the test:

Decision At the 0.05 significance level, since $d(\tilde{D}_0, \tilde{0}) > d(\tilde{D}_1, \tilde{0})$, we do not reject the null hypothesis H_0.

Interpretation We do not reject the null hypothesis that the average of the present state of businesses is fairly good or near the value 4, implicitly mentioned in the triangular representations of $\tilde{H}_0^T$ and $\tilde{H}_1^T$. In the same way, we tend to reject the alternative hypothesis that this average is away from 4.

From the calculations of both FCI, one could remark that the support and the core sets of the LR-FCI are slightly smaller than the ones of the traditional FCI. This fact confirms the output obtained from the simulation study of Sect. 6.3.2. Regarding the decisions based on these FCI, it is clear to see that both procedures of estimation of FCI give the same decisions with very close degrees of rejection or acceptance of a particular hypothesis. This result is consequently considered as appeasing.

6.5.2 Application 2

For the second application, we consider another modelling context with the same data set. We will fuzzify the variable "gross profit compared to the last 12 months" by triangular fuzzy numbers. The support set of the fuzzy observations is generated randomly, while their core sets are fixed. In this case, the core sets are reduced to one element only. They are nothing but the real variable itself. We proceed by fixing the difference between the core from one side, and the minimum and maximum values of the support set to 0.5. Thus, the support set of a given fuzzy number could be of width 1. For instance, if a given firm answered by 2, we could model this response by the tuple $(1.82, 2, 2.031)$ for example.

We would like to test whether we reject the following fuzzy hypotheses related to "the average of the gross profit compared to the last 12 months":

$$\tilde{H}_0^T = (2, 3, 4) \quad \text{vs.} \quad \tilde{H}_1^{OR} = (3, 3, 5),$$

where the α-cuts of the fuzzy null hypothesis are given by

$$(\tilde{H}_0^T)_\alpha = \begin{cases} (\tilde{H}_0^T)_\alpha^L = 2 + \alpha; \\ (\tilde{H}_0^T)_\alpha^R = 4 - \alpha. \end{cases} \tag{6.122}$$

We would like to calculate the fuzzy p-value related to this test. In our situation, the rejection region is defined related to the test statistic seen in Eq. (6.8).

First of all, let us calculate the fuzzy sample mean $\overline{\tilde{X}}$. It is given by the tuple $\overline{\tilde{X}} = (3, 3.265, 3.5)$ with its corresponding left and right α-cuts $\overline{\tilde{X}}_\alpha^L$ and $\overline{\tilde{X}}_\alpha^R$. Furthermore, since the distribution of the variable is supposed to be normal with a standard deviation $\sigma = 0.796$, we could calculate the α-cuts of the fuzzy p-value where the functions $\theta_L(\alpha)$ and $\theta_R(\alpha)$ are the bounds of the integrals for which the integrands are the probability density function related to the normal distribution. The functions $\theta_L(\alpha)$ and $\theta_R(\alpha)$ are expressed by

$$\theta_L(\alpha) = \frac{\overline{\tilde{X}}_\alpha^R - (\tilde{H}_0^T)_\alpha^L}{\sigma / \sqrt{n}} = \frac{(3.5 - 0.235\alpha) - (2 + \alpha)}{0.796 / \sqrt{234}} = 28.826 - 23.734\alpha,$$

and $\theta_R(\alpha) = \dfrac{\overline{\tilde{X}}_\alpha^L - (\tilde{H}_0^T)_\alpha^R}{\sigma / \sqrt{n}} = \dfrac{(3 + 0.265\alpha) - (4 - \alpha)}{0.796 / \sqrt{234}} = -19.217 + 24.310\alpha.$

Consequently, the α-cuts of the fuzzy p-value $\tilde{p}$ associated to the density function of the normal distribution are written by

$$\tilde{p}_\alpha = \left[\int_{\theta_L(\alpha)}^{+\infty} (2\pi)^{-\frac{1}{2}} \exp(\frac{-u^2}{2}) du, \int_{\theta_R(\alpha)}^{+\infty} (2\pi)^{-\frac{1}{2}} \exp(\frac{-u^2}{2}) du \right]. \tag{6.123}$$

The obtained fuzzy p-value is shown in Fig. 6.12. It is clear that a decision cannot be made since $\tilde{p}$ is very fuzzy and it overlaps with the significance level. Therefore, we would like to defuzzify the fuzzy p-value to get a clearer decision. We could use the SGD as instance, and we get the following defuzzified p-value:

$$d(\tilde{p}, \tilde{0}) = \frac{1}{2} \int_0^1 \left(P_{\theta_L}(T \geq \tilde{t}_\alpha^R) + P_{\theta_R}(T \geq \tilde{t}_\alpha^L) \right) d\alpha$$

$$= \frac{1}{2} \int_0^1 \left(\int_{28.826 - 23.734\alpha}^{+\infty} (2\pi)^{-\frac{1}{2}} \exp(\frac{-u^2}{2}) du + \int_{-19.217 + 24.310\alpha}^{+\infty} (2\pi)^{-\frac{1}{2}} \exp(\frac{-u^2}{2}) du \right) d\alpha$$

$$= 0.395.$$

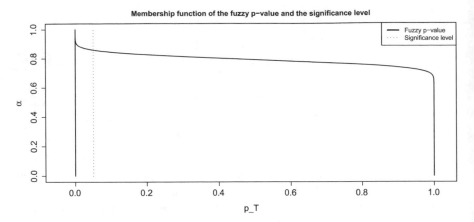

Fig. 6.12 Fuzzy p-value of the test—Sect. 6.5.2

For the GSGD with $\theta^\star = 1$, the defuzzified p-value is 0.559. Thus, for both tests, the decision related to the defuzzified p-values is written in the following manner:

Decision The defuzzified p-value 0.395 (respectively, 0.559 for the d_{GSGD} distance) is higher than the significance level $\delta = 0.05$. Thus, we tend to not reject the null hypothesis with the degree of 0.395 (0.559 with d_{GSGD}).

From another side, we would like to explore the influence of the shape of the fuzzy hypotheses on the decisions made by the fuzzy p-value approach. In other words, we would be interested to know whether the obtained decision depends on the amount of fuzziness contained in the fuzzy hypotheses, and if it is true, to know in which direction regarding rejecting or not the null hypothesis. In Berkachy and Donzé (2017a), we treated a similar case. One of our main findings was that the shape and spread of the alternative hypothesis do not influence the final decision. This hypothesis is only responsible of determining which form of the rejection region R has to be considered. Thus, considering a less or more imprecise alternative hypothesis has no influence. However, the variation of the fuzzy null hypothesis could change drastically the decision taken. To validate and expand this result, we re-consider the previous test, and we vary the support set of the fuzzy null hypothesis. Table 6.6 gives a review of different considered tests where the null hypothesis changes, and where we consider $\tilde{H}_1$ to be the same such that $\tilde{H}_1^{OR} = (3, 3, 5)$.

It is clear to see that intersections between the obtained fuzzy p-values and the significance level exist. We deduce that no decision can be made. For this purpose, we express the defuzzified p-values by the SGD and the GSGD. The decisions are accordingly made. The obtained fuzzy p-values are shown in Fig. 6.13.

From Table 6.6 and Fig. 6.13, we can directly see that the defuzzified p-value is sensitive to the variation of the support set of the fuzzy null hypothesis. In addition, it is clear that with respect to both distances, the highest defuzzified p-

Table 6.6 Testing different fuzzy hypotheses on the significance level $\delta = 0.05$—Sect. 6.5.2

Test	Fuzzy null hypothesis	$d_{SGD}(\tilde{p}, \tilde{0})$	And decision	$d_{GSGD}(\tilde{p}, \tilde{0})$	And decision ($\theta^\star = \frac{1}{3}$)	$d_{GSGD}(\tilde{p}, \tilde{0})$	And decision ($\theta^\star = 1$)
1	$\tilde{H}_0^T = (2, 3, 4)$	0.395	Not reject $\tilde{H}_0$	0.456	Not reject $\tilde{H}_0$	0.559	Not reject $\tilde{H}_0$
2	$\tilde{H}_0^T = (2.5, 3, 3.5)$	0.326	Not reject $\tilde{H}_0$	0.376	Not reject $\tilde{H}_0$	0.461	Not reject $\tilde{H}_0$
3	$\tilde{H}_0^T = (2.8, 3, 3.2)$	0.213	Not reject $\tilde{H}_0$	0.246	Not reject $\tilde{H}_0$	0.301	Not reject $\tilde{H}_0$
4	$\tilde{H}_0^T = (2.9, 3, 3.1)$	0.134	Not reject $\tilde{H}_0$	0.155	Not reject $\tilde{H}_0$	0.190	Not reject $\tilde{H}_0$
5	$\tilde{H}_0^T = (2.9, 3, 3.05)$	0.083	Not reject $\tilde{H}_0$	0.096	Not reject $\tilde{H}_0$	0.118	Not reject $\tilde{H}_0$
6	$\tilde{H}_0^T = (2.9, 3, 3.0162)$	**0.05**	Not reject nor reject $\tilde{H}_0$	0.058	Slightly not reject $\tilde{H}_0$	0.07	Not reject $\tilde{H}_0$
7	$\tilde{H}_0^T = (2.9, 3, 3.0085)$	0.0433	Reject $\tilde{H}_0$	**0.05**	Not reject nor reject $\tilde{H}_0$	0.06	Not reject $\tilde{H}_0$
8	$\tilde{H}_0^T = (3, 3, 3)$	0.0365	Reject $\tilde{H}_0$	0.0421	Reject $\tilde{H}_0$	0.052	Not reject nor reject $\tilde{H}_0$

The values in bold are the significance levels

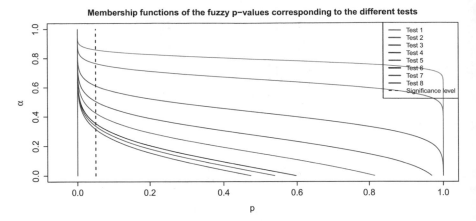

Fig. 6.13 Fuzzy p-values of the different tests given—Table 6.6

value corresponds to the fuzzy p-value for which the support set is the largest. It is the opposite for the ones having the lowest spreaded fuzzy p-value. Moreover, in terms of the differences between the signed distance and the generalized signed distance, we could say that by the generalized one we tend more strongly to not reject the null hypothesis, since the defuzzified p-value is higher. Contrariwise, since the defuzzified p-value by the GSGD distance is higher as well, we tend less strongly to reject the null hypothesis.

For these right one-sided tests, if we look closely to the tests 4–7, we can remark that the left hand side of the null hypothesis is fixed. Thus, the left hand side has no influence on the defuzzified p-value. The variation depends rather on the right hand side of the fuzzy null hypothesis. Note that for a left one-sided test, the left hand side of the null hypothesis has the influence instead of the right one. In addition, we mention that we intentionally ordered the tests in the following order: the fuzziest null hypothesis to the most precise one. We can clearly see that the defuzzified p-values have a decreasing order as well, such that the most we get closer to a precise hypothesis, the more we tend to reject the null hypothesis.

Furthermore, we can find the significance level $\delta = 0.05$ as the defuzzified p-value for the test 6 with respect to d_{SGD}, and for the test 7 with respect to d_{GSGD} when $\theta^{\star} = \frac{1}{3}$. One could deduce that the right hand sides of the fuzzy null hypothesis related to these tests are considered to be thresholds of rejection. Note that the fuzzy null hypotheses of both tests are slightly different. We add that for the case where $\theta^{\star} = 1$, we could not find any null hypothesis induced by the calculations to get exactly the threshold of significance. Moreover, if we compare the two cases where $\theta^{\star} = \frac{1}{3}$ and $\theta^{\star} = 1$, even though the calculations are overall the same, we can expect to get different decisions. It is the effect of the weight $\theta^{\star}$ one puts on the shape of the involved fuzzy numbers through the calculation of the distance that matters in such situations.

We could conclude that the defuzzified p-value seems to be a pertinent indicator of the fuzziness contained in the null hypothesis. Therefore, in order to make the most suitable decision, one has to model prudently the required fuzzy null hypothesis.

As a final remark about the above-described applications, we have almost every time assumed that the distributions are normally distributed to simplify the cases. In a more realistic situation where we do not actually know in advance the theoretical distribution of the set, we would recommend to use the empirical density function instead. Thus, the outcome of the integration over this function on the boundaries of the fuzzy null hypothesis would also be empirical.

6.5.3 Classical Versus Fuzzy Hypotheses Testing

For the final section, it is interesting to compare the classical and fuzzy approaches of testing hypotheses in terms of the decisions made. For this purpose, we perform exactly the same tests described in the previous two subsections, with the data set, but in the classical traditional context. Thus, we consider the data and the hypotheses as precise. For the Sects. 6.5.1 and 6.5.2, we obtained, respectively, the results shown in Tables 6.7 and 6.8. We remind that similarly to the aforementioned approaches, there exist two means of deciding in the classical theory: we could compare the test statistic T to the quantiles of the corresponding distribution, or to compare the p-value with the significance level.

Both approaches summarized in Tables 6.7 and 6.8 showed that for the treated cases, the classical and fuzzy approaches gave different decisions, where by the classical we tend to reject more strongly the null hypothesis at the significance level $\delta = 0.05$, while with the fuzzy one, we have a different margin of acceptability of the null hypothesis.

For instance, for the first case (Table 6.7), we expressed the fuzzy decisions with their respective defuzzification by the generalized signed distance. Following, we tend to not reject the null hypothesis with a degree of conviction of 0.7311 in the fuzzy approach. However, by the classical test, we totally reject it. In addition, we

Table 6.7 The results of the crisp and fuzzy tests corresponding to the variable "The present state of business"—Sect. 6.5.1

The present state of the business	The test results	The decision
Crisp two-sided test $H_0: \mu = 4$ vs. $H_1: \mu \neq 4$	$\|T\| = \|12.0404\| \notin (t_a; t_b) = (-1.9702; 1.9702)$	H_0 is totally rejected
Fuzzy two-sided test $\tilde{H}_0: \mu$ is close to 4 vs. $\tilde{H}_1: \mu$ is away from 4	$\tilde{D}_0 = (0.567, 0.757, 0.831)$, with $d_{GSGD}(\tilde{D}_0, \tilde{0}) = 0.7311$ $\tilde{D}_1 = (0, 0.243, 0.474)$, with $d_{GSGD}(\tilde{D}_1, \tilde{0}) = 0.2676$	$\tilde{H}_0$ is not rejected with a degree of conviction 0.7311.

Table 6.8 The results of the crisp and fuzzy tests corresponding to the variable "The gross profit compared to the last 12 months"—Sect. 6.5.2

The gross profit compared to the last 12 months	The p-values	The decision
Crisp right-sided test H_0: $\mu = 3$ vs. H_1: $\mu \geq 3$	$3.64e^{-07}$	H_0 is strongly rejected
Fuzzy right-sided test $\tilde{H}_0$: μ is approximately 3 vs. $\tilde{H}_1$: μ is approximately bigger than 3	$d_{SGD} = 0.395$ $d_{GSGD} = 0.559$	$\tilde{H}_0$ is rejected with the degree 0.605 $\tilde{H}_0$ is not rejected with the degree 0.559

have to remark that the obtained fuzzy decisions have a relatively high amount of fuzziness. Moreover, the decision $\tilde{D}_0$ has a support set close to the value 0.5, i.e. an evident threshold of comparison between the two possible decisions. It is then more difficult to truly assume whether to reject or not the null hypothesis. Therefore, even though the defuzzification operation is seen as a loss of information, by this tool, we get a better view of the possible decisions to make. They are reflected in the produced degrees of conviction. We remind that the idea behind these degrees has effectively been introduced to better interpret the fuzzy decisions. Note that these degrees are mainly related to the no rejection of a given hypothesis.

For the second case (Table 6.8 and Fig. 6.12), it is easy to see that the significance level and the fuzzy p-value overlap. No clear decision can be made. A defuzzification is then needed. Table 6.8 shows that at the significance level $\delta = 0.05$, since 0.395 (0.559 with d_{GSGD}) $> \delta$, then we could not reject the null hypothesis $\tilde{H}_0$. However, the fuzzy p-value clearly contains a big amount of fuzziness, and the decision made could be judged as not very realistic because of the loss of information provoked by the defuzzification. Then, since the overlapping is a typical case in similar contexts, one could imagine the following supplementary decision rule for such cases:

We propose to compare the defuzzified p-value related to the rejection of the null hypothesis H_0 written by $d(\tilde{p}, \tilde{0})$ with a given distance d, to the threshold 0.5. This value has been chosen because of the implicit idea of equally weighting the two possible decisions: rejection of H_0 or no rejection of H_0. In this case, the supplementary decision rule will be expressed by

- If $d(\tilde{p}, \tilde{0}) < 0.5$, we tend to reject the null hypothesis with a degree of conviction $1 - d(\tilde{p}, \tilde{0})$.
- If $d(\tilde{p}, \tilde{0}) > 0.5$, we tend to not reject the null hypothesis with a degree of conviction $d(\tilde{p}, \tilde{0})$.
- If $d(\tilde{p}, \tilde{0}) = 0.5$, one should decide whether or not to reject the null hypothesis.

Thus, by this new decision rule, the last decision made will be partially revoked, and since by the signed distance $0.395 < 0.5$, we will consequently tend to slightly reject the null hypothesis with the degree of conviction $1 - 0.395 = 0.605$. However, by the generalized signed distance, we have a defuzzified p-value of 0.559, which means that we slightly tend to not reject the null hypothesis with the degree 0.559.

Following this added decision rule, Berkachy and Donzé (2019c) asserted that when the considered parameter is positioned in the no-rejection region, the fuzzy approach is seen as more pessimistic than the classical one. In other words, the classical p-value tends to stronger not reject the null hypothesis compared to the fuzzy one. However, for the rejection region in the present case (Table 6.8), we can see that by the classical approach, we strongly reject the null hypothesis. This conjunction is less strong in the fuzzy approach. One could say that in the rejection region, the fuzzy approach gives a less pessimistic decision than the classical one, showing a given amount of flexibility.

Although generally the interpretations of the fuzzy decisions by the confidence intervals of a given test are somehow different between both approaches, one can deduce that the classical tests are seen as the extremes of the fuzzy ones. Thus, this latter is in some sense a sort of generalization of the classical cases: the more the fuzziness tends to be absent, the more we are closer to the results and decisions of the classical theory. For the concept of the p-value, the interpretations of the decisions are different since for the fuzzy context, if we adopt the supplementary decision rule, the p-value is said to be a degree of acceptability related to the null hypotheses. These definitions gave us a general idea on how to choose the better hypotheses testing approach fitting the data.

To sum up and following the discussion on the comparison between the classical and fuzzy approaches, it is important to provide some pros and cons of the use of each one of them. Giving some corresponding guidelines is absolutely an asset. Berkachy and Donzé (2019c) displayed a brief list of their advantages and disadvantages.

Even though the decisions of both approaches can be in many cases the same, a first main difference is in the nature and the interpretation of the obtained decisions. Thus, in the classical theory, the decision is reduced to one of both possible decisions: reject the null hypothesis or not reject it. Yet, by the fuzzy approach, the decision is precisely interpreted by a degree of conviction or a degree of acceptability. This fact is clearly seen in the applications of Sects. 6.5.1 and 6.5.2. In other words, the decision by the fuzzy approach is not restrictive to one decision only. Instead, we would be able to reject or not the null hypothesis at a given degree of conviction. At the same time, we would reject or not the alternative hypothesis at a "complementary" degree of conviction.

Furthermore, we can say that the classical approach is seen as the extreme case regarding the fuzzy one, since the classical approach tends more strongly to reject or not a given null hypothesis. This fact can be seen as an absolute advantage in the cases where the involved hypothesis is actually true. When it is the opposite, this particularity of the classical approach is considered to be a disadvantage. Thus, the decision made will be far from the reality. Similarly, when the hypothesis is actually true, the fuzzy approach is disadvantaging.

On a more global level, an utter difference is that by the fuzzy approach, the uncertainty and subjectivity of the hypotheses and the data sets can now on be taken into account. In addition, despite the fact that the defuzzification induces a loss of information, we managed to produce a crisp interpretable decision, similarly to the classical approach. Therefore, being able to provide a decision—fuzzy or defuzzified—associated to a vague context is an asset. Moreover, since the shape and spread of the fuzzy hypotheses and the fuzzy data influence the decisions made by the tests or the p-values, we saw that the defuzzified p-value can be seen as a pertinent indicator of fuzziness of the hypothesis. A convenient decision can consequently be made. To conclude, it is clear that the question about which of both procedures modelling the uncertainty to choose arises strongly in such contexts.

References

Berkachy, R., & Donzé, L. (2017a). Defuzzification of a fuzzy hypothesis decision by the signed distance method. In *Proceedings of the 61st World Statistics Congress, Marrakech, Morocco*.

Berkachy, R., & Donzé, L. (2017b). Testing fuzzy hypotheses with fuzzy data and defuzzification of the fuzzy p-value by the signed distance method. In *Proceedings of the 9th International Joint Conference on Computational Intelligence (IJCCI 2017)* (pp. 255–264). ISBN: 978-989-758-274-5.

Berkachy, R., & Donzé, L. (2019a). Defuzzification of the fuzzy p-value by the signed distance: Application on real data. In *Computational intelligence. Studies in computational intelligence* (Vol. 829.1, pp. 77–97). Springer International Publishing. ISSN: 1860-949X. https://doi.org/10.1007/978-3-030-16469-0

Berkachy, R., & Donzé, L. (2019b). Fuzzy confidence interval estimation by likelihood ratio. In *Proceedings of the 2019 Conference of the International Fuzzy Systems Association and the European Society for Fuzzy Logic and Technology (EUSFLAT 2019)*.

Berkachy, R., & Donzé, L. (2019c). Testing hypotheses by fuzzy methods: A comparison with the classical approach. In A. Meier, E. Portmann, & L. Terán (Eds.), *Applying fuzzy logic for the digital economy and society* (pp. 1–22). Cham: Springer International Publishing. ISBN: 978-3-030-03368-2. https://doi.org/10.1007/978-3-030-03368-2_1.

Couso, I., & Sanchez, L. (2011). Inner and outer fuzzy approximations of confidence intervals. *Fuzzy Sets and Systems, 184*(1), 68–83. ISSN: 0165-0114. https://doi.org/10.1016/j.fss.2010.11.004. http://www.sciencedirect.com/science/article/pii/S0165011410004550

DeGroot, M. H., & Schervish, M. J. (2012). *Probability and statistics* (4th ed.). Pearson Education, Inc.

Denoeux, T. (2011). Maximum likelihood estimation from fuzzy data using the EM algorithm. *Fuzzy Sets and Systems, 183*(1), 72–91. ISSN: 0165-0114. https://doi.org/10.1016/j.fss.2011.05.022

Efron, B. (1979). Bootstrap methods: Another look at the jackknife. *The Annals of Statistics, 7*(1), 1–26. ISSN: 0090-5364. http://www.jstor.org/stable/2958830

Everitt, B. S. (2006). *The Cambridge dictionary of statistics* (3rd ed.). Cambridge University Press. ISBN: 978-0-511-24323-3.

Filzmoser, P., & Viertl, R. (2004). Testing hypotheses with fuzzy data: The fuzzy p-value. *Metrika, Springer-Verlag, 59*, 21–29. ISSN: 0026-1335. https://doi.org/10.1007/s001840300269

Gil, M. A., & Casals, M. R. (1988). An operative extension of the likelihood ratio test from fuzzy data. *Statistical Papers, 29*(1), 191–203. ISSN: 1613-9798. https://doi.org/10.1007/BF02924524

Gonzalez-Rodriguez, G., et al. (2006). Bootstrap techniques and fuzzy random variables: Synergy in hypothesis testing with fuzzy data. *Fuzzy Sets and Systems, 157*(19), 2608–2613. ISSN: 0165-0114. https://doi.org/10.1016/j.fss.2003.11.021. http://www.sciencedirect.com/science/article/pii/S0165011406002089

Grzegorzewski, P. (2000). Testing statistical hypotheses with vague data. *Fuzzy Sets and Systems, 112*(3), 510. ISSN: 0165-0114. http://dx.doi.org/10.1016/S0165-0114(98)00061-X. http://www.sciencedirect.com/science/article/pii/S016501149800061X

Grzegorzewski, P. (2001). Fuzzy tests - defuzzification and randomization. *Fuzzy Sets and Systems, 118*(3), 437–446. ISSN: 0165–0114. http://dx.doi.org/10.1016/S0165-0114(98)00462-X. http://www.sciencedirect.com/science/article/pii/S016501149800462X

Hryniewicz, O. (2018). Statistical properties of the fuzzy p-value. *International Journal of Approximate Reasoning, 93*, 544–560. ISSN: 0888-613X. https://doi.org/10.1016/j.ijar.2017.12.003. http://www.sciencedirect.com/science/article/pii/S0888613X17307296

Kahraman, C., Otay, I., & Öztaysi, B. (2016). Fuzzy extensions of confidence intervals: Estimation for μ, $\sigma 2$, and p. In C. Kahraman & Ö. Kabak (Eds.), *Fuzzy statistical decision-making: Theory and applications* (pp. 129–154). Cham: Springer International Publishing. ISBN: 978-3-319-39014-7. https://doi.org/10.1007/978-3-319-39014-7_9

Kruse, R., & Meyer, K. D. (1987). *Statistics with vague data* (Vol. 6). Springer Netherlands.

Montenegro, M., et al. (2004). Asymptotic and Bootstrap techniques for testing the expected value of a fuzzy random variable. *Metrika, 59*(1), 31–49. ISSN: 1435-926X. https://doi.org/10.1007/s001840300270

Neyman, J., & Pearson, E. S. (1933). The testing of statistical hypotheses in relation to probabilities a priori. *Mathematical Proceedings of the Cambridge Philosophical Society, 29*(04), 492–510. ISSN: 1469-8064. https://doi.org/10.1017/S030500410001152X. http://journals.cambridge.org/article_S030500410001152X

Parchami, A. (2018). *EM.Fuzzy: EM Algorithm for Maximum Likelihood Estimation by Non-Precise Information, R package.* https://CRAN.R-project.org/package=EM.Fuzzy

Parchami, A., Nourbakhsh, M., & Mashinchi, M. (2017). Analysis of variance in uncertain environments. *Complex & Intelligent Systems, 3*(3), 189–196

Parchami, A., Taheri, S. M., & Mashinchi, M. (2010). Fuzzy p-value in testing fuzzy hypotheses with crisp data. *Statistical Papers, 51*(1), 209–226. ISSN: 0932-5026. https://doi.org/10.1007/s00362-008-0133-4

Viertl, R. (2011). *Statistical methods for fuzzy data.* John Wiley and Sons, Ltd.

Viertl, R., & Yeganeh, S. M. (2016). Fuzzy confidence regions. In C. Kahraman & Ö. Kabak (Eds.), *Fuzzy statistical decision-making: Theory and applications* (pp. 119–127). Cham: Springer International Publishing. ISBN: 978-3-319-39014-7. https://doi.org/10.1007/978-3-319-39014-7_8.

Zadeh, L. A. (1965). Fuzzy sets. *Information and Control, 8*(3), pp. 338–353. ISSN: 0019-9958. http://dx.doi.org/10.1016/S0019-9958(65)90241-X. http://www.sciencedirect.com/science/article/pii/S001999586590241X

Zadeh, L. A. (1968). Probability measures of fuzzy events. *Journal of Mathematical Analysis and Applications, 23*(2), 421–427. ISSN: 0022-247X. https://doi.org/10.1016/0022-247X(68)90078-4

Conclusion Part I

This part is composed of 5 chapters intended to propose novel methods of hypotheses testing procedures in the fuzzy context. To do so, we expressed in the first chapter the foundations of the fuzzy set theory, such as the concepts and definitions of fuzzy sets, fuzzy numbers, the extension principle, and the different logical and arithmetic fuzzy operations.

We briefly exposed in Chap. 3 fundamental notions of the fuzzy rule-based systems, where we clearly stated our interest to the defuzzification step. In Chap. 4, we listed several known metrics between fuzzy numbers with their properties, including the signed distance operator. Following, we introduced two novel L_2 metrics, i.e. the $d_{SGD}^{\theta^\star}$ and the generalized signed distance d_{GSGD}. Their construction is centered on the definition of the signed distance, but where we add a newer component taking into consideration the shape of the fuzzy numbers. We highlight that the generalized signed distance preserves the directional property of the signed distance. Thereafter, Chap. 5 aims to recall two main approaches of defining random variables in the fuzzy environment, i.e. the Kwakernaak–Kruse and Meyer and the Féron–Puri and Ralescu approaches. By the Kwakernaak–Kruse and Meyer method, a fuzzy random variable is defined as the fuzzy perception of a traditional random variable. The outcome of such methods is then fuzzy. The Féron–Puri and Ralescu approach relies on the levelwise extension of the random sets in real spaces to the spaces of fuzzy sets. It models then uncertain random variables for which this uncertainty is by nature. It is clear that conceptual differences exist between both approaches. However, in a one-dimensional setting, both approaches coincide. We add that for the Féron–Puri and Ralescu methodology, a metric is needed to solve the problem induced by the fuzzy difference operation. The generalized signed distance or other metrics can consequently be used. We exposed in this chapter also some interesting statistical measures based on both approaches, such as the expectation and the variance, and the notion of point estimators. We closed this chapter by several simulations where we randomly constructed uncertain data sets with various sizes. The aim was to explore the influence of the sample size and the main differences between the use of the previously listed distances in terms of statistical measures. We were particularly interested in the sample moments and, consequently by the variance, the skewness and the excess of kurtosis. From these computations, we found that the sample size does not affect the value of the different measures. According to the use of all the distances in such contexts, we saw that no particular interpretation can be done on the variance and the excess of kurtosis. However, for the skewness, we remarked that by the directional distances, we tend to get a null skewness. The skewness measures of the data sets using the non-directional distances are very close and farther from 0 than the directional ones. This finding can be an indication that by a directional distance, we might be close to a normal distribution.

For the last and most important chapter of this part, we introduced a fuzzy hypotheses test by fuzzy confidence intervals, where we suppose that the fuzziness is a matter of data and/or hypotheses. For the calculation of the fuzzy confidence intervals, we proposed a practical procedure of construction based on the likelihood ratio method. The great advantage of this method is that all types of parameters can be used, without the need of a priori defining a corresponding theoretical distribution. We remind that our test statistic gives fuzzy decisions related to the null and the alternative hypotheses. Since these decisions might overlap, a defuzzification could be advantageous in order to obtain an interpretable decision. The defuzzified values are said to be the degrees of conviction of not rejecting a given hypothesis. An inference test can also be done by calculating a p-value and comparing it to the significance level. We expressed a fuzzy p-value where the data and/or the hypotheses are considered as uncertain, as well as its decision rule. A defuzzification of this fuzzy p-value could be useful since a crisp decision might be needed. This operation could eventually be done by a suitable operator, such as the original signed distance or the generalized one. We showed after two detailed applications using the Finanzplatz data set, where we computed the abovementioned approaches. By these examples, we wanted to investigate the influence of the shape of the fuzzy hypotheses on the decisions made by the fuzzy p-value approach. We found that the alternative hypothesis has no influence on the decision made. This hypothesis is only responsible for defining the rejection region. However, variating the shape and the spread of the fuzzy null hypothesis could influence drastically the decision. We finally performed the same tests but in the classical theory. The aim is to compare the fuzzy and the classical cases. We consequently saw that the classical approach seems to be a particular case of the fuzzy one. Thus, the more the fuzziness tends to be absent, the more we get closer to the results and obviously to the decisions of the classical analysis.

To sum up, we could state that by the proposed procedures, new ways of "treating" the uncertainty and subjectivity contained in the hypotheses or the data sets are introduced. We add that even though the defuzzification operation provokes a loss of information, obtaining a crisp decision is in many situations an asset, specifically in terms of interpretations.

Part II
Applications

Chapter 7
Evaluation of Linguistic Questionnaire

Linguistic questionnaires have gained lots of attention in the last decades. They are a prominent tool used in many fields to convey the opinion of people on different subjects. For instance, they are often used in satisfaction surveys, in politics and surveys related to voting, in media and ratings, etc. They are then seen as a way of expressing the feelings of individuals based on their answers to the different questions. These latter are usually coded using a Likert scale. While these feelings are expected to be precise, this fact is not actually realistic since human perception often waves between different ideas at the same time. In the spirit of precisely measuring the perception of people, one could suggest to increase the number of categories by question in order to give a wider panoply of possibilities. This increase could eventually induce an increase of the variability in the considered variable. An ideal situation would then be to increase endlessly this number of categories to achieve a certain continuity. However, this outcome cannot be described in natural language. Fuzzy tools are efficient to solve such drawbacks. By these tools, one could take into consideration the uncertainty expressed in the answers.

In many cases, an overall assessment of the questionnaire can be of good use. This evaluation encompasses the general point of view of a given respondent, or the set of all the respondents. In this context, de la Rosa de Saa et al. (2015) proposed a fuzzy rating method. They asserted that their method gives a precise idea of the subjectivity and uncertainty of the human perception. Gil et al. (2015) introduced a fuzzy rating scale-based approach applied in psychology to analyze the answers of 9-year-old children. Wang and Chen (2008) proposed an assessment for the performance of high school teachers using fuzzy arithmetics. Bourquin (2016) constructed fuzzy poverty measures to analyze the poverty in Switzerland based on the linguistic answers of members of Swiss households. Lin and Lee (2009) and Lin and Lee (2010) proposed a method based on the evaluation of the questionnaire, so-called aggregative assessment, where the fuzziness in the answers is taken into account. This latter is particularly introduced to assess the questionnaire at a global level. As for us, we are not only interested in the assessment of the whole questionnaire, but also in the assessment at an individual level. In a list of papers Berkachy and Donzé (2015), Berkachy and Donzé (2016a), Berkachy and Donzé (2016b), we presented a model based on the individual and the global evaluations of a linguistic questionnaire, where we considered sampling weights under the assumption of complete cases. The obtained evaluations give another way of interpreting the answers for a given questionnaire. Nevertheless, we showed that the aggregative assessment is nothing but the weighted mean of the individual ones. We highlight that contrariwise to the traditional fuzzy systems based on the three-steps process composed of the fuzzification, the IF-THEN rules, and the defuzzification as described in Chap. 3, our approach consists of aggregating the records of an observation intermediately using a convenient distance. Then, a sort of score is constructed based of these partial aggregations. As instance, we illustrated our approach in Donzé and Berkachy (2019) by an empirical study applied in the context of measuring the poverty in Swiss households.

Moreover, we know that real life linguistic questionnaires are often exposed to missingness. This latter is seen as a drawback for the above presented approaches. Note that the missingness can be intentional, or not intentional due to any inconvenience in the collection of opinion for an example. For this purpose, we proposed in Berkachy and Donzé (2016d) an update of our model in order to be able to consider the missing values in the calculations. Thereafter, we suggested to proceed by readjustment of weights. In survey statistics defined in the classical environment, researchers introduced indicators related to non-response, such as Rubin (1987) who proposed the "fraction of information about missing due to non-response." Inspired by this idea, we provided in Berkachy and Donzé (2016d) the so-called indicators of information rates at the individual and the global levels. These indicators give us information about the influence of the amount of missingness contained in the answers of a given observation regarding the weights chosen for the model. For such evaluation models, fuzzy arithmetics are needed in the calculations. Thus, we propose to use a chosen distance between fuzzy numbers. As such, in further researches, we proposed to use the signed distance, known to be an efficient ranking tool.

The outcome of the defended model is a distribution of the individual evaluations defined in the crisp environment. Therefore, one could treat this distribution exactly the same as any traditional distribution in the classical statistical approach. In addition, since we are mainly interested in the statistical characteristics of such distributions, we showed in Berkachy and Donzé (2016b) a study on some of them. We saw that by the signed distance, we get close enough to a normal distribution. From another perspective, in Berkachy and Donzé (2017), we performed, using simulations, the calculations by fuzzy rule-based systems as described in Chap. 3. For this purpose, we used different defuzzification operators. The aim was to analyze the robustness of our method, and the sensitivity of the distributions using the signed distance. In addition, we wanted to compare the statistical characteristics of distributions obtained by these systems, with the ones obtained from the individual evaluations. Interesting remarks arose from this comparison: although both approaches offer the opportunity of characterizing the units by their overall assessments while taking into account the uncertainty, they give different interpretations in terms of statistical measures.

Inspired by Lin and Lee (2009), we will then recall in Sect. 7.1 the settings of the proposed model describing the global and individual evaluations where we suppose that the sample weights and the missingness are both allowed. For the problem of missingness, we show in Sect. 7.2 a method based of the readjustment of weights. We should clarify that the proposed approach is not a correction for the missingness in the sample as widely known in survey statistics. In Sect. 7.3, we give the expressions of the individual and global assessments, followed by the description of the indicators of information rate related to missing answers in Sect. 7.4. The model is after illustrated by a numerical application related to the Finanzplatz data set, given in Sect. 7.5. The objective of this empirical application is to clearly see that the obtained individual evaluations can be treated similarly to any data set in the classical theory. In addition, we will empirically remark that the obtained distributions tend to be normally distributed. Afterward, in Sect. 7.6, we perform different analyses by simulations on the statistical measures of these distributions. The aim of this section is to compare the individual evaluations with respect to all the distances of Chap. 4, and to see the influence of the symmetry of the modelling fuzzy numbers and the sample sizes on different statistical measures. We close the section by a comparison between the evaluations by the defended model, and the ones obtained through a usual fuzzy system using different defuzzification operators. Interesting findings of these analyses are that corresponding statistical measures are independent from the sample sizes, and that the use of the traditional fuzzy rule-based systems is not always the most convenient tool when non-symmetrical modelling shapes are used, contrariwise to the defended approach. Our approach by the individual evaluations is in such situations suggested.

7.1 Questionnaire in Fuzzy Terms

Lin and Lee (2009) and Lin and Lee (2010) stated that the linguistic questionnaires similar to the one given in Table 7.1 can be evaluated at a global level. Thus, they exposed an approach for the complete questionnaire. In Berkachy and Donzé (2015), Berkachy and Donzé (2016a) and Berkachy and Donzé (2016b), we asserted that a method based on the two levels "global" and "individual" could be eventually compelling to explore and develop. For our model, we propose to use the same reasoning of Lin and Lee (2009) and Lin and Lee (2010) expressed in the following manner.

Consider the linguistic questionnaire proposed in Table 7.1. We define a so-called main-item denoted by B_j, $j = 1, \ldots, r$ as a collection of sub-items, i.e. questions in our case. Each main-item has a weight given by b_j, such that $0 \leq b_j \leq 1$ and $\sum_{j=1}^{r} b_j = 1$. Indeed, a given questionnaire is said to have r main-items. We mention that in this case, the weight b_j related to a given main-item j is a priori equal for all the observations of a sample. In the same way, a sub-item B_{jk}, $k =$

Table 7.1 Example of a linguistic questionnaire

Please select your answer:				
1.		General overview of the thesis		
	1.1	Do you find this thesis boring?	☐ Yes, extremely	
			☐ Yes, a little	
			☐ No, not really	
			☐ No, not at all	
	1.2	How interested are you in this topic?	☐ Very interested	
			☐ Somewhat interested	
			☐ Not interested nor interested	
			☐ Not very interested	
			☐ Not interested at all	
	$\vdots$			
	$1.m_1$		☐ Yes, extremely	L_1
			$\vdots$	$L_{q_{1m_1}}$
			☐ No, not at all	$L_{Q_{1m_1}}$
2.		About the structure of the thesis		
	2.1	...		
	2.2	...		
	$\vdots$			
$\vdots$	...			
j.				
	j.1	...		
	$\vdots$			
	j.k	...		
	$\vdots$			
	$j.m_j$	...		
$\vdots$	...			
r.	...			

$1, \ldots, m_j$, is a given "question" of the main-item B_j composed of q_{jk} linguistic terms. Under the main-item B_j, each sub-item B_{jk} has a weight b_{jk} with $0 \leq b_{jk} \leq 1$ and $\sum_{k=1}^{m_j} b_{jk} = 1$, for which $j = 1, \ldots, r$, $k = 1, \ldots, m_j$. Similarly to the weight of the main-items, we suppose that the weights of a sub-item k at a main-item j is the same for all the observations. In addition, let $L_{q_{jk}}, q_{jk} = 1, 2, \ldots, Q_{jk}$ be the linguistic terms of a sub-item B_{jk}. These terms are assumed to be fuzzy and are modelled by their corresponding fuzzy homologues $\tilde{L}_{q_{jk}}, q_{jk} = 1, 2, \ldots, Q_{jk}$. We give as well their respective distances to the fuzzy origin written as $d(\tilde{L}_{q_{jk}}, \tilde{0})$.

For the setups of our model, consider a sample of N units. Several assumptions must be taken into consideration. Suppose the following ones:

- The distances (the signed distances for example) between the fuzzy numbers $\tilde{L}_1, \ldots, \tilde{L}_{q_{jk}}, \ldots, \tilde{L}_{Q_{jk}}$ from one side, and the fuzzy origin from another one are linearly ordered, i.e. $d(\tilde{L}_1, \tilde{0}) < d(\tilde{L}_2, \tilde{0}) < \ldots < d(\tilde{L}_{Q_{jk}}, \tilde{0})$;
- The interviewees are allowed to choose one linguistic term only per sub-item;
- Missing values are allowed;
- If a given sample unit i is weighted by a sampling weight denoted by w_i, $i = 1, \ldots, N$, we consider a weighting parameter ξ_i such that $0 \leq \xi_i \leq 1$ and $\sum_{i=1}^{N} \xi_i = 1$. By default, this parameter is defined by the following expression:

$$\xi_i = \frac{w_i}{\sum_{i=1}^{N} w_i}.$$

- If no sampling weights are envisaged in the concerned data set, we suppose that the weighting parameter ξ_i is equal to the value 1 for every unit of the sample, i.e. $\xi_i = 1, \forall i$.

We denote by δ_{jkqi} the indicator of an answer at the term $L_{q_{jk}}$, for the unit i, $i = 1, \ldots, N$. It is written as

$$\delta_{jkqi} = \begin{cases} 1 & \text{if the observation } i \text{ has an answer for the linguistic } L_{q_{jk}}; \\ 0 & \text{otherwise.} \end{cases} \tag{7.1}$$

Under the sub-item B_{jk}, consider the total weighted number of answers at a given linguistic $L_{q_{jk}}$, denoted by $n_{jkq\bullet}$, and written as

$$n_{jkq\bullet} = \sum_{i=1}^{N} \xi_i \delta_{jkqi}. \tag{7.2}$$

Analogously, we define $n_{jk\bullet\bullet}$ to be the total weighted number of answers at a given sub-item B_{jk}, such that

$$n_{jk\bullet\bullet} = \sum_{i=1}^{N} \sum_{q=1}^{Q_{jk}} \xi_i \delta_{jkqi}, \tag{7.3}$$

with $j = 1, \ldots, r$ and $k = 1, \ldots, m_j$. Note that the fuzzy entity $\frac{n_{jkq\bullet}}{n_{jk\bullet\bullet}} \tilde{L}_{q_{jk}}$ is a weighted fuzzy number.

For a particular case, suppose that no sampling weights are envisaged, i.e. the sample units are not weighted and $\xi_i = 1$ for every unit i. Additionally, if we consider that no missing values exist in the data set, i.e. $\delta_{jkqi} = 1, \forall j, k, i$, one could easily remark that

$$n_{jk\bullet i} = \sum_{q=1}^{Q_{jk}} \xi_i \delta_{jkqi} = 1.$$

A direct consequence of these conditions is to get the following expression:

$$n_{jk\bullet\bullet} = \sum_{i=1}^{N} \sum_{q=1}^{Q_{jk}} \xi_i \delta_{jkqi} = N. \tag{7.4}$$

Similarly, if the condition on the non-existence of the missingness is only applied, i.e. $\delta_{jkqi} = 1$, if the sample units are weighted, i.e. $\xi_i \neq 1$, $\forall i$, we could write that $n_{jk\bullet\bullet} = \sum_{i=1}^{N} \xi_i$, since one answer only is allowed per sub-item.

All the abovementioned information are explicited in the annotated questionnaire given in Table 7.2. Note that this Table 7.2 corresponds exactly to the Table 7.1.

However, for the proposed setups, a main problem of assessment arises: the missingness in the data. A method solving this problem should be displayed.

Table 7.2 The annotated questionnaire corresponding to Table 7.1

Please select your answer:			
1.	General overview of the thesis Main-item B_1 with weight b_1		
1.1	Do you find this thesis boring?	☐ Yes, extremely	$\tilde{L}_1$ $n_{111\bullet}$
	Sub-item B_{11} with weight b_{11}	☐ Yes, a little	$\tilde{L}_2$ $n_{112\bullet}$
		☐ No, not really	$\tilde{L}_3$ $n_{113\bullet}$
		☐ No, not at all	$\tilde{L}_4$ $n_{114\bullet}$
1.2	How interested are you in this topic?	☐ Very interested	$\tilde{L}_1$ $n_{121\bullet}$
	Sub-item B_{12} with weight b_{12}	☐ Somewhat interested	$\tilde{L}_2$ $n_{122\bullet}$
		☐ Not interested nor interested	$\tilde{L}_3$ $n_{123\bullet}$
		☐ Not very interested	$\tilde{L}_4$ $n_{124\bullet}$
		☐ Not interested at all	$\tilde{L}_5$ $n_{125\bullet}$
⋮ ...	Sub-item B_{1m_j} with weight b_{1m_j}		
j. ...	Main-item B_j with weight b_j		
...	Sub-item B_{jk} with weight b_{jk}		
r. ...	Main-item B_r with weight b_r		

7.2 Readjustement of Weights

When missingness in a given observation of the data set intervenes, researchers often tended to suppress the record related to this observation. In the opposite side, if one would like to keep these observations in the data, methods "treating" this missingness should be known. For our context, if for a given sub-item an answer is missing, we propose to re-adjust the weights attributed to each component of the questionnaire. Indeed, we proceed by revisiting iteratively the weights of each main-item and sub-item.

As such, let us first introduce the indicator of no-missingness, i.e. the existence of an answer:

For a sub-item B_{jk} under the main-item B_j, we denote by Δ_{jki}, an indicator of no-missingness of the answer related to the individual i, written as

$$\Delta_{jki} = \begin{cases} 0 & \text{if the observation i is missing;} \\ 1 & \text{otherwise.} \end{cases} \tag{7.5}$$

The idea is then to correct the weights given to each sub-item and main-item by the corresponding weights based on the presence of missingness at a given level. For instance, consider the individual i. We correct the weight of the sub-item B_{jk}, written as $b_{jk} \equiv b_{jki}$, by the weight b^*_{jki} depending on i, such that

$$b^*_{jki} = \frac{\Delta_{jki} b_{jk}}{\sum_{k=1}^{m_j} \Delta_{jki} b_{jk}}. \tag{7.6}$$

In the same way, we could correct the weight of the main-item B_j, written as the weight $b_j \equiv b_{ji}$, by b^*_{ji} indexed additionally by the unit i, using the Eq. 7.6. This weight can be given in the following manner:

$$b^*_{ji} = \frac{\left(\sum_{k=1}^{m_j} b^*_{jki} \right) b_j}{\sum_{j=1}^{r} b_j \left(\sum_{k=1}^{m_j} b^*_{jki} \right)}. \tag{7.7}$$

Furthermore, we remind that a necessary condition on such weights is to have their total sum equal to one. In fact, we know by hypotheses that $\sum_{j=1}^{r} b_j = 1$ and $\sum_{k=1}^{m_j} b_{jk} = 1$. Since the total sums of weights in each main-item and sub-item are conserved, then we get that $\sum_{j=1}^{r} b^*_{ji} = 1$ and $\sum_{k=1}^{m_j} b^*_{jki} = 1$.

We underline that our method is only designed to "treat" the missingness partially existing in the data. Thus, if all the answers of a given observation are missing, we suggest to suppress the record of this observation.

To sum up, by the described method based on readjusting the weights, we tended to redistribute the weights on the "complete" main-items or sub-items without penalizing the concerned main or sub-item for not being fully "complete."

Accordingly, we are now able to propose the assessments related to the individual and the global levels.

7.3   Global and Individual Evaluations

Consider a distance d on a given metric space. We define d_{jkq} to be the distance of the fuzzy number $(n_{jkq\bullet}/n_{jk\bullet\bullet})\tilde{L}_{q jk}$ to the fuzzy origin. This entity could be written using Eqs. 7.2 and 7.3, as

$$d_{jkq} = d\left(\frac{n_{jkq\bullet}}{n_{jk\bullet\bullet}}\tilde{L}_{q jk}, \tilde{0}\right) = \frac{n_{jkq\bullet}}{n_{jk\bullet\bullet}}d(\tilde{L}_{q jk}, \tilde{0}) = \frac{\sum_{i=1}^{N}\xi_i\delta_{jkqi}}{\sum_{i=1}^{N}\sum_{q=1}^{Q_{jk}}\xi_i\delta_{jkqi}}d(\tilde{L}_{q jk}, \tilde{0}). \tag{7.8}$$

By combining the information of Eqs. 7.6, 7.7, and 7.8, we expose the global assessment of a whole linguistic questionnaire denoted by $\mathcal{P}$, by the following expression:

$$\mathcal{P} = \sum_{i=1}^{N}\sum_{j=1}^{r} b_{ji}^* \sum_{k=1}^{m_j} b_{jki}^* \sum_{q=1}^{Q_{jk}} \delta_{jkqi}\, d_{jkq}$$

$$= \sum_{i=1}^{N}\sum_{j=1}^{r} b_{ji}^* \sum_{k=1}^{m_j} b_{jki}^* \sum_{q=1}^{Q_{jk}} \frac{n_{jkq\bullet}}{n_{jk\bullet\bullet}} \delta_{jkqi}\, d(\tilde{L}_{q jk}, \tilde{0}). \tag{7.9}$$

On an individual level, it is as well important to display the assessment of the questionnaire related to a single individual i. The outcome of such evaluations is in some sense a score resuming all the information contained in the answers of the individual i. Let us first define the individual evaluation of the main-item B_j for the observation i. It is given by $\mathcal{P}_i^{(j)}$ such that

$$\mathcal{P}_i^{(j)} = \sum_{k=1}^{m_j} b_{jki}^* \sum_{q=1}^{Q_{jk}} \delta_{jkqi} d(\tilde{L}_{q jk}, \tilde{0}). \tag{7.10}$$

This expression leads to the individual assessment of the questionnaire for the individual i denoted by $\mathcal{P}_i$. This assessment is then expressed by:

$$\mathcal{P}_i = \sum_{j=1}^{r} b_{ji}^* \mathcal{P}_i^{(j)} = \sum_{j=1}^{r} b_{ji}^* \sum_{k=1}^{m_j} b_{jki}^* \sum_{q=1}^{Q_{jk}} \delta_{jkqi} d(\tilde{L}_{q jk}, \tilde{0}). \tag{7.11}$$

When missingness is not depicted in a collection of answers, the global evaluation of a questionnaire is often seen as the weighted arithmetic mean of the
individual evaluations of the involved sample. This idea is exposed in the following
proposition:

Proposition 7.3.1 *Under the assumption of no missing values in the data, the
global evaluation of a questionnaire denoted by $\mathcal{P}$ is nothing but the weighted
arithmetic mean of the individual evaluations $\mathcal{P}_i, i = 1, \dots, N$.*

Proof The weighted arithmetic mean denoted by $\overline{\mathcal{P}_\bullet}$ of the individual evaluations
can be written as

$$\overline{\mathcal{P}_\bullet} = \frac{\sum_{i=1}^{N} \xi_i \mathcal{P}_i}{\sum_{i=1}^{N} \xi_i},$$

where ξ_i is the weight assigned to the observation i, with $0 \le \xi_i \le 1$.
   From the expression of the individual evaluation given in Eq. 7.11, and under the
assumption of no-missingness in the data, i.e. $b_{ji}^* \equiv b_j$ and $b_{jki}^* \equiv b_{jk}, \forall i$, we write
$\overline{\mathcal{P}_\bullet}$ as follows:

$$\frac{1}{\sum_{i=1}^{N} \xi_i} \sum_{i=1}^{N} \xi_i \sum_{j=1}^{r} b_{ji}^* \sum_{k=1}^{m_j} b_{jki}^* \sum_{q=1}^{Q_{jk}} \delta_{jkqi} d(\tilde{L}_{q_{jk}}, \tilde{0}) = \sum_{j=1}^{r} b_j \sum_{k=1}^{m_j} b_{jk} \sum_{q=1}^{Q_{jk}} \frac{\sum_{i=1}^{N} \xi_i \delta_{jkqi}}{\sum_{i=1}^{N} \xi_i} d(\tilde{L}_{q_{jk}}, \tilde{0}).$$

The idea is to show that

$$\frac{n_{jkq\bullet}}{n_{jk\bullet\bullet}} = \frac{\sum_{i=1}^{N} \xi_i \delta_{jkqi}}{\sum_{l=1}^{N} \xi_i}.$$

Yet, we know from Eqs. 7.2 and 7.3, that

$$\frac{n_{jkq\bullet}}{n_{jk\bullet\bullet}} = \frac{\sum_{i=1}^{N} \xi_i \delta_{jkqi}}{\sum_{i=1}^{N} \sum_{q=1}^{m} \xi_i \delta_{jkqi}}.$$

Then, we only have to prove that

$$\sum_{i=1}^{N} \sum_{q=1}^{Q_{jk}} \xi_i \delta_{jkqi} = \sum_{i=1}^{N} \xi_i. \tag{7.12}$$

   Since δ_{jkqi} is the indicator function of an answer, we know that for every positive
answer for a particular linguistic L_q, we have $\delta_{jkqi} = 1$. It is equal to 0 otherwise.
Therefore, by definition of the function δ_{jkqi} given in Eq. 7.1, then Eq. 7.12 is
true. $\square$

   Analogously, one could also prove that if no sampling weights are applied in
the data set, i.e. all the sample units have exactly the same weight, then the global
evaluation is equal to the arithmetic mean of the individual evaluations.

The distance d exposed in the calculations of these evaluations are supposed to be generic. In other words, we could propose to use any of the described distances of Chap. 4. Furthermore, some distances might be equivalent in terms of the distributions of the individual evaluations, such as the Bertoluzza and the mid/spr metrics, as seen in the following proposition:

Proposition 7.3.2 *In a one-dimensional setting, the individual evaluations with respect to the Bertoluzza distance $d_{Bertoluzza}$ are equivalent to the ones with respect to the mid/spr one $d_{Mid/Spr}$.*

Proof We know from Remarks 4.4.1 that in a one-dimensional setting, the Bertoluzza distance is equivalent to the mid/spr metric. This latter induces directly that the distributions of the individual evaluations related to both distances are equivalent as well.   □

Moreover, our previous research showed that the SGD could be advantageous in such methodologies. At this level, we are not only interested in using it but also in using its generalized form in the purpose of comparing them. One could easily see that the individual evaluations by the $d_{SGD}^{\theta^*}$ and the d_{GSGD} distances are equivalent. As such, we have the following proposition:

Proposition 7.3.3 *The individual evaluations with respect to the $d_{SGD}^{\theta^*}$ distance are equivalent to the ones with respect to the generalized signed distance d_{GSGD}.*

Proof The proof of this proposition is direct, by construction of d_{GSGD}.   □

7.4 Indicators of Information Rate of a Data Base

For cases where missing values are present, it could be very disadvantageous for a practitioner to naively exclude the record of the observations containing missingness. For this reason, we propose to calculate the so-called indicator of information rate of the whole questionnaire at a global level. We could also introduce an indicator based on an individual level. The purpose is then to get an idea of the "quantity" of the information obtained despite the presence of missing values.

We construct an indicator of information rate of the full questionnaire denoted by R, $0 \leq R \leq 1$ and based on the weighted number of missing values in all the answers, such that

$$R = 1 - \frac{1}{\sum_{i=1}^{N} \sum_{j=1}^{r} \sum_{k=1}^{m_j} \xi_i b_{jk}} \sum_{i=1}^{N} \xi_i \sum_{j=1}^{r} \sum_{k=1}^{m_j} (b_{jk} - b_{jki}^*) \Delta_{jki} \cdot \frac{M_{ji}}{m_j},$$

$$(7.13)$$

where ξ_i is the sampling weight of the individual i, b_{jk} is the unadjusted weight of the sub-item B_{jk}, b_{jki}^* is the adjusted one, Δ_{jki} is the indicator of no-missingness of a given answer, M_{ji} is the total number of missing values occurring in a main-item j for i such that $M_{ji} = \sum_{k=1}^{m_j} \Delta_{jki}$, and m_j is the total number of sub-items per main-item j. In addition, a value 1 of R means that there is no missing values in the data set. If $R = 0$, it is the opposite. The indicator R can then be interpreted as a measure of the presence of missing values in the data base.

In the same way, we denote by R_i the indicator of information rate of the answers of the individual i with $0 \le R_i \le 1$. The indicator R_i can then be written as

$$R_i = 1 - \frac{1}{\sum_{j=1}^{r} \sum_{k=1}^{m_j} b_{jk}} \sum_{j=1}^{r} \sum_{k=1}^{m_j} (b_{jk} - b_{jki}^*) \Delta_{ijk} \cdot \frac{M_{ji}}{m_j}. \tag{7.14}$$

It is clear, by analogy to the global and individual evaluations, that the indicator R is seen as a sort of weighted mean of the indicator R_i. By these indicators, two main conclusions on both levels can be deduced. The first one is related to the relevance of the data set when missingness in the answers intervenes. Thus, an indicator of the whole questionnaire as R could be useful to draw the attention of the practitioner on whether to consider the missing values or not. Analogously, the indicator R_i could be an ideal indication of stating whether to exclude or not the answers of a given observation of the data set depending on the amount of contained missingness.

The computation of the global and individual evaluations can be summarized in the following algorithm:

Algorithm:

1. Consider a linguistic questionnaire with its collection of questions.
2. Define the main-items B_j, $j = 1, \ldots, r$ with their weights b_j, $0 \le b_j \le 1$ and $\sum_{j=1}^{r} b_j = 1$.
3. For each main-item B_j, define the sub-items B_{jk} with their weights b_{jk}, $j = 1, \ldots, r$ and $k = 1, \ldots, m_j$, such that $0 \le b_{jk} \le 1$ and $\sum_{k=1}^{m_j} b_{jk} = 1$.
4. Fuzzify the data set using the fuzzy numbers modelling the linguistic terms $L_1, \ldots, L_{q_{jk}}$ corresponding to each sub-item B_{jk}.
5. Calculate the distance of each fuzzy linguistic to the fuzzy origin $d(\tilde{L}_{q_{jk}}, \tilde{0})$.
6. Calculate $n_{jkq\bullet}$ and $n_{jk\bullet\bullet}$—Eqs. 7.2 and 7.3.
7. If missing values are present in the data set, calculate for each individual i, the readjusted weights b_{jki}^* and b_{ji}^*—Eqs. 7.6 and 7.7. Otherwise, go to 9.
8. Eventually calculate the global and individual indicators of information rate R and R_i—Eqs. 7.13 and 7.14.
9. Calculate the global and individual evaluations $\mathcal{P}$ and $\mathcal{P}_i$—Eqs. 7.9 and 7.11.

In order to clarify the computations needed for such procedures, the afore-mentioned algorithm is illustrated by the following application on the Finanzplatz questionnaire.

7.5   Application: Finanzplatz

For the application of the global and individual evaluations, we use the Finanzplatz data set described in Appendix B. This data set is composed of 234 firms, for which we chose to analyze their answers to 15 linguistic questions divided into three main groups. These questions are coded by 5 linguistics each, going from 1 (bad) to 5 (good). Every unit should choose exclusively one linguistic only. Suppose that the fuzziness in such answers intervenes. Therefore, we propose to model the involved linguistics by fuzzy numbers. For sake of simplicity, we propose to use triangular and trapezoidal fuzzy numbers only. Note that the considered data base includes the information about the company sizes (small, medium, or big) and the company branches (banks or insurances, etc.). In addition, a sampling weight is exposed.

We would like to assess our questionnaire at the two levels: individual and global. The objective is to adopt the aforementioned methodology of modelling the uncertainty and produce these evaluations. However, our data set of 234 observations includes missing values. Therefore, we proceed by readjustment of weights as given in Sect. 7.3.

Let us first fuzzify the data set. We suppose that we model the linguistics 1 to 5 denoted by L_1, L_2, L_3, L_4, and L_5, by their corresponding fuzzy numbers $\tilde{L}_1$, $\tilde{L}_2$, $\tilde{L}_3$, $\tilde{L}_4$, and $\tilde{L}_5$, written as

$$\tilde{L}_1 = (1, 1, 1.2);$$

$$\tilde{L}_2 = (1, 1.5, 2.4, 3.1);$$

$$\tilde{L}_3 = (2.2, 2.6, 3.4, 4.2);$$

$$\tilde{L}_4 = (3.2, 3.7, 4.2, 4.5);$$

$$\tilde{L}_5 = (4.8, 5, 5, 5).$$

We mention that for this case, we chose arbitrarily to model the same linguistics by the same fuzzy numbers. In contrary, in some other contexts, one could expect that the considered firm would eventually choose by itself the corresponding fuzzy number modelling its perception.

According to the procedure presented in Sect. 7.1, we consider three main-items B_1, B_2, and B_3, such that their weights are respectively b_1, b_2, and b_3 with $b_1 + b_2 + b_3 = 1$. In addition, each main-item is composed of sub-items defined by the 5 linguistics $\tilde{L}_1$ to $\tilde{L}_5$.

Since missingness exists in the data set, then for each observation i, we have to consider the readjusted weights of sub- and main-items given in Sect. 7.2. A detailed example of calculation for the observation 17 is given in Table 7.3. This latter shows the three main-items where we supposed that they are equally weighted $b_1 = b_2 = b_3 = \frac{1}{3}$. The sub-items are also at the first stage equally weighted. For this example, we used the SGD given in Sect. 4.6.

Once we perform the same procedure on the answers of the 234 observations, we get a new sample of evaluations where each of them refers to the overall overview of the answers of a given observation. In addition, one could calculate the global evaluation of the complete data set, where the sampling weights are taken into account. For our present case, the global evaluation is

$$\mathcal{P} = 3.5339.$$

Furthermore, we are interested in calculating the indicator of information rate as given in Eqs. 7.13 and 7.14. For the entire data set containing missing values, the indicator denoted by R is

$$R = 0.9781.$$

This latter of around 97.81% indicates that even though missingness exists, we can rely on our data set since the amount of missingness is apparently limited regarding the number of existing answers and their weights. In the same way, one could compute the indicator of information rate related to a particular observation. Consider for example the observation 17 for which the answers were described in Table 7.3. The corresponding indicator of information rate is

$$R_{17} = 0.9948.$$

This indicator is very close to the value 1, which means that the effect of the missingness on the evaluation of the answers of the observation 17 is slight. Thus, this observation should not be excluded from the data set.

The produced vector of individual evaluations can be treated similarly to any other data set in the classical statistical theory. In Berkachy and Donzé (2016c), we presented a study on the different statistical measures obtained by such procedures. The aim of this contribution was to compare the individual evaluations to distributions obtained by a traditional fuzzy rule-based system. By that, we mean that they were calculated by the three-steps process (fuzzification, IF-THEN rules, defuzzification), as given in Chap. 3. Our purpose is then to investigate the characteristics of the underlying distributions seen as traditional vectors of data, in order to be able to use them in the modelling process using appropriate classical statistical methods. In addition, it is important to know if this distribution is as instance symmetric, or a fortiori tends to be normal. The Figs. 7.1 and 7.2 show the distributions of the evaluated data set. A clear graphical remark of these figures is that the obtained data set tends to be normally distributed. Another empirical way

Table 7.3 Application of the individual evaluation using the signed distance on Finanzplatz: answers of the observation 17—Sect. 7.5

B_j	b_j	B_{jk}	b_{jk}	L_q	δ_{jkqi}	b^*_{jki}	$d(\bar{L}_q,\bar{0})$	$\mathcal{P}_i$
1	$\frac{1}{3}$	1.1 The present state of business is	0.5	Bad	0	0.5	3.9	3.930238
				L_2	0			
				Fair	0			
				L_4	1			
				Excellent	0			
		1.2 The expected state of business in 12 months is	0.5	Worst	0	0.5	4.95	
				L_2	0			
				Same	0			
				L_4	0			
				Better	1			
2	$\frac{1}{3}$	2.1 The demand for our services/products compared to the last 12 months is	0.125	L_4	1	0.143	3.9	
		2.2 The demand of domestic private customers for our products compared to the last 12 months is	0.125	L_4	1	0.143	3.9	
		2.3 The demand from foreign private clients for services/products compared to the last 12 months is	0.125	L_4	1	0.143	3.9	
		2.4 The demand of corporate/institutional clients for our services/products compared to the last 12 months is	0.125	L_3	1	0.143	3.1	
		2.5 The expected demand for our services/products in the next 12 months is	0.125	L_4	1	0.143	3.9	
		2.6 The expected demand of domestic private customers for our products in the next 12 months is	0.125	L_4	1	0.143	3.9	
		2.7 The expected demand from foreign private clients for services/products in the next 12 months is	0.125	NA	NA	0	0	
		2.8 The expected demand of corporate/institutional clients for our services / products in the next 12 months is	0.125	L_4	1	0.143	3.9	
3	$\frac{1}{3}$	3.1 The gross profit compared to business in 12 months is	0.2	L_4	1	0.2	3.9	
		3.2 The employment compared to the last 12 months is	0.2	L_3	1	0.2	3.1	
		3.3 The expected gross profit in the next 12 months is	0.2	L_4	1	0.2	3.9	
		3.4 The expected employment in the next 12 months is	0.2	L_3	1	0.2	3.1	
		3.5 The expected use of technical and personal capacities in the next 12 months is	0.2	L_4	1	0.2	3.9	

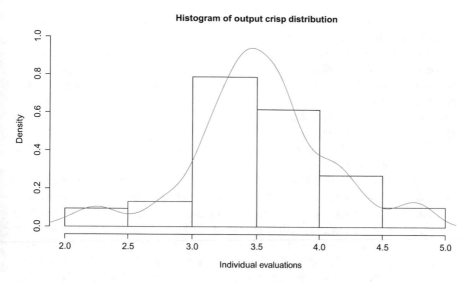

Fig. 7.1 The histogram of the distribution of the weighted individual evaluations of the Finanz-platz linguistic questionnaire—Sect. 7.5

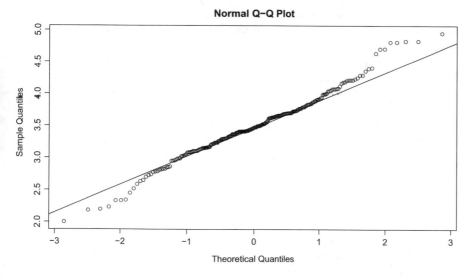

Fig. 7.2 The QQ-plot of the distribution of the weighted individual evaluations of the Finanzplatz linguistic questionnaire—Sect. 7.5

to verify the assumption of normality is to calculate the asymmetry measures. Thus, for this case, we have a skewness of -0.0601 and an excess of kurtosis of 0.7192. These results show a symmetric distribution and a possible normal distribution induced from the evaluations of our questionnaire. Last for now, different normality tests exist in statistics. If we use the Chi-squared, the Shapiro–Wilk, the Anderson–Darling, and the Lilliefors tests,[1] we could not reject the null hypothesis that the calculated distribution may represent a normal distribution, or is close enough.

We saw that the distributions of the individual evaluations tended empirically to be normal. Yet, theoretically, to prove that the produced distribution is potentially normal, one could imagine to reason by postulating a multinomial distribution related to the set of variables of the questionnaire, and therefore, to deduce a tendency of the individual evaluations to be asymptotically drawn from a univariate normal distribution. A further research should be done on this purpose to validate (or negate) this point of view.

From another side, to illustrate the usability of the individual evaluations as a usual data set, we propose to analyze another facet of it. Since our original data set contains additional information such as the company size and branch, based on the obtained evaluations, we could analyze the difference between these groups by means of statistical measures. The Tables 7.5 and 7.6 give a summary of the different statistical measures calculated with the corresponding sampling weights. If we look closely to the mean of the individual assessments, we can see that the evaluations of the small sized companies are higher than the medium and the big ones. By that, one could conclude that small businesses expect more improvement than the medium and big ones. This result can be seen in the boxplot given in Fig. 7.3. Analogously, by classifying the set with respect to the branches of the firms, the Financial services companies have the highest scores among others. The Fig. 7.4 gives a clear view of such conclusions.

It is an advantage to be able to use the observations where missingness occurs. We would like to compare the processed case with a complete one. If we decide to exclude every block of records related to a given unit, containing at least one missing value, we should expect to have a big loss of observations. However, since the missing values are concentrated in 6 variables only, we propose to take into account the variables with no missingness only. Thus, 9 variables left will be studied. Similarly to the last case, we expose in Table 7.7 the statistical measures of the evaluations with and without "NA." By this table, we can say that the measures of both data sets are comparable. We additionally draw a QQ-plot between these distributions in order to see if a given potential relation exists between them. This is seen in the Fig. 7.5. It is consequently interesting to see that a linear relation exists between both distributions. By this conclusion, we could say that the presence of the missing values as in our data set does not penalize the results of the evaluations. Thus, we can clearly see the relevance of our proposed model with "NA" compared to the complete case. This latter is seen as a less realistic case in terms of collecting data. To sum up, in real life situations, when such evaluations are needed, with respect to our methodology, we are a fortiori now on able confidently to treat the missing values in order to avoid the high loss of information due to the exclusion of observations.

---

[1]These tests are described further respectively in Pearson (1900), Fisher (1922) and Fisher (1924) for the Chi-squared test, Shapiro and Wilk (1965) for the Shapiro–Wilk test, Anderson and Darling (1952) for the Anderson–Darling test, and Lilliefors (1967) for the Lilliefors test.

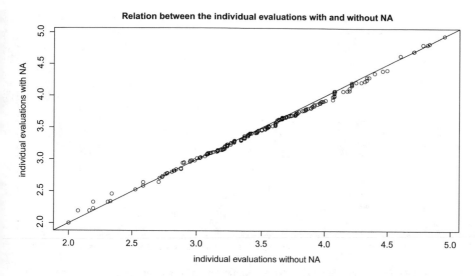

Fig. 7.3 Boxplot of the weighted individual evaluations classified by company size—Sect. 7.5

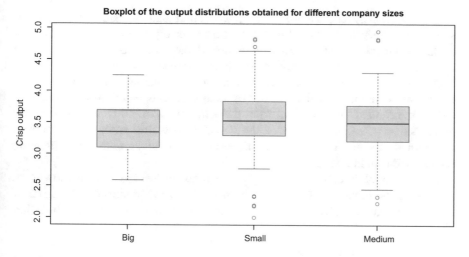

Fig. 7.4 Boxplot of the weighted individual evaluations classified by company branch—Sect. 7.5

7.6 Analyses by Simulations on the Individual Evaluations

The idea now is to investigate different statistical characteristics of the distribution of individual evaluations induced from our methodology of assessment. Similar work has been done in Berkachy and Donzé (2017) and Berkachy and Donzé (2016a). In this section, we elaborate the results obtained in our previous papers in order to go further in the generalization of our interpretation of such characteristics.

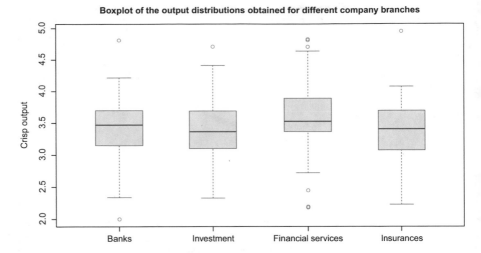

Boxplot of the output distributions obtained for different company branches

Fig. 7.5 QQ-plot of the distributions of the weighted individual evaluations with and without NA—Sect. 7.5

From Berkachy and Donzé (2017), we found that for small data sets of around 20 observations, the statistical measures corresponding to the distribution of the individual evaluations are seen as non-stable. The reason is obviously the inadequate number of observations. We found as well that from a threshold of 1000 observations, the measures become much more stable. This result confirms the known assumption of the stability of such measures based on increasing the size of the studied sample drawn from the same distribution. However, from our previous researches, we saw that all the statistical measures are independent from sample sizes. We note that when sizes increased, almost all presented an asymptotic stability with slight fluctuations.

In our present case, we perform by simulations several tests which allow us to understand the influence of different components on the produced distribution of individual evaluations. The objectives of our tests are then given by the following ideas:

- To understand the influence of the sample size;
- To understand the influence of the asymmetry in the shapes of the modelling membership functions;
- To understand the influence of the choice of defuzzification method, and if a distance is required, to test the influence of the chosen distances.

As such, the setups of the simulations are as follows:

We consider a questionnaire of 4 equally weighted main-items. Each main-item has respectively 2 sub-items. The sub-items in each main-item are equally weighted as well. All the sub-items are composed of 5 linguistics $L_1, \ldots, L_5$. We would like to take into consideration the fuzziness existing in the answers of a given observation. We model then each linguistic by its corresponding fuzzy number

$\tilde{L}_1, \ldots, \tilde{L}_5$. We chose arbitrarily these linguistics. We add that for the same data set, we considered the two cases of symmetrical and non-symmetrical triangles and trapezoids in order to understand the difference between them. For the symmetrical triangular shapes, we used the following fuzzy numbers:

$$\tilde{L}_1 = (1, 1, 2);$$
$$\tilde{L}_2 = (1, 2, 3);$$
$$\tilde{L}_3 = (2, 3, 4);$$
$$\tilde{L}_4 = (3, 4, 5);$$
$$\tilde{L}_5 = (4, 5, 5);$$

and for the non-symmetrical shapes, we used the fuzzy linguistics shown in Sect. 7.5 and given by:

$$\tilde{L}_1 = (1, 1, 1.2);$$
$$\tilde{L}_2 = (1, 1.5, 2.4, 3.1);$$
$$\tilde{L}_3 = (2.2, 2.6, 3.4, 4.2);$$
$$\tilde{L}_4 = (3.2, 3.7, 4.2, 4.5);$$
$$\tilde{L}_5 = (4.8, 5, 5, 5).$$

Note that for the present case, we chose to model the same linguistics by the same fuzzy number. However, in further studies, one could suggest to ask each unit to answer by its proper fuzzy number for example. In addition, no sampling weights or missing values are allowed in our situation, the reason is obviously to avoid the possible bias that might be induced.

Our first attempt is to investigate the influence of the sample size of the data set on the statistical measures of the distribution of the produced individual evaluations. For this purpose, we generated random data sets of different sizes starting from the threshold found in Berkachy and Donzé (2017), i.e. 1000 observations, to 10000 observations with a step of 100. A first finding is that the measures of central tendency, location, dispersion, and asymmetry are independent of sample sizes. This result confirms the statement showed in Berkachy and Donzé (2017). Thus, to use a data set of 1000 observations is enough to get stable statistical measures. It can also be seen in Table 7.8. For the rest of these analyses, we show the cases of 1000 and 10000 observations only.

It is interesting now to interpret the results given in Table 7.8. For instance, if we would like to compare the distances in terms of the obtained measures. We can clearly see that the variances and consequently the standard deviations of all the distributions are very similar. It is the same for the excess of kurtosis. Thus, no particular conclusion can be done on this purpose. All the distributions have

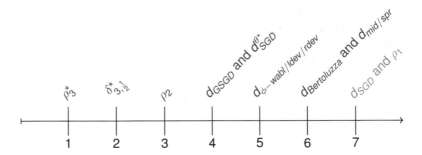

Fig. 7.6 Order of the skewness of the distributions of the individual evaluations with respect to the different distances (the red ones are the L_3 group, the blue ones are the L_2 group, and the gray one is the L_1 group)

very close means except the one with respect to the distance $\rho_3^\star$. It is the same for the quantiles of the distribution. One could deduce that the observations in all the distributions except the $\rho_3^\star$ ones are quite closely located.

However, for the skewness, the case is different. It appears that an order for the skewness of the different distributions exists, as seen in Fig. 7.6. We have to mention that the skewness of the distributions of the defended distances d_{GSGD} and $d_{SGD}^{\theta^\star}$ are between all, from the closest to 0. A possible interpretation of this order is that the distributions are grouped by their L_p spaces. In other terms, we can directly recognize that the L_3 metrics are the closest to 0, followed by the group of L_2 metrics, and finally by the L_1 ones. The closeness to the 0 level skewness is maybe due to the higher exponent of the expressions of the different distances. We highlight that in the group of L_2 metrics, the generalized signed distance presents measures which are very close to the ones of the ρ_2 metric. This statement confirms the topological equivalence that we intended to expose in Proposition 4.7.4 of Chap. 4. Nevertheless, if we look carefully to both asymmetry measures, we can remark that they are close to the ones of the normal distribution. This remark signalizes thus an eventual relation with the standard normal distribution.

For the comparison between the cases related to the symmetry of the modelling fuzzy numbers, we found that almost all the measures are very close, except the skewness, the minimum, and the maximum. The relatively big difference in the minimum and the maximum measures is maybe due to the weight that the adopted method puts on the edges of the distributions. Contrariwise, it is interesting to see that the quantiles 1, 2, and 3 are very close between both shapes. By choosing non-symmetric shapes for the modelling numbers, we tend to stretch the range of the distributions, more than with symmetric shapes, signalizing a larger dispersion of observations.

For the skewness and concerning the size of the sample, one could expect to get a closest skewness to 0 with a higher sample size. On the other hand, modelling with non-symmetric fuzzy numbers gave closer values to 0 than the symmetric ones. It is the case of all the distances except the $\rho_3^\star$ and the $\delta_{3,\frac{1}{2}}^\star$ ones. For these latter, choosing symmetric shapes provides closer skewness to 0. Finally, we have to mention that the ρ_2 metric gave a 0 skewness. For our elaborated version of the signed distance, we get a skewness of 0.002 for non-symmetrical fuzzy numbers. From these findings, one could clearly deduce that the defended model can be well adapted to non-symmetrical cases. The distributions obtained from such calculations are not penalized by the irregularities of the modelling shapes.

The distances displayed in these calculations are used in some sense as a defuzzification operator. We propose now to perform the procedure described in Chap. 3 on the same data sets with the same setups and compare them with the individual evaluations using the different distances. The aim is to be able to recognize the possible drawbacks of each of both approaches. Since we are particularly interested in the defuzzification step, we apply all the defuzzification operators presented in Sect. 3.3. The statistical measures obtained from the produced data sets are given in Table 7.9. We mention that for the inference rules of the involved systems, we used the t-norm and t-conorm min and max.

The idea now is to interpret the results of Table 7.9 and compare them with the results of the individual evaluations. From this table, we overall found that the distributions of the centroid, the bisector, and the MOM are relatively similar in terms of statistical measures, notably the mean, the median, the skewness, and the quantiles. However, the distributions using the SOM and LOM present extreme measures, which is due to their definitions.

On another level, for the dispersion measures, the distributions induced from the fuzzy inference systems give lower variances and standard deviations than the ones by the different distances. Thus, the evaluations through the fuzzy system are more concentrated than the evaluations by the defended model. The variances resulting from the distributions of the SOM and LOM are the highest between all defuzzification operators. This fact shows a disadvantage of these operators regarding the others.

According to the calculated asymmetry measures, the distributions by these systems are not close enough to suspect their relation with the normal distribution. Thus, the excess of kurtosis is relatively higher, and the skewness is farther from 0 than the distributions of the individual evaluations. We add that the distributions by the SOM and LOM display respectively extreme positive and negative skewness. Therefore, one could conclude that the use of the SOM and LOM is not efficient to preserve the symmetry of distributions.

For the comparison between the shapes of the modelling fuzzy numbers, we can mainly say that the interpretations of the differences between both cases are similar to the ones of the individual evaluations of Table 7.8. Moreover, the skewness of the distributions by the centroid, the bisector, and the MOM methods is very sensible to the irregularities in the modelling membership functions. Thus, by adopting a non-symmetric fuzzy model, we lost in the symmetry of the output distributions. This result could be due to the inference rules adopted for the computations. Therefore, if a questionnaire consists of many questions composed of multiple answers, refining these rules can be seen as a complicated task. It is evident thus to conclude that in terms of symmetry, if non-symmetric membership functions are required in the modelling procedure, then the use of our procedure of evaluation with a distance as a defuzzification operator is more convenient. Our approach of calculations produces distributions close to be drawn from a normal distribution, and this fact is seen as an important advantage in terms of hypothesis of many statistical analysis.

7.7   Conclusion

To conclude, we proposed an approach of aggregating a linguistic questionnaire at an individual and global levels where we supposed that uncertainty is embedded in the answers. This imprecision is not only taken into account, but treated in situations where sampling weights and missing values are considered. In the situation where the missingness occurs in data sets, we introduced two indicators of information rate related respectively to the individual and the global levels. Their aim is to indicate whether a given record is relevant in terms of the presence of missing values, or not. We illustrated our methodology by multiple empirical applications, having each particular objectives to investigate.

In a general manner, we found empirically that the calculated individual evaluations tend to be drawn from a standard normal distribution. We also saw that a linear relation exists between the evaluations with missing values from one side, and the evaluations without missing values from another side. A conclusion is that the presence of missing values and the way we treated them did not affect the results of the assessment.

To sum up, if we would like to perform a similar operation in the classical statistical theory, one could rather proceed by calculating the mean of means or constructing a given score, which works as well as an aggregation of the answers. This measure is advantageous in case we want to broadly describe the information of the questionnaire. However, if missingness occurs, the mean of means could be strongly affected by the non-response bias. The estimates will consequently be biased. If imprecision arises in a linguistic questionnaire, our approach of assessment is suggested specifically since the missing values can be taken into account.

Appendix

The first section is devoted to the description of the Finanzplatz data set.

B Description of the Finanzplatz Data Base

The questionnaire, the so-called Finanzplatz Zurich: Umfrage 2010 described in BAK (2010) is a survey of the financial place of Zurich, Switzerland. On behalf of the Department of Economic Affairs of the Canton of Zurich, the ASAM (Applied Statistics And Modelling) group was in charge of the design plan and the data analysis of this survey in 2010. The chosen sample is mainly composed of financial enterprises, i.e. banks or insurances of the canton of Zurich and the surrounding. The aim of this survey is to investigate the situation of the financial market of the canton of Zurich, by answering to a written questionnaire about their actual situation and their views for the future. As such, it intends to understand the expectations of the demand, the income, and the employment of the different firms for the foreseeable future.

This questionnaire consists of 21 questions from which 19 are linguistic ones with five Likert scale terms each. This survey is then recognized as a linguistic one. These questions are divided into 4 groups. The data base is composed of 245 observations, i.e. firms. Variables related to company size and branch are also present, as well as the sampling weights. From this questionnaire, we consider 18 variables only, including 15 linguistic ones divided into 3 groups. The Table 7.4 gives a detailed description of the considered attributes.

Note that non-response exists in this data base. We excluded the records for which a particular firm did not answer to any of the questions of this questionnaire. The size of the sample becomes then 234 observations.

Table 7.4 The variables of the Finanzplatz data set—Sect. B

Variables	Possible answers	Description
Group 1		
$F2.1$	1 (bad), 2, 3 (fair), 4, 5 (excellent)	The present state of business
$F2.2$	1 (worst), 2, 3 (same), 4, 5 (better)	The expected state of business in the next 12 months
Group 2		
$F3.1$	1 (less), 2, 3 (equal), 4, 5 (greater)	The demand for our services/products compared to the last 12 months
$F3.4$	1 (less), 2, 3 (equal), 4, 5 (greater)	The demand of domestic private customers for our products compared to the last 12 months
$F3.5$	1 (less), 2, 3 (equal), 4, 5 (greater)	The demand from foreign private clients for services/products compared to the last 12 months
$F3.6$	1 (less), 2, 3 (greater), 4, 5 (greater)	The demand of corporate/institutional clients for our services/products compared to the last 12 months
$F3.7$	1 (worst), 2, 3 (same), 4, 5 (better)	The expected demand for our services/products in the next 12 months
$F3.10$	1 (worst), 2, 3 (same), 4, 5 (better)	The expected demand of domestic private customers for our products in the next 12 months
$F3.11$	1 (worst), 2, 3 (same), 4, 5 (better)	The expected demand from foreign private clients for services/products in the next 12 months
$F3.12$	1 (worst), 2, 3 (same), 4, 5 (better)	The expected demand of corporate/institutional clients for our services/products in the next 12 months
Group 3		
$F4.1$	1 (less), 2, 3 (equal), 4, 5 (greater)	The gross profit compared to the last 12 months
$F4.2$	1 (less), 2, 3 (equal), 4, 5 (greater)	The employment compared to the last 12 months
$F4.4$	1 (worst), 2, 3 (same), 4, 5 (better)	The expected gross profit in the next 12 months
$F4.5$	1 (worst), 2, 3 (same), 4, 5 (better)	The expected employment in the next 12 months
$F4.6$	1 (worst), 2, 3 (same), 4, 5 (better)	The expected use of technical and personal capacities in the next 12 months
$w12$	A numerical positive value	The sampling weights
size	1 (small), 2 (medium), 3 (big)	The size of the firms
Branch	1 (bank), 2 (investment), 3 (financial services), 4 (insurance)	The branch of the firms

C   Results of the Applications of Chap. 7

This Annex provides the detailed tables of results of the studies of Chap. 7. Tables 7.5 and 7.6 show the summary of the weighted statistical measures of the individual evaluations using the signed distance, classified by company size and company branch correspondingly. Table 7.7 shows the weighted statistical measures of the distributions obtained from individual evaluations using the signed distance, for cases with and without "NA". Finally, Table 7.8 gives the summary of the statistical measures of the distributions obtained from individual evaluations using all distances, while Table 7.9 provides the measures of the output distribution of the fuzzy rule-based system with different defuzzification operators.

Table 7.5 Summary of the weighted statistical measures of the individual evaluations using the signed distance, classified by company size—Sect. 7.5

Statistical measure	Company size		
	Small	Medium	Big
Mean	3.573	3.494	3.370
Median	3.527	3.495	3.348
Variance	0.333	0.237	0.161
Standard deviation	0.577	0.481	0.402
Skewness	−0.192	0.084	0.268
Excess of kurtosis	0.692	0.779	−0.596
Quantiles			
$P = 0$	2	2.229	2.587
$P = 0.25$	3.293	3.207	3.1
$P = 0.5$	3.527	3.495	3.348
$P = 0.75$	3.835	3.77	3.696
$P = 1$	4.827	4.95	4.25

Table 7.6 Summary of the weighted statistical measures of the individual evaluations using the signed distance, classified by company branch—Sect. 7.5

| Statistical measure | Company branch | | | |
	Banks	Financial services companies	Investment companies	Insurance companies
Mean	3.422	3.624	3.456	3.367
Median	3.473	3.527	3.367	3.407
Variance	0.287	0.281	0.340	0.226
Standard deviation	0.536	0.530	0.583	0.476
Skewness	−0.633	−0.006	0.228	0.742
Excess of kurtosis	0.574	1.106	−0.284	2.325
Quantiles				
$P = 0$	2	2.181	2.33	2.229
$P = 0.25$	3.150	3.367	3.104	3.077
$P = 0.5$	3.473	3.527	3.367	3.407
$P = 0.75$	3.7	3.888	3.690	3.696
$P = 1$	4.81	4.827	4.705	4.95

Table 7.7 Summary of the weighted statistical measures of the distributions obtained from individual evaluations using the signed distance, for cases with and without NA—Sect. 7.5

Statistical measure	Case with NA	Case without NA
Mean	3.534	3.565
Median	3.495	3.542
Variance	0.295	0.308
Standard deviation	0.543	0.555
Skewness	−0.060	−0.153
Excess of kurtosis	0.719	0.489
Quantiles		
$P = 0$	2.000	2.000
$P = 0.25$	3.233	3.260
$P = 0.5$	3.495	3.542
$P = 0.75$	3.809	3.8888
$P = 1$	4.950	4.950

Table 7.8 Summary of the statistical measures of the distribution obtained from the individual evaluations for all distances—Sect. 7.6

Distances	Statistical measures	Symmetrical MF		Non-symmetric MF	
		1000	10000	1000	10000
ρ_2	Mean	3.044	3.048	3.051	3.053
	Median	3.042	3.044	3.055	3.057
	Variance	0.200	0.199	0.233	0.230
	Standard deviation	0.447	0.446	0.483	0.480
	Skewness	0.134	0.023	0.104	0
	Excess of kurtosis	−0.198	−0.182	−0.182	−0.162
	Minimum	1.855	1.489	1.678	1.325
	Maximum	4.458	4.458	4.602	4.602
	Quartile 1	2.716	2.732	2.704	2.707
	Quartile 3	3.361	3.363	3.407	3.404
$d_{Bertoluzza}$ and $d_{Mid/Spr}$	Mean	3.015	3.018	3.017	3.019
	Median	3.016	3.016	3.012	3.022
	Variance	0.202	0.201	0.236	0.234
	Standard deviation	0.450	0.449	0.486	0.484
	Skewness	0.143	0.030	0.126	0.018
	Excess of kurtosis	−0.195	−0.182	−0.178	−0.163
	Minimum	1.831	1.459	1.668	1.300
	Maximum	4.445	4.445	4.592	4.592
	Quartile 1	2.677	2.705	2.665	2.665
	Quartile 3	3.329	3.330	3.374	3.374
$\delta^{\star}_{3,\frac{1}{2}}$	Mean	3.088	3.091	3.097	3.098
	Median	3.083	3.085	3.094	3.094
	Variance	0.196	0.194	0.230	0.227
	Standard deviation	0.442	0.441	0.480	0.476
	Skewness	0.123	0.015	0.072	−0.026
	Excess of kurtosis	−0.202	−0.183	−0.185	−0.160
	Minimum	1.899	1.539	1.693	1.356
	Maximum	4.477	4.477	4.615	4.615
	Quartile 1	2.772	2.775	2.758	2.758
	Quartile 3	3.407	3.410	3.448	3.430
	Mean	3.425	3.428	3.408	3.409
	Median	3.402	3.405	3.395	3.428
	Variance	0.201	0.200	0.237	0.232
	Standard deviation	0.449	0.447	0.487	0.482

(continued)

Table 7.8 (continued)

Distances	Statistical measures	Symmetrical MF		Non-symmetric MF	
		1000	10000	1000	10000
$\rho_3^\star$	Skewness	0.056	−0.039	−0.114	−0.175
	Excess of kurtosis	−0.220	−0.194	−0.18	−0.162
	Minimum	2.170	1.798	1.848	1.518
	Maximum	4.755	4.759	4.771	4.771
	Quartile 1	3.094	3.094	3.089	3.089
	Quartile 3	3.727	3.713	3.763	3.761
$d_{\phi-wabl/ldev/rdev}$	Mean	3.014	3.017	3.016	3.019
	Median	3.014	3.015	3.012	3.022
	Variance	0.203	0.202	0.236	0.234
	Standard deviation	0.450	0.449	0.486	0.484
	Skewness	0.142	0.029	0.126	0.018
	Excess of kurtosis	−0.195	−0.182	−0.178	−0.163
	Minimum	1.827	1.455	1.668	1.300
	Maximum	4.444	4.444	4.592	4.592
	Quartile 1	2.676	2.702	2.664	2.664
	Quartile 3	3.329	3.330	3.374	3.374
d_{SGD} and ρ_1	Mean	2.998	3.002	2.999	3.001
	Median	3	3	2.994	3
	Variance	0.204	0.203	0.238	0.236
	Standard deviation	0.452	0.451	0.488	0.486
	Skewness	0.145	0.032	0.137	0.027
	Excess of kurtosis	−0.194	−0.181	−0.175	−0.164
	Minimum	1.812	1.438	1.663	1.287
	Maximum	4.438	4.438	4.587	4.588
	Quartile 1	2.656	2.688	2.644	2.644
	Quartile 3	3.312	3.312	3.350	3.350
$d_{SGD}^{\theta^\star}$ and d_{GSGD}	Mean	3.031	3.035	3.048	3.050
	Median	3.030	3.031	3.053	3.055
	Variance	0.201	0.200	0.233	0.231
	Standard deviation	0.449	0.448	0.483	0.480
	Skewness	0.135	0.024	0.106	0.002
	Excess of kurtosis	−0.198	−0.182	−0.181	−0.162
	Minimum	1.840	1.471	1.676	1.322
	Maximum	4.451	4.451	4.601	4.601
	Quartile 1	2.701	2.718	2.701	2.703
	Quartile 3	3.349	3.350	3.405	3.402

Table 7.9 Summary of the statistical measures of the output distribution of the fuzzy rule-base system with different defuzzification operators—Sect. 7.6

Operators	Statistical measures	Symmetrical MF		Non-symmetric MF	
		1000	10000	1000	10000
Centroid	Mean	2.997	3.002	2.824	2.836
	Median	3	3	2.821	2.821
	Variance	0.033	0.037	0.044	0.047
	Standard deviation	0.180	0.192	0.209	0.217
	Skewness	0.130	0.144	0.510	0.601
	Excess of kurtosis	3.522	4.942	2.392	2.100
	Minimum	2.320	1.820	2.160	1.950
	Maximum	3.680	4.180	3.533	3.767
	Quartile 1	2.820	2.820	2.740	2.740
	Quartile 3	3.001	3.180	2.873	2.873
Bisector	Mean	2.990	2.998	2.787	2.803
	Median	3	3	2.800	2.800
	Variance	0.068	0.071	0.071	0.078
	Standard deviation	0.261	0.267	0.266	0.279
	Skewness	−0.034	0.098	0.664	0.876
	Excess of kurtosis	3.196	3.295	2.177	2.915
	Minimum	2.120	1.840	2	1.960
	Maximum	3.880	4.160	3.760	4.800
	Quartile 1	2.840	2.840	2.560	2.560
	Quartile 3	3.120	3.160	2.840	2.840
MOM	Mean	3.002	3.002	2.817	2.834
	Median	3	3	2.813	2.813
	Variance	0.116	0.121	0.077	0.082
	Standard deviation	0.340	0.347	0.277	0.287
	Skewness	0.124	0.066	0.771	0.835
	Excess of kurtosis	0.655	1.139	2.343	2.063
	Minimum	2	1.5	2.028	1.899
	Maximum	4	4.5	3.832	4.034
	Quartile 1	2.750	2.750	2.707	2.707
	Quartile 3	3.250	3.250	2.845	2.845

(continued)

Table 7.9 (continued)

Operators	Statistical measures	Symmetrical MF		Non-symmetric MF	
		1000	10000	1000	10000
SOM	Mean	1.195	1.188	1.110	1.109
	Median	1	1	1	1
	Variance	0.187	0.195	0.071	0.082
	Standard deviation	0.433	0.441	0.267	0.286
	Skewness	2.080	2.359	3.171	3.642
	Excess of kurtosis	3.570	5.355	12.642	16.980
	Minimum	1	1	1	1
	Maximum	3	4	2.600	3.720
	Quartile 1	1	1	1	1
	Quartile 3	1	1	1	1
LOM	Mean	4.823	4.816	4.858	4.853
	Median	5	5	5	5
	Variance	0.172	0.184	0.110	0.119
	Standard deviation	0.415	0.429	0.332	0.344
	Skewness	−2.226	−2.315	−2.226	−2.359
	Excess of kurtosis	4.250	5.131	4.250	5.658
	Minimum	3	2	3.400	2.360
	Maximum	5	5	5	5
	Quartile 1	5	5	5	5
	Quartile 3	5	5	5	5

References

Anderson, T. W., & Darling, D. A. (1952). Asymptotic theory of certain "Goodness of Fit" criteria based on stochastic processes. *Annals of Mathematical Statistics, 23*, 193–212. https://doi.org/10.1214/aoms/1177729437

BAK. (2010). *FINANZPLATZ ZÜRICH 2010, Monitoring, Impact-Analyse und Ausblick*. Tech. Rep. Amt fur Wirtschaft und Arbeit (AWA) des Kantons Zürich.

Berkachy, R., & Donzé, L. (2015). Linguistic questionnaire evaluation: Global and individual assessment with the signed distance defuzzification method. In *Advances in Computational Intelligence Proceedings of the 16th International Conference on Fuzzy Systems FS'15, Rome, Italy* (vol. 34, pp. 13–20).

Berkachy, R., & Donzé, L. (2016a). Individual and global assessments with signed distance defuzzification, and characteristics of the output distributions based on an empirical analysis. In *Proceedings of the 8th International Joint Conference on Computational Intelligence - Volume 1: FCTA* (pp. 75–82). ISBN: 978-989-758-201-1. https://doi.org/10.5220/006036500750082

Berkachy, R., & Donzé, L. (2016b). Linguistic questionnaire evaluation: An application of the signed distance de- fuzzification method on different fuzzy numbers. The impact on the skewness of the output distributions. *International Journal of Fuzzy Systems and Advanced Applications, 3*, 12–19.

Berkachy, R., & Donzé, L. (2016c). Linguistic questionnaire evaluations using the signed distance defuzzification method: Individual and global assessments in the case of missing values. In *Proceedings of the 13th Applied Statistics 2016 International Conference, Ribno, Slovenia*.

Berkachy, R., & Donzé, L. (2016d). Statistical characteristics of distributions obtained using the signed distance defuzzification method compared to other methods. In *Proceedings of the International Conference on Fuzzy Management Methods ICFMSquare Fribourg, Switzerland* (pp. 48–58).

Berkachy, R., & Donzé, L. (2017). Statistical characteristics of distributions obtained using the signed distance defuzzification method compared to other methods. In A. Meier et al. (Eds.), *The Application of Fuzzy Logic for Managerial Decision Making Processes: Latest Research and Case Studies* (pp. 35–45). Cham: Springer. ISBN: 978-3-319-54048-1. https://doi.org/10.1007/9783319540481_4

Bourquin, J. (2016). *Mesures floues de la pauvreté. Une application au phénomène des working poor en Suisse*. MA Thesis. University of Fribourg, Switzerland.

de Saa, R., & de la, S., et al. (2015). Fuzzy rating scale-based questionnaires and their statistical analysis. *IEEE Transactions on Fuzzy Systems, 23*(1), 111–126. ISSN: 1063-6706. https://doi.org/10.1109/TFUZZ.2014.2307895

Donzé, L., & Berkachy, R. (2019). Fuzzy individual and global assessments and FANOVA. Application: Fuzzy measure of poverty with Swiss data. In *Proceedings of the 62st World Statistics Congress, Kuala Lumpur, Malaysia*.

Fisher, R. A. (1922). On the interpretation of χ^2 from contingency tables and the calculation of P. *Journal of the Royal Statistical Society, 85*(1), 87–94. ISSN: 09528385. http://www.jstororg/stable/2340521

Fisher, R. A. (1924). The conditions under which χ^2 measures the discrepancy between observation and hypothesis. *Journal of the Royal Statistical Society, 87*, 442–450. http://puhep1.princeton.edu/~mcdonald/examples/statistics/fisher_jrss_87_442_24.pdf

Gil, M. Á., Lubiano, M. A., de la Rosa de Sáa, S., & Sinova, B. (2015). Analyzing data from a fuzzy rating scale-based questionnaire a case study. *Psicotherma, 27*(2), 182–191.

Lilliefors, H. W. (1967). On the Kolmogorov-Smirnov test for normality with mean and variance unknown. *Journal of the American Statistical Association, 62*(318), 399–402. ISSN: 01621459. http://www.jstor.org/stable/2283970

Lin, L., & Lee, H.-M. (2009). Fuzzy assessment method on sampling survey analysis. *Expert Systems with Applications, 36*(3), 5955–5961. https://doi.org/10.1016/j.eswa.2008.07.087. http://dx.doi.org/10.1016/j.eswa.2008.07.087

Lin, L., & Lee, H.-M. (2010). Fuzzy assessment for sampling survey defuzzification by signed distance method. *Expert Systems with Applications, 37*(12), 7852–7857. http://dx.doi.org/10.1016/j.eswa.2010.04.052

Pearson, K. (1900). On the criterion that a given system of deviations from the probable in the case of a correlated system of variables is such that it can be reasonably sup-posed to have arisen from random sampling. *The London, Edinburgh, and Dublin Philosophical Magazine and Journal of Science, 50*(302), 157–175. https://doi.org/10.1080/14786440009463897

Rubin, D. B. (1987). *Multiple imputation for nonresponse in surveys*. Wiley series in probability and mathematical statistics. ISBN: 0-471-08705-X.

Shapiro, S. S., & Wilk, M. B. (1965). An analysis of variance test for normality. *Biometrica, 52*, 591–611. https://doi.org/10.1093/biomet/52.3-4.591

Wang, C.-H., & Chen, S.-M. (2008). Appraising the performance of high school teachers based on fuzzy number arithmetic operations. *Soft Computing, 12*(9), 919–934. ISSN: 1433-7479. https://doi.org/10.1007/s00500-007-0240-5

Chapter 8
Fuzzy Analysis of Variance

The multi-ways analysis of variance is a well-known statistical tool. It is based on inference tests having the aim of comparing sample means of different populations clustered by groups. This procedure is directly related to a regression model, particularly where the independent variables are categorical, the so-called factors. The terminology "multi-ways" is due to the fact that such models may be composed of multiple factors. If a model consists of one factor only, it will be denoted by the one-way analysis of variance (ANOVA). The multi-ways analysis of variance extends the one-way ANOVA by taking into account multiple independent variables to be combined in a regression model, the purpose being to explain the difference of means of the dependent variable related to the different factor levels. We remind that the computational process of such approaches relies on calculating the sums of squares related to the decomposition of the variance on the different components of the model, and constructing test-based statistics and thereafter corresponding decision rules.

In an ANOVA model, we often assume that the considered data are precise. When we decide to treat the fuzziness contained in the data set, we realize that using the conventional multi-factors ANOVA approach, the situation becomes complicated. This approach presents somehow a lack in the flexibility since it could not model the dependence between the difference of means of the different populations in a way taking into account at most the uncertain situation. One has to find a way to overcome this problem. This latter can eventually be solved by fuzzy tools. We propose a fuzzy version of the Mult-ANOVA that we consequently denote by Mult-FANOVA. One relates the obtained fuzzy model to a fuzzy regression one, similarly to the classical theory. Following this reasoning, two main approaches of regression models in fuzzy contexts have been particularly introduced. The different concepts of fuzzy regression models are due to the beliefs of the researchers about the sources of fuzziness of their models. We recall two of them. It is clear that one could distinguish much more approaches to fuzzy regression. For example, Tanaka et al. (1982) who proposed a procedure of fuzzy linear regression models

© The Author(s), under exclusive license to Springer Nature Switzerland AG 2021
R. Berkachy, *The Signed Distance Measure in Fuzzy Statistical Analysis*,
Fuzzy Management Methods, https://doi.org/10.1007/978-3-030-76916-1_8

believed that the fuzziness is a matter of model structure. However, Celmins (1987) introduced an approach based on the least-squares method, where they asserted that the fuzziness in their case is a matter of data. In our Mult-FANOVA model, our independent variables are factors. Thereafter, we suppose that the fuzziness in our case is a matter of the parameters related to the considered Mult-FANOVA model (resulting from a fuzzy regression model), and of the dependent variable.

In the context of ANOVA in a univariate or a multivariate environment, various researchers have already worked on similar topics, such as Kang and Han (2004) who introduced very briefly a multivariate analysis of variance for fuzzy data. Analogously but in a univariate context, Wu (2007) proposed an ANOVA in the fuzzy environment. The author also gave the notions of rejection of the null and the alternative hypotheses of the statistical test by the so-called pessimistic and optimistic degrees. Parchami et al. (2017) exposed an approach of analysis of variance in uncertain environments. For their approach, they proposed different computational ways to proceed, using different types of fuzzy numbers. In the same way, Jiryaei et al. (2013) discussed a one-way analysis of variance in similar contexts where triangular and exponential fuzzy numbers were used to model the required parameters.

In order to encounter the computational difficulty due to the difference and the product between fuzzy numbers, multiple metrics have been previously used in literature in the approaches of fuzzy analysis of variance. We note, for instance, Gonzalez-Rodriguez et al. (2006) who developed independent and paired tests of equality of means based on extending the Student t-tests to the fuzzy environment. Gil et al. (2006) proposed similar tests but in a multi-sample context, where they used the Bertoluzza distance. For the univariate case, Montenegro et al. (2004) and Gonzalez-Rodriguez et al. (2012), and others, proposed a univariate one-way ANOVA with random fuzzy variables, using different metrics. In the same way, we showed in Berkachy and Donzé (2018) a one-way ANOVA model using the signed distance. However, these simple ANOVA models are seen as limited since we cannot include more than one factor only. A generalization to the multi-factors case is then needed.

The aim of this chapter is to expose a Mult-FANOVA approach when the fuzziness is taken into consideration, based on a fuzzy regression model. As such, we expose two exploratory approaches: the first one by preserving the fuzzy nature of the calculated sums of squares using fuzzy approximations and afterward by defining a related heuristic decision rule to be able to decide, and the second one using a particular metric all along the calculations, namely the signed distance and the generalized one. We highlight that the purpose of this chapter is to show the usefulness of the signed distance measure as a ranking tool in the context of the fuzzy analysis of variance.

Applying a distance in the process of calculating the sums of squares leads to a set of crisp sums. An interesting finding of our approaches is that using the SGD, if the data set is modelled by symmetrical fuzzy numbers, we reproduce the analysis by the classical approach. Thus, we can say that the classical approach can be seen as a particular case of the fuzzy approach in the described context. Consequently, the Mult-FANOVA seems to be a reproduction of the classical case, but with more flexibility in the process of modelling the uncertainty contained in the data. These results are clearly confirmed by the study Berkachy (2018) done on a data set from Nestlé beverages, where we performed using the SGD multiple Mult-FANOVA tests on several sets of membership functions, and different variants of estimated models.

For sake of illustration, we give two detailed empirical applications of our approach, one on the data set from Nestlé beverages, and the other one on the Finanzplatz data set. Both models are composed by multiple factors. By these applications, the objective is to present the Mult-FANOVA model where the GSGD is used and to compare it with the analysis using the same setups but in the classical environment. One of our main findings is that in both cases we get the same decision of rejecting or not the hypothesis of equality of the population mean vectors, but the decomposition of the sums is different. For the analysis on the residuals of the constructed model, the normality of their distributions is rejected, but stronger in the classical approach. Berkachy and Donzé (2018) reported that a possible cause is the presence of a greater lack of adjustment of the fitting model by the Mult-ANOVA approach than the Mult-FANOVA one. This result can be seen as a disadvantage of the classical approach, and by consequence strengthen our support to the fuzzy one. For the tests of the homogeneity of variances between groups, the analysis by the classical analysis tends to reject more often than the fuzzy one the hypothesis of homogeneity, violating by that a basic postulated assumption of typical Mult-ANOVA models.

This chapter is organized as follows. In Sect. 8.1, we recall the basic concepts of the classical Mult-ANOVA, followed by a brief introduction of fuzzy regression in Sect. 8.2. We expose in Sect. 8.3 the fuzzy multi-ways analysis of variance Mult-FANOVA, where we propose two approaches of decision making: a heuristic one based on the obtained fuzzy sums, and another one totally based on a chosen metric. We give after a particular case of Mult-FANOVA by the signed distance, where we show that using symmetrical shapes of modelling fuzzy numbers, we return exactly the classical approach. The section is closed by a procedure of calculating sequential sums of squares, when the ordering effect intervenes. This is caused by the unbalance in the studied design. Finally, the Sect. 8.4 is devoted for two empirical applications where we provide multiple tests in order to investigate the influence of the choice of distance, and of shapes of the modelling fuzzy numbers, on Mult-FANOVA models.

8.1   Classical Multi-Ways Analysis of Variance

It is easy to see that the multi-ways ANOVA or Mult-ANOVA can be done by a regression model with categorical independent variables. The terminology of the multi-ways ANOVA approach relates to the fact that multiple categorical variables are considered to explain one response variable only.

Based on a regression model, let us briefly recall the Mult-ANOVA in the classical theory. To simplify the case, we propose a model based on two factors only with no interactions. The multiple case, with eventually adding the interactions, can analogously be conceived.

In a given experiment, consider a sample X of N observations, and 2 factors. We denote by $k = 1, 2$ the index of a given factor in the set of factors. Consider $i_k = 1, \ldots, g_k$ to be the index of a given level related to the factor k, with n_{i_k} the corresponding number of observations. In addition, let $j_k = 1, \ldots, n_{i_k}$ be the index of a given observation in the set of observations corresponding to level i_k of the factor k, such that $N = \sum_{i_k=1}^{g_k} n_{i_k}$ for every value of $k = 1, 2$. We suppose in a first step that our design is balanced, i.e. the number of observations is the same across all the factor levels. Therefore, we get that for every cell related simultaneously to the level i_1 of factor 1 and the level i_2 of factor 2, the number of observations is equal to n. As such, we have that $j_k = j$.

The approach of the multi-factors ANOVA relies on testing the null hypothesis H_0 that the means related to all the levels of a factor are equal in the experiment, against the alternative one H_1 that at least one pair of means is not equal, at a significance level δ. The null and alternative hypotheses H_0 and H_1 can then be written as follows:

$$H_0 : \mu_1 = \ldots = \mu_{i_k} = \ldots = \mu_{g_k}, \quad \text{against} \quad H_1 : \text{not all the means are equal,}$$

$$(8.1)$$

where μ_{i_k} is the mean of the population at the level i_k related to the factor k. In such hypotheses tests, the following conditions are generally assumed:

- The population variance of the response variable denoted by σ^2 is the same across all the observations;
- The observations are supposed to be independent;
- Each observation is drawn from a normal distribution.

For the last condition on the normality of the distribution of the data set, the central limit theorem is often used to asymptotically validate this assumption in case of large samples. Consequently, it would be useful to investigate after whether the distribution of residuals of the considered model tends to be normal or close enough.

Let us expose the decomposition of the data set of the N observations on 2 factors in the following manner:

		Factor 2				
		1	2	$\dots i_2$		$\dots g_2$
	1	X_{11}	X_{12}	$\dots$	X_{1i_2}	$\dots$ X_{1g_2}
	2	X_{21}	X_{22}	$\dots$	X_{2i_2}	$\dots$ X_{2g_2}
	$\vdots$					
Factor 1	i_1	X_{i_11}	X_{i_12}	$\dots$	$X_{i_1i_2}$	$\dots$ $X_{i_1g_2}$
	$\vdots$					
	g_1	X_{g_11}	X_{g_12}	$\dots$	$X_{g_1i_2}$	$\dots$ $X_{g_1g_2}$

where $X_{i_1i_2} = \left(X_{i_1i_21}, \dots, X_{i_1i_2j}, \dots, X_{i_1i_2n}\right)'$ is the vector of observations belonging to the level i_1 of factor 1, and i_2 of factor 2 at the same time.

Let us now expose the multi-ways model explaining the variable $X_{i_1i_2j}$. It is given as follows:

$$X_{i_1i_2j} = \mu + \tau_{i_1} + \beta_{i_2} + \epsilon_{i_1i_2j}, \qquad (8.2)$$

where μ is the overall sample mean, τ_{i_1} is the i_1-th treatment effect of the first factor with $\sum_{i_1=1}^{g_1} \tau_{i_1} = 0$, and β_{i_2} is the i_2-th treatment effect of the second factor with $\sum_{i_2=1}^{g_2} \beta_{i_2} = 0$, and $\epsilon_{i_1i_2j}$ is an independent random error drawn from the normal distribution with a mean 0 and a variance σ^2.

Nevertheless, since the aim is to test whether the mean of the dependent variable is statistically different across all the levels, or not, it would be interesting to express the multi-factors ANOVA model in terms of the expectation of the response variable analogously to the model described in Berkachy and Donzé (2018) for the univariate case. Thus, for $i_1 = 1, \dots, g_1$, $i_2 = 1, \dots, g_2$ and $j = 1, \dots, n$, the multi-factors ANOVA model can then be written as follows:

$$\underbrace{\mathbb{E}(X_{i_1i_2j})}_{\text{mean response}} = \underbrace{\mu}_{\text{overall expectation}} + \underbrace{\tau_{i_1}}_{\substack{\text{treatment effect of} \\ \text{factor 1}}} + \underbrace{\beta_{i_2}}_{\substack{\text{treatment effect of} \\ \text{factor 2}}},$$

$$(8.3)$$

where the sample analogue $\overline{X}_{\bullet\bullet\bullet}$ of the overall expectation μ is given by

$$\overline{X}_{\bullet\bullet\bullet} = \frac{1}{g_1g_2n} \sum_{i_1=1}^{g_1} \sum_{i_2=1}^{g_2} \sum_{j=1}^{n} X_{i_1i_2j}. \qquad (8.4)$$

One could also calculate the mean of the i_k-th population of the factor k. As instance, we denote by $\overline{X}_{i_1\bullet\bullet}$, the mean of the observations of level i_1 related to the factor 1.

It is expressed by

$$\overline{X}_{i_1 \bullet \bullet} = \frac{1}{g_2 n} \sum_{i_2=1}^{g_2} \sum_{j=1}^{n} X_{i_1 i_2 j}. \qquad (8.5)$$

The mean $\overline{X}_{i_1 \bullet \bullet}$ is the empirical analogue of the expectation of the i_1-th population denoted by μ_{i_1}. In the same way, the i_2-th population mean of the factor 2 can be written as:

$$\overline{X}_{\bullet i_2 \bullet} = \frac{1}{g_1 n} \sum_{i_1=1}^{g_1} \sum_{j=1}^{n} X_{i_1 i_2 j}. \qquad (8.6)$$

The model given in Eq. 8.3 leads to the following decomposition of a given observation:

$$\underbrace{X_{i_1 i_2 j}}_{\text{observation}} = \underbrace{\overline{X}_{\bullet\bullet\bullet}}_{\substack{\text{overall sample} \\ \text{mean}}} + \underbrace{\left(\overline{X}_{i_1 \bullet \bullet} - \overline{X}_{\bullet\bullet\bullet}\right)}_{\substack{\text{estimated treatments or} \\ \text{hypotheses effect for} \\ \text{factor 1}}}$$

$$+ \underbrace{\left(\overline{X}_{\bullet i_2 \bullet} - \overline{X}_{\bullet\bullet\bullet}\right)}_{\substack{\text{estimated treatments or} \\ \text{hypotheses effect for} \\ \text{factor 2}}} + \underbrace{\left(\overline{X}_{i_1 i_2 j} - \overline{X}_{i_1 i_2 \bullet}\right)}_{\text{residuals}}, \qquad (8.7)$$

where

$$\overline{X}_{i_1 i_2 \bullet} = \frac{1}{n} \sum_{j=1}^{n} X_{i_1 i_2 j},$$

is the mean of the observations belonging at the same time to the level i_1 from the factor 1 and the level i_2 from the factor 2. By decomposing $X_{i_1 i_2 j}$, the sum $\overline{X}_{i_1 i_2 \bullet}$ can also be written as $\overline{X}_{i_1 i_2 \bullet} = \overline{X}_{i_1 \bullet \bullet} + \overline{X}_{\bullet i_2 \bullet} - \overline{X}_{\bullet\bullet\bullet}$.

Let us now square and sum the deviations given by $X_{i_1 i_2 j} - \overline{X}_{\bullet\bullet\bullet}$ seen in Eq. 8.7, over i_1, i_2, and j. We get the following:

$$\sum_{i_1=1}^{g_1} \sum_{i_2=1}^{g_2} \sum_{j}^{n} \left(X_{i_1 i_2 j} - \overline{X}_{\bullet\bullet\bullet}\right)^2 = \sum_{i_1=1}^{g_1} \sum_{i_2=1}^{g_2} \sum_{j}^{n} \left(\overline{X}_{i_1 \bullet \bullet} - \overline{X}_{\bullet\bullet\bullet}\right)^2 + \sum_{i_1=1}^{g_1} \sum_{i_2=1}^{g_2} \sum_{j}^{n} \left(\overline{X}_{\bullet i_2 \bullet} - \overline{X}_{\bullet\bullet\bullet}\right)^2$$

$$+ \sum_{i_1=1}^{g_1} \sum_{i_2=1}^{g_2} \sum_{j}^{n} \left(\overline{X}_{i_1 i_2 j} - \overline{X}_{i_1 i_2 \bullet}\right)^2,$$

$$= \sum_{i_1=1}^{g_1} ng_2\left(\overline{X}_{i_1\bullet\bullet} - \overline{X}_{\bullet\bullet\bullet}\right)^2 + \sum_{i_2=1}^{g_2} ng_1\left(\overline{X}_{\bullet i_2\bullet} - \overline{X}_{\bullet\bullet\bullet}\right)^2$$

$$+ \sum_{i_1=1}^{g_1}\sum_{i_2=1}^{g_2}\sum_{j}^{n} \left(\overline{X}_{i_1 i_2 j} - \overline{X}_{i_1 i_2\bullet}\right)^2. \tag{8.8}$$

We denote, respectively, by T the total sum of squares, by B_1 the treatment sum of squares related to the factor 1, by B_2 the treatment sum of squares related to the factor 2, and by E the residual sum of squares. These sums are written as follows:

$$T = \sum_{i_1=1}^{g_1}\sum_{i_2=1}^{g_2}\sum_{j}^{n} \left(X_{i_1 i_2 j} - \overline{X}_{\bullet\bullet\bullet}\right)^2, \tag{8.9}$$

$$B_1 = \sum_{i_1=1}^{g_1} ng_2\left(\overline{X}_{i_1\bullet\bullet} - \overline{X}_{\bullet\bullet\bullet}\right)^2, \tag{8.10}$$

$$B_2 = \sum_{i_2=1}^{g_2} ng_1\left(\overline{X}_{\bullet i_2\bullet} - \overline{X}_{\bullet\bullet\bullet}\right)^2, \tag{8.11}$$

$$E = \sum_{i_1=1}^{g_1}\sum_{i_2=1}^{g_2}\sum_{j}^{n} \left(X_{i_1 i_2 j} - \overline{X}_{i_1 i_2\bullet}\right)^2, \tag{8.12}$$

such that

$$T = B_1 + B_2 + E.$$

We add that the decomposition in sums of squares given in Eq. 8.8 is associated with a decomposition in the degrees of freedom as follows:

$$g_1 g_2 n - 1 = (g_1 - 1) + (g_2 - 1) + \left(g_1 g_2 n - (g_1 + g_2 - 1)\right), \tag{8.13}$$

where $g_1 g_2 n - 1$, $g_1 - 1$, $g_2 - 1$, and $g_1 g_2 n - (g_1 + g_2 - 1)$ are the degrees of freedom of T, B_1, B_2, and E, respectively.

Different methods of testing can be adopted in this context. Researchers used to calculate the ratio of the treatment mean sum of squares to the error mean sum of squares. This ratio serves afterward as a test statistic. According to the decompositions given in Eqs. 8.8 and 8.13, the ratio F_1 related to the first factor of the model can be expressed by

$$F_1 = \frac{MB_1}{ME} = \frac{B_1/df_{B_1}}{E/df_E}, \tag{8.14}$$

where MB_1 is the treatment mean sum of squares related to the factor 1, ME is the error mean sum of squares, with $df_{B_1} = g_1 - 1$ and $df_E = \left(g_1 g_2 n - (g_1 + g_2 - 1)\right)$ the degrees of freedom of B_1 and E, respectively. The ratio F_2 related to the second factor can be in the same way calculated. This ratio can then be written by

$$F_2 = \frac{MB_2}{ME} = \frac{B_2/df_{B_2}}{E/df_E}, \tag{8.15}$$

where MB_2 is the treatment mean sum of squares related to the factor 2 with $df_{B_2} = g_2 - 1$.

A decision rule is needed to be able to decide whether to reject or not the equality of means (hypothesis given in Eq. 8.1). Under the classical assumptions of the analysis of variance model, one could prove that the test statistic is taken from a Fisher-Snedecor distribution. Namely, the previously calculated statistic F_1 related to the factor 1 is taken from the Fisher-Snedecor distribution with df_{B_1} and df_E degrees of freedom. The corresponding decision rule is then given by:

Decision Rule The statistic F_1 will be compared to the quantile of the Fisher-Snedecor distribution with df_{B_1} and df_E degrees of freedom at the significance level δ, i.e. $F_{(1-\delta;df_{B_1},df_E)}$. The rule is given by:

- if $F_1 > F_{(1-\delta;df_{B_1},df_E)}$, we reject the null hypothesis H_0 at the significance level δ;
- Otherwise, we do not reject H_0.

The decision rule of the test statistic F_2 related to the factor 2 can be similarly conceived.

We mention that in a multivariate context, a wide range of sophisticated test statistics are constructed for similar decision making problems, such as the Wilks Lambda, the Hotelling-Lawley Trace or the Pillai Trace, etc.[1] In addition, some researchers used also the so-called Bonferroni correction given in Bonferroni (1935) to measure the effects of belonging to a given level.

8.2 Introduction to Fuzzy Regression

Based on the regression, known to be a performant tool in statistical analysis, and in the spirit of developing new ways of modelling data, researchers tended to design new approaches of analysis. When some of them started to take into account the uncertainty contained in the data they collect, the need for regression models defined in a fuzzy environment became subject to discussion.

---

[1] The detailed references can be found, respectively, in Wilks (1932), Wilks (1946), Hotelling (1951), Lawley (1938), and Pillai (1965).

Researchers often explained that fuzzy regression models are a fuzzy variant of the classical regression models, in which specific elements have a fuzzy nature. The fuzziness may affect a given model by having uncertain data and/or model parameters. These entities can consequently be modelled by fuzzy sets. It can be due to the fuzziness of the dependent variables, the independent variable or the regression coefficients of the model. Thus, for a model explaining one response variable only, researchers roughly reduced the possible scenarios to the following ones:

1. the response variable is fuzzy, but the explanatory variables and the parameters are crisp. The model can then be written by

$$\mathbb{E}(\tilde{Y}) = \beta_0 + \beta_1 X_1 + \ldots + \beta_p X_p, \tag{8.16}$$

where $\tilde{Y}$ is the fuzzy response variable, $X_1, \ldots, X_p$ are the crisp explanatory variables, and $\beta_0, \ldots, \beta_p$ are the crisp parameters of the regression model;
2. the response variable and the explanatory variables are crisp, but the parameters are fuzzy, such that

$$\mathbb{E}(Y) = \tilde{\beta}_0 \oplus \tilde{\beta}_1 X_1 \oplus \ldots \oplus \tilde{\beta}_p X_p; \tag{8.17}$$

3. the response variable and the explanatory variables are fuzzy, but the parameters are crisp, such that

$$\mathbb{E}(\tilde{Y}) = \beta_0 \oplus \beta_1 \tilde{X}_1 \oplus \ldots \oplus \beta_p \tilde{X}_p; \tag{8.18}$$

4. the response variable and the parameters are fuzzy, and the explanatory variables are crisp. Then, we get the following model:

$$\mathbb{E}(\tilde{Y}) = \tilde{\beta}_0 \oplus \tilde{\beta}_1 X_1 \oplus \ldots \oplus \tilde{\beta}_p X_p; \tag{8.19}$$

5. all the entities are fuzzy. The model is then given such that

$$\mathbb{E}(\tilde{Y}) = \tilde{\beta}_0 \oplus (\tilde{\beta}_1 \otimes \tilde{X}_1) \oplus \ldots \oplus (\tilde{\beta}_p \otimes \tilde{X}_p). \tag{8.20}$$

Since in some of these cases, the fuzzy multiplication and fuzzy addition are needed, then we can remark that a fuzzy type prediction is not always ensured. We add that in regression models, the uncertainty can be due to many factors: the data, the model structure, the ignorance of the relation between the independent and dependent variables, etc. Some authors might not believe that the fuzziness is only contained in the data set or in the analysis, but considered otherwise that the fuzziness of the model is rather a consequence of the way the data are collected. The uncertainty and the randomness are present in such methodologies. This idea was less studied in literature than the one related to the fuzziness of the data and/or the parameters.

For such models, two main approaches were often discussed in literature: the Tanaka's fuzzy linear regression models, and the fuzzy least-squares regression models. The first one is based on the so-called minimum fuzziness criterion. This approach was firstly described by Tanaka et al. (1982) who introduced a definition of the concept of fuzzy linear regression models. In this contribution, the parameters were considered as fuzzy sets, particularly fuzzy numbers, and the dependent variable as interval or fuzzy-valued. In this approach, the authors asserted that the fuzziness is a matter of model structure. Therefore, Tanaka et al. (1982) and Guo and Tanaka (2006) stated that the deviations between the predicted and the observed entities of their model depend on the indefiniteness of the structure of the system. In addition, in the theory of Tanaka et al. (1982), the possibilistic models were estimated by means of linear or quadratic programming algorithms relying on the principle of "minimum fuzziness" or "maximum possibility" under the studied parameters. Accordingly, the idea of the authors was to determine the fuzzy coefficients by minimizing the so-called index of fuzziness of the possibilistic model. Several researchers have extended the idea of Tanaka et al. (1982) and paved the way to other regression-based models. As instance, Buckley and Jowers (2008) provided an estimation of the model parameters using Monte Carlo calculations. In addition, they stated that if the fuzzy response variable is modelled by triangular fuzzy numbers, then the "best" model fitting the data will be the one for which the fuzzy linear regression parameters are triangular fuzzy numbers.

For the fuzzy least-squares approach, the aim of all the researches was to find the "best" fitting model in a given metric space and where the response variable is supposed to be modelled by L-R fuzzy numbers. For this case, the fuzziness is assumed to be a matter of data. Celmins (1987) introduced a methodology of estimation based on the least-squares method, where the objective was to minimize a given least-squares function. Diamond (1988) introduced the Hausdorff metric in the procedure of estimation in order to define a criterion function based on the least-squares. This function has to be afterward minimized, as in the classical theory. Inspired by this work, Nather (2006) embedded the notion of random fuzzy sets in the regression model, and by that he designed a model estimating these parameters. A wide number of researchers has worked on this specific approach since it is seen as a direct extension of the classical one. It is obvious to see that in similar contexts, the problem of estimation can be solved by using an adequate metric. In other words, the objective in this situation is to minimize the deviation between the fuzzy prediction and the fuzzy output with respect to particular distances between fuzzy numbers.

We are mainly interested in modelling a Mult-ANOVA based on a regression model. Indeed, we model the response variable as a function of factors. In our case, the dependent variable is considered to be uncertain and will be modelled by fuzzy numbers. The objective is then to test whether the means of the observations related to the different levels are equal or not. In this perspective, it is natural to see that our model will be similar to the one given in Case 4. We would like to consider the Aumann-type expectation of the fuzzy response variable shown in Definition 5.4.4. Therefore, following the above description of the two fuzzy regression approaches, we will particularly proceed by the least-squares one. The idea is then to prudently choose a metric to be used in the process. We propose to use the metrics introduced in Chap. 4, in order to investigate their utility in such contexts.

8.3 Fuzzy Multi-Ways Analysis of Variance

By analogy to the classical statistical theory, the Mult-ANOVA in the fuzzy context can be seen as a particular case of a fuzzy regression model. In other terms, we will consider a set of crisp independent factors, and we would like to explain a given fuzzy dependent variable. The considered model will be similar to the one shown in Case 4 of Sect. 8.2. We then expect the parameters of the model to be fuzzy as well as the response variable. By the Mult-FANOVA and accordingly to the test given in Eq. 8.1 in the fuzzy environment, we intend to verify whether the fuzzy means corresponding to all the groups of a given factor are equal, or not, at the significance level δ. If at least 2 means are not equal, then the hypothesis of equality will be rejected. Otherwise, this hypothesis will not be rejected. In the same setups of the multi-ways ANOVA model proposed in Sect. 8.1, the fuzzy version of this model can be expressed in the following sense.

Let X be a sample of N observations, and 2 factors. Consider again the following indices:

- $k = 1, 2$ is the index of a factor in the set;
- $i_k = 1, \ldots, g_k$ is the index of a level related to factor k;
- and $j_k = 1, \ldots, n_{i_k}$ is the index of a given observation corresponding to the level i related to factor k, for which $N = \sum_{i_k=1}^{g_k} n_{i_k}$. Similarly to the classical case, we suppose that the design of the subsequent model is balanced. In other words, the number of observations related any cell is the same. It is then equal to n. Therefore, there is no need to index j. Thus, we get that $j = 1, \ldots, n$.

We would like to extend the classical approach of multi-ways ANOVA to the fuzzy environment. Two possible scenarios can be applied on the fuzziness contained in the response variable X: the collected data set could be rather given by fuzzy numbers by nature, or given by crisp numbers which will be afterward modelled by corresponding sets of membership functions. In both cases, the fuzzy variable will be written by $\tilde{X}$. We note that in such models, the same conditions of the classical model have also to be fulfilled in the fuzzy context.

By generalizing the classical approach to the fuzzy sets, the sample containing the fuzzy variable $\tilde{X}$ related to both factors can be arranged as follows:

		Factor 2			
	1	2	... i_2	... g_2
	1	$\tilde{X}_{11}$	$\tilde{X}_{12}$	... $\tilde{X}_{1i_2}$	... $\tilde{X}_{1g_2}$
	2	$\tilde{X}_{21}$	$\tilde{X}_{22}$	... $\tilde{X}_{2i_2}$	... $\tilde{X}_{2g_2}$
	⋮				
Factor 1 i_1		$\tilde{X}_{i_11}$	$\tilde{X}_{i_12}$	... $\tilde{X}_{i_1i_2}$	... $\tilde{X}_{i_1g_2}$
	⋮				
	g_1	$\tilde{X}_{g_11}$	$\tilde{X}_{g_12}$	... $\tilde{X}_{g_1i_2}$	... $\tilde{X}_{g_1g_2}$

where $\tilde{X}_{i_1i_2} = \left(\tilde{X}_{i_1i_21}, \ldots, \tilde{X}_{i_1i_2j}, \ldots, \tilde{X}_{i_1i_2n}\right)'$ is the vector of fuzzy observations $\tilde{X}_{i_1i_2j}$ belonging to the i_1-th level of factor 1 and the i_2-th level of factor 2.

In compliance with the model given in Case 4 of Sect. 8.2 and with the model of the classical theory shown in Eq. 8.3, the corresponding Mult-FANOVA model in terms of the expectation of the fuzzy response variable can be written in the following manner:

$$\tilde{\mathbb{E}}(\tilde{X}_{i_1i_2j}) = \tilde{\mu} + \tilde{\tau}_{i_1} + \tilde{\beta}_{i_2}, \tag{8.21}$$

where $\tilde{\mathbb{E}}(\tilde{X}_{i_1i_2j})$ is the fuzzy expectation of the population, $\tilde{\mu}$ is the fuzzy overall expectation, $\tilde{\tau}_{i_1}$ is the i_1-th population fuzzy treatment effect for the factor 1, and $\tilde{\beta}_{i_2}$ is the i_2-th population fuzzy treatment effect for the factor 2. We note that in the fuzzy environment, the sample estimator of the fuzzy expectation $\tilde{\mu}$ denoted by $\overline{\tilde{X}}_{\bullet\bullet\bullet}$ is expressed by

$$\overline{\tilde{X}}_{\bullet\bullet\bullet} = \frac{1}{g_1g_2n} \sum_{i_1=1}^{g_1} \sum_{i_2=1}^{g_2} \sum_{j=1}^{n} \tilde{X}_{i_1i_2j}. \tag{8.22}$$

We denote by $\tilde{\mu}_{i_k}$ the fuzzy expectation related to the level i_k of the factor k. This expectation $\tilde{\mu}_{i_k}$ can be estimated by a fuzzy sample mean for each of the corresponding levels i_k of the factor k. For a given level i_1 related to the first factor of our model, the fuzzy sample mean written as $\overline{\tilde{X}}_{i_1\bullet\bullet}$ can be expressed by

$$\overline{\tilde{X}}_{i_1\bullet\bullet} = \frac{1}{g_2n} \sum_{i_2=1}^{g_2} \sum_{j=1}^{n} \tilde{X}_{i_1i_2j} = \frac{1}{g_2n} \left[\left(\sum_{j=1}^{n} \tilde{X}_{i_11j} \right) \oplus \ldots \oplus \left(\sum_{j=1}^{n} \tilde{X}_{i_1g_2j} \right) \right]. \tag{8.23}$$

It is the same for a particular level i_2 related to the second factor. Thus, the fuzzy sample mean for the level i_2 written as $\overline{\tilde{X}}_{\bullet i_2 \bullet}$ can be expressed by

$$\overline{\tilde{X}}_{\bullet i_2 \bullet} = \frac{1}{g_1 n} \sum_{i_1=1}^{g_1} \sum_{j=1}^{n} \tilde{X}_{i_1 i_2 j} = \frac{1}{g_1 n} \left[\left(\sum_{j=1}^{n} \tilde{X}_{1 i_2 j} \right) \oplus \ldots \oplus \left(\sum_{j=1}^{n} \tilde{X}_{g_1 i_2 j} \right) \right]. \quad (8.24)$$

The Mult-FANOVA model intends to compare the different populations' means. For models described in the context of the multivariate analysis of variance, Kang and Han (2004) asserted that in the fuzzy environment, a given dependent variable can be partitioned into different sums, as well as the partition done in the classical theories. If we reason similarly to Kang and Han (2004) regarding our Mult-FANOVA model, and to the decomposition given in Eq. 8.8 for the classical approach, we consider the following decomposition into fuzzy sums:

$$\tilde{X}_{i_1 i_2 j} = \underbrace{\overline{\tilde{X}}_{\bullet\bullet\bullet}}_{\substack{\text{fuzzy overall} \\ \text{sample mean}}} \oplus \underbrace{\left(\overline{\tilde{X}}_{i_1 \bullet\bullet} \ominus \overline{\tilde{X}}_{\bullet\bullet\bullet} \right)}_{\substack{\text{fuzzy estimated treat-} \\ \text{ments or hypotheses ef-} \\ \text{fect for factor 1}}} \oplus \underbrace{\left(\overline{\tilde{X}}_{\bullet i_2 \bullet} \ominus \overline{\tilde{X}}_{\bullet\bullet\bullet} \right)}_{\substack{\text{fuzzy estimated treat-} \\ \text{ments or hypotheses ef-} \\ \text{fect for factor 2}}} \oplus \underbrace{\left(\tilde{X}_{i_1 i_2 j} \ominus \overline{\tilde{X}}_{i_1 i_2 \bullet} \right)}_{\text{fuzzy residuals}},$$

(8.25)

where $\overline{\tilde{X}}_{i_1 i_2 \bullet}$ is the fuzzy mean of the observations which belong simultaneously to the level i_1 from factor 1, and to the level i_2 from factor 2. It can be expressed as follows:

$$\overline{\tilde{X}}_{i_1 i_2 \bullet} = \frac{1}{n} \sum_{j=1}^{n} \tilde{X}_{i_1 i_2 j}. \quad (8.26)$$

This sum can also be written[2] as

$$\overline{\tilde{X}}_{i_1 i_2 \bullet} = \overline{\tilde{X}}_{i_1 \bullet\bullet} \oplus \overline{\tilde{X}}_{\bullet i_2 \bullet} \ominus \overline{\tilde{X}}_{\bullet\bullet\bullet}. \quad (8.27)$$

Following this decomposition and the procedure of the classical Mult-ANOVA approach, one could eventually define the fuzzy total sum of squares $\tilde{T}$ given by

$$\tilde{T} = \sum_{i_1=1}^{g_1} \sum_{i_2=1}^{g_2} \sum_{j}^{n} \left(\tilde{X}_{i_1 i_2 j} \ominus \overline{\tilde{X}}_{\bullet\bullet\bullet} \right)^{\textcircled{2}}. \quad (8.28)$$

---

[2]Despite the complexity induced by the difference arithmetic in the decomposition of $\tilde{X}_{i_1 i_2 j}$, remark that $\overline{\tilde{X}}_{i_1 i_2 \bullet}$ is expressed by this form in accordance with the expression of the classical theory.

The fuzzy sum $\tilde{T}$ is partitioned into three different fuzzy sums:

- the fuzzy treatment sum of squares for factor 1 denoted by $\tilde{B}_1$, also called the fuzzy hypothesis sum of squares for factor 1;
- the fuzzy treatment sum of squares for factor 2 denoted by $\tilde{B}_2$, also called the fuzzy hypothesis sum of squares for factor 2;
- the fuzzy residual sum of squares denoted by $\tilde{E}$, also called fuzzy error sum of squares.

The fuzzy sum $\tilde{T}$ can then be written as:

$$\tilde{T} = \tilde{B}_1 \oplus \tilde{B}_2 \oplus \tilde{E}, \tag{8.29}$$

such that the fuzzy sums $\tilde{B}_1$, $\tilde{B}_2$ and $\tilde{E}$ can be expressed as follows:

$$\tilde{B}_1 = \sum_{i_1=1}^{g_1} n g_2 \left(\overline{\overline{X}}_{i_1 \bullet \bullet} \ominus \overline{\overline{X}}_{\bullet \bullet \bullet} \right)^{\textcircled{2}}, \tag{8.30}$$

$$\tilde{B}_2 = \sum_{i_2=1}^{g_2} n g_1 \left(\overline{\overline{X}}_{\bullet i_2 \bullet} \ominus \overline{\overline{X}}_{\bullet \bullet \bullet} \right)^{\textcircled{2}}, \tag{8.31}$$

$$\tilde{E} = \sum_{i_1=1}^{g_1} \sum_{i_2=1}^{g_2} \sum_{j}^{n} \left(\overline{\overline{X}}_{i_1 i_2 j} \ominus \overline{\overline{X}}_{i_1 i_2 \bullet} \right)^{\textcircled{2}}. \tag{8.32}$$

The described decomposition of the Mult-FANOVA model with the corresponding degrees of freedom is summed up in Table 8.1.

Nevertheless, it is expected in such situations to calculate a test statistic able to serve in the process of decision making. In our fuzzy situation, it is clear that the calculated sums of squares will be fuzzy. However, since the intermediate fuzzy operations (difference, product, and consequently the square) are expensive in terms

Table 8.1 The Mult-FANOVA table

Mult-FANOVA		
Source of variation	df	Fuzzy sum of squares
Treatment effect—Factor 1	$g_1 - 1$	$\tilde{B}_1$
Treatment effect—Factor 2	$g_2 - 1$	$\tilde{B}_2$
Residual—Error	$g_1 g_2 n - (g_1 + g_2 - 1)$	$\tilde{E}$
Total	$g_1 g_2 n - 1$	$\tilde{T}$

of computations, this task is seen as laborious. Yet, two possible ideas can be given:

- **Method 1.** Based on preserving the fuzziness in the calculations using fuzzy approximations.
- **Method 2.** Based on using a specific metric in the calculation procedure.

Method 1

The first one is a preliminary method. From our point of view, this idea is interesting to explore at this stage since it is directly inspired from ranking fuzzy numbers by the signed distance measure. The method intends to preserve the fuzzy nature of these sums and construct decision rules based on the fuzzy entities. Yet, a typical drawback for such methodologies is the computational burden that can be induced by the calculations of the difference and the product operations in the fuzzy context. For this purpose, we propose to calculate numerically the fuzzy sums by using convenient fuzzy approximations to estimate the fuzzy arithmetic operations. The problems can be overcome in specific conditions of the shapes of the modelling fuzzy numbers only and, thus, cannot be generalized to all possible cases. Thus, in our approximations, we consider the cases of trapezoidal and triangular fuzzy numbers only. By this exploratory method, we get decisions in the same spirit as the outcome of the tests by fuzzy confidence intervals shown in Sect. 6.4.1. In other terms, using some particular fuzzy approximations of these operations, a new set of fuzzy sums of squares could be constructed. We could then benefit from a decision produced on the basis of fuzzy entities.

Ideally, we would be interested in calculating a fuzzy test statistic. However, the ratio between fuzzy sets is challenging as well. Therefore, we propose to compare the different means of the fuzzy sums of squares instead of calculating their ratio. Since no fuzzy ordering exists between fuzzy numbers and in the idea of ranking fuzzy numbers, a reasoning by defuzzification similar to the one in Sect. 6.4.2 can be envisaged for the comparison between the fuzzy sums of squares.

At this stage, the decision rule is required. From a practical point of view, we propose a heuristic reasoning based on calculating the surface of the concerned fuzzy numbers. For a particular factor k, consider $\mathrm{Surf}(\overline{\tilde{B}_k})$ and $\mathrm{Surf}(\overline{\tilde{E}})$ to be the surfaces under the membership functions of the fuzzy numbers $\overline{\tilde{B}_k}$ and $\overline{\tilde{E}}$ denoting the respective means of the treatment and the residuals fuzzy sums of squares. In the case of L-R fuzzy sums of squares, it is easy to see that these surfaces are nothing but two times the signed distances of the corresponding fuzzy numbers measured from the fuzzy origin $\tilde{0}$. In other terms, they can be expressed as follows:

$$\mathrm{Surf}(\overline{\tilde{B}_k}) = \int_0^1 \left(\overline{\tilde{B}_{k_\alpha}}^L(\alpha) + \overline{\tilde{B}_{k_\alpha}}^R(\alpha) \right) d\alpha = 2 \times d_{SGD}(\overline{\tilde{B}_k}, \tilde{0}), \qquad (8.33)$$

$$\mathrm{Surf}(\overline{\tilde{E}}) = \int_0^1 \left(\overline{\tilde{E}_\alpha}^L(\alpha) + \overline{\tilde{E}_\alpha}^R(\alpha) \right) d\alpha = 2 \times d_{SGD}(\overline{\tilde{E}}, \tilde{0}). \qquad (8.34)$$

In the conventional case, we used to compare the ratio of the mean sums of squares of the treatment and the residuals to a quantile of the Fisher-Snedecor distribution. In the fuzzy context, another important drawback appears at this stage. It is mainly because of the ignorance of the theoretical distribution of the corresponding test statistic. Thus, consider T_{Surf} to be a test statistic expressed by

$$T_{\text{Surf}} = \frac{\text{Surf}(\overline{\tilde{B}_k})}{\text{Surf}(\overline{\tilde{E}})} = \frac{d_{SGD}(\overline{\tilde{B}_k}, \tilde{0})}{d_{SGD}(\overline{\tilde{E}}, \tilde{0})}. \tag{8.35}$$

The knowledge about the theoretical distribution of such tests is not evident and have to be investigated. This theoretical distribution could be empirically estimated, for example, by bootstrap techniques and Monte Carlo simulations.

Based on an initial data set, we propose to construct bootstrap statistics to estimate T_{Surf}. We would like then to generate D bootstrap samples. This task can be done by simply drawing with replacement a given number of observations.

Another idea to accomplish this task can be done using Proposition 4.7.6. In other terms, one could generate bootstrap samples for which every observation corresponds to a given unit of the set of the nearest symmetrical trapezoidal fuzzy numbers. Therefore, by this step, we intend to preserve the couple of characteristics of location and dispersion (s_0, ϵ) of each fuzzy number constructing our primary data set.

To sum up, for such constructions two algorithms can be eventually used. The first one consists on simply generating random bootstrap samples as follows:

Algorithm 1:

1. Compute the value of the test statistic T_{Surf} related to the considered original fuzzy sample.
2. From the primary data set, draw randomly with replacement a set of fuzzy observations to construct the bootstrap data set.
3. Calculate the bootstrap statistic $T_{\text{Surf}}^{\text{boot}}$.
4. Recursively repeat the Steps 2 and 3 a large number D of times to get a set of D values constructing the bootstrap distribution of test statistic.
5. Find κ the $1 - \delta$-quantile of the bootstrap distribution of test statistic.

In the same way, the (s_0, ϵ)-based algorithm can be written in the following manner:

Algorithm 2:

1. Consider a fuzzy sample. Calculate for each observation the couple of characteristics (s_0, ϵ).
2. From the calculated set of couples (s_0, ϵ), draw randomly with replacement and with equal probabilities a sample of couples (s_0, ϵ). Construct the bootstrap sample.
3. Calculate the bootstrap statistic $T_{\text{Surf}}^{\text{boot}}$.
4. Recursively repeat the Steps 2 and 3 a large number D of times to get a set of D values constructing the bootstrap distribution of test statistic.
5. Find κ the $1 - \delta$-quantile of the bootstrap distribution of test statistic.

Once κ the $1 - \delta$-quantile of the bootstrap distribution of the test statistic $T_{\text{Surf}}^{\text{boot}}$ is estimated, one could deduce the following decision rule:

Decision Rule For a particular factor k, by comparing the test statistic T_{Surf} of the original data set to the quantile κ of the bootstrap distribution of test statistics at the significance level δ, the decision rule could be given by

- If $T_{\text{Surf}} = \dfrac{\text{Surf}(\overline{\overline{B_k}})}{\text{Surf}(\overline{\overline{E}})} > \kappa$, the null hypothesis is rejected.
- Otherwise, the null hypothesis is not rejected.

Furthermore, since it is difficult to precisely calculate the value of the p-value of such test statistic, it would be important to conveniently estimate the bootstrap p-values related to the distribution of test statistics. This p-value denoted by p^{boot} can be written as follows:

$$p^{\text{boot}} = \frac{1 + \#(T_{\text{Surf}}^{\text{boot}} \geq T_{\text{Surf}})}{D + 1}. \tag{8.36}$$

Another way to decide whether to reject or not the null hypothesis in terms of the surfaces under the membership functions could also be established. Let κ be the $1 - \delta$-quantile of the bootstrap distribution of test statistic and by $\overline{\overline{E}}^{*}$ the weighted fuzzy number expressed by

$$\overline{\overline{E}}^{*} = \overline{\overline{E}} \times \kappa, \tag{8.37}$$

such that its surface is given by the following expression:

$$\text{Surf}(\overline{\overline{E}}^{*}) = \text{Surf}(\overline{\overline{E}} \times \kappa) = 2 \times d_{SGD}(\overline{\overline{E}}^{*}, \tilde{0}). \tag{8.38}$$

It is then obvious that if we would like to compare $\mathrm{Surf}(\overline{\tilde{B}_k})$ to $\mathrm{Surf}(\overline{\tilde{E}}^*)$, we could eventually compare $d_{SGD}(\overline{\tilde{B}_k}, \tilde{0})$ to $d_{SGD}(\overline{\tilde{E}}^*, \tilde{0})$. Concerning the decision rule, we have to compare the constructed test statistic T_{Surf} given in Eq. 8.35 to the quantile κ. This latter is equivalent to comparing the signed distance of $\overline{\tilde{E}}^*$ to the signed distance of $\overline{\tilde{B}_k}$, regarded as ranking the fuzzy numbers $\overline{\tilde{E}}^*$ and $\overline{\tilde{B}_k}$. Thus, we could write the subsequent decision rule:

Decision Rule By comparing $d_{SGD}(\overline{\tilde{E}}^*, \tilde{0}) = d_{SGD}(\overline{\tilde{E}} \times \kappa, \tilde{0})$ to $d_{SGD}(\overline{\tilde{B}_k}, \tilde{0})$ at the significance level δ, the decision rule can be written as follows:

- If $d_{SGD}(\overline{\tilde{E}}^*, \tilde{0}) \leq d_{SGD}(\overline{\tilde{B}_k}, \tilde{0})$, the null hypothesis H_0 is rejected;
- If $d_{SGD}(\overline{\tilde{E}}^*, \tilde{0}) > d_{SGD}(\overline{\tilde{B}_k}, \tilde{0})$, the null hypothesis H_0 is not rejected.

An additional argument to strengthen our support to this procedure is that if $\overline{\tilde{B}_k}$ and $\overline{\tilde{E}}^*$ overlap partially or totally, the case is considered to be critical. No decision can be made since rejecting or not the null hypothesis is not completely clear. A reasoning on the surfaces under the curves as above described, seen as a defuzzification by the SGD operator, is most probably needed.

If we want to reason similarly to the hypotheses test by confidence interval given in Sect. 6.4, we could additionally translate (in the sense of "standardizing") the fuzzy numbers $\overline{\tilde{B}_k}$ and $\overline{\tilde{E}} \times \kappa$ to the interval $[0; 1]$ of the x-axis, in order to get fuzzy numbers close in terms of properties and interpretations to the ones of the fuzzy decisions, and to probably obtain a degree of "rejection" of the null hypothesis by defuzzifying these fuzzy numbers.

Finally, if one would like to enhance the decision regarding the different hypotheses, it would be interesting to define as well the so-called contribution of the error and the treatment sums of squares to the total sum of squares. They are written as the following ratios:

$$C(\overline{\tilde{B}_k}) = \frac{\mathrm{Surf}(\overline{\tilde{B}_k})}{\mathrm{Surf}(\overline{\tilde{B}_k}) + \mathrm{Surf}(\overline{\tilde{E}}^*)}, \tag{8.39}$$

$$C(\overline{\tilde{E}}^*) = \frac{\mathrm{Surf}(\overline{\tilde{E}}^*)}{\mathrm{Surf}(\overline{\tilde{B}_k}) + \mathrm{Surf}(\overline{\tilde{E}}^*)}, \tag{8.40}$$

where $C(\overline{\tilde{B}_k})$ and $C(\overline{\tilde{E}}^*)$ are, respectively, the contributions of the treatment and of the error sums of squares to the model. They are contained in the interval $[0; 1]$ such that $C(\overline{\tilde{B}_k}) + C(\overline{\tilde{E}}^*) = 1$. These contributions are written in terms of the weighted mean of the fuzzy residual sum of squares $\overline{\tilde{E}}^*$ since the objective of constructing such indicators is to complement the decision made. However, it would also make sense to write them using the mean of the fuzzy residual sums of squares $\overline{\tilde{E}}$ if the aim is to compare both means of the fuzzy sums without the need of a threshold of comparison (obviously the quantile κ in our case).

A final remark is that a further investigation of this approach is crucial. Actually, the choice of convenient fuzzy numerical approximations of the arithmetics should be argumented or criticized, as well as the decisions obtained by the aforementioned procedure. In addition, a further investigation of the theoretical distribution of the corresponding test statistic has to be performed.

Method 2

The second approach is to consider the Aumann-type expectation of the fuzzy dependent variable as shown in Definition 5.4.4. It could be done by choosing a convenient metric to model the distance between fuzzy numbers. We know that the multi-ways case extends the one-way model by considering a set of factors instead of only one. In Berkachy and Donzé (2018), we showed a univariate case of fuzzy one-way ANOVA where we proposed to use the SGD between fuzzy numbers. A generalization to the multi-factors case is thoroughly conceivable. Hence, the Chap. 4 giving a list of possible choices of distances can be of good use in this purpose.

Consider a corresponding metric denoted by d. We could re-write the sums given in 8.28, 8.30–8.32 with respect to this distance d. We highlight that the obtained sums will be crisp entities. They can be written as follows:

$$T_d = \sum_{i_1=1}^{g_1} \sum_{i_2=1}^{g_2} \sum_{j=1}^{n} d\left(\tilde{X}_{i_1 i_2 j}, \overline{\tilde{X}}_{\bullet\bullet\bullet}\right) \times d\left(\tilde{X}_{i_1 i_2 j}, \overline{\tilde{X}}_{\bullet\bullet\bullet}\right) = \sum_{i_1=1}^{g_1} \sum_{i_2=1}^{g_2} \sum_{j=1}^{n} \left(d\left(\tilde{X}_{i_1 i_2 j}, \overline{\tilde{X}}_{\bullet\bullet\bullet}\right)\right)^2,$$
(8.41)

$$B_{1d} = \sum_{i_1=1}^{g_1} n g_2 d\left(\overline{\tilde{X}}_{i_1 \bullet\bullet}, \overline{\tilde{X}}_{\bullet\bullet\bullet}\right) \times d\left(\overline{\tilde{X}}_{i_1 \bullet\bullet}, \overline{\tilde{X}}_{\bullet\bullet\bullet}\right) = \sum_{i_1=1}^{g_1} n g_2 \left(d\left(\overline{\tilde{X}}_{i_1 \bullet\bullet}, \overline{\tilde{X}}_{\bullet\bullet\bullet}\right)\right)^2,$$
(8.42)

$$B_{2d} = \sum_{i_2=1}^{g_2} n g_1 d\left(\overline{\tilde{X}}_{\bullet i_2 \bullet}, \overline{\tilde{X}}_{\bullet\bullet\bullet}\right) \times d\left(\overline{\tilde{X}}_{\bullet i_2 \bullet}, \overline{\tilde{X}}_{\bullet\bullet\bullet}\right) = \sum_{i_2=1}^{g_2} n g_1 \left(d\left(\overline{\tilde{X}}_{\bullet i_2 \bullet}, \overline{\tilde{X}}_{\bullet\bullet\bullet}\right)\right)^2,$$
(8.43)

$$E_d = \sum_{i_1=1}^{g_1} \sum_{i_2=1}^{g_2} \sum_{j=1}^{n} d\left(\tilde{X}_{i_1 i_2 j}, \overline{\tilde{X}}_{i_1 i_2 \bullet}\right) \times d\left(\tilde{X}_{i_1 i_2 j}, \overline{\tilde{X}}_{i_1 i_2 \bullet}\right) = \sum_{i_1=1}^{g_1} \sum_{i_1=1}^{g_2} \sum_{j=1}^{n} \left(d\left(\tilde{X}_{i_1 i_2 j}, \overline{\tilde{X}}_{i_1 i_2 \bullet}\right)\right)^2.$$
(8.44)

We have to verify now that the following decomposition equation

$$T_d = B_{1d} + B_{2d} + E_d$$
(8.45)

is true. The proof is given as follows:

Proof From Eqs. 8.41–8.44, and for a well pre-defined associative and distributive metric d, the sum T_d can be decomposed as follows:

$$\mathrm{T}_d = \sum_{i_1=1}^{g_1}\sum_{i_2=1}^{g_2}\sum_{j=1}^{n}\left(d\big(\tilde{X}_{i_1i_2j},\overline{\overline{\tilde{X}}}_{\bullet\bullet\bullet}\big)\right)^2,$$

$$= \sum_{i_1=1}^{g_1}\sum_{i_2=1}^{g_2}\sum_{j=1}^{n}\left[d\big(\tilde{X}_{i_1i_2j},\overline{\tilde{X}}_{i_1i_2\bullet}\big) + d\big(\overline{\tilde{X}}_{i_1i_2\bullet},\overline{\overline{\tilde{X}}}_{\bullet\bullet\bullet}\big)\right]^2,$$

since

$$\left(d\big(\tilde{X}_{i_1i_2j},\overline{\overline{\tilde{X}}}_{\bullet\bullet\bullet}\big)\right)^2 = \left(d\big(\tilde{X}_{i_1i_2j},\tilde{0}\big) - d\big(\overline{\overline{\tilde{X}}}_{\bullet\bullet\bullet},\tilde{0}\big)\right)^2$$

$$= \left(d\big(\tilde{X}_{i_1i_2j},\tilde{0}\big) - d\big(\overline{\overline{\tilde{X}}}_{\bullet\bullet\bullet},\tilde{0}\big) + d\big(\tilde{X}_{i_1i_2\bullet},\tilde{0}\big) - d\big(\tilde{X}_{i_1i_2\bullet},\tilde{0}\big)\right)^2$$

$$= \left[d\big(\tilde{X}_{i_1i_2j},\overline{\tilde{X}}_{i_1i_2\bullet}\big) + d\big(\overline{\tilde{X}}_{i_1i_2\bullet},\overline{\overline{\tilde{X}}}_{\bullet\bullet\bullet}\big)\right]^2.$$

Then, by developing $\left[d\big(\tilde{X}_{i_1i_2j},\overline{\tilde{X}}_{i_1i_2\bullet}\big) + d\big(\overline{\tilde{X}}_{i_1i_2\bullet},\overline{\overline{\tilde{X}}}_{\bullet\bullet\bullet}\big)\right]^2$, we get the following decomposition of T_d:

$$\mathrm{T}_d = \sum_{i_1=1}^{g_1}\sum_{i_2=1}^{g_2}\sum_{j=1}^{n}\left[\left(d\big(\tilde{X}_{i_1i_2j},\overline{\tilde{X}}_{i_1i_2\bullet}\big)\right)^2 + \left(d\big(\overline{\tilde{X}}_{i_1i_2\bullet},\overline{\overline{\tilde{X}}}_{\bullet\bullet\bullet}\big)\right)^2\right.$$

$$\left. +2d\big(\tilde{X}_{i_1i_2j},\overline{\tilde{X}}_{i_1i_2\bullet}\big)\,d\big(\overline{\tilde{X}}_{i_1i_2\bullet},\overline{\overline{\tilde{X}}}_{\bullet\bullet\bullet}\big)\right]$$

$$= \mathrm{E}_d + \sum_{i_1=1}^{g_1}\sum_{i_2=1}^{g_2}\sum_{j=1}^{n}\left[\left(d\big(\overline{\tilde{X}}_{i_1i_2\bullet},\overline{\overline{\tilde{X}}}_{\bullet\bullet\bullet}\big)\right)^2 + 2d\big(\tilde{X}_{i_1i_2j},\overline{\tilde{X}}_{i_1i_2\bullet}\big)\,d\big(\overline{\tilde{X}}_{i_1i_2\bullet},\overline{\overline{\tilde{X}}}_{\bullet\bullet\bullet}\big)\right].$$

Our purpose is then to prove that

$$\sum_{i_1=1}^{g_1}\sum_{i_2=1}^{g_2}\sum_{j=1}^{n}\left(d\big(\overline{\tilde{X}}_{i_1i_2\bullet},\overline{\overline{\tilde{X}}}_{\bullet\bullet\bullet}\big)\right)^2 = \mathrm{B}_{1d} + \mathrm{B}_{2d}, \tag{8.46}$$

and that

$$\sum_{i_1=1}^{g_1}\sum_{i_2=1}^{g_2}\sum_{j=1}^{n}2d\big(\tilde{X}_{i_1i_2j},\overline{\tilde{X}}_{i_1i_2\bullet}\big)\,d\big(\overline{\tilde{X}}_{i_1i_2\bullet},\overline{\overline{\tilde{X}}}_{\bullet\bullet\bullet}\big) = 0. \tag{8.47}$$

It can be done in the following manner:

1. At a first stage, let us prove the Eq. 8.46. From Eq. 8.27, we know that $\overline{\tilde{X}}_{i_1 i_2 \bullet} = \overline{\tilde{X}}_{i_1 \bullet \bullet} \oplus \overline{\tilde{X}}_{\bullet i_2 \bullet} \ominus \overline{\tilde{X}}_{\bullet \bullet \bullet}$. Then, if we would like to calculate the distance of $\overline{\tilde{X}}_{i_1 i_2 \bullet}$ measured from the fuzzy origin $\tilde{0}$, we could write the following:

$$d(\overline{\tilde{X}}_{i_1 i_2 \bullet}, \tilde{0}) = d(\overline{\tilde{X}}_{i_1 \bullet \bullet}, \tilde{0}) + d(\overline{\tilde{X}}_{\bullet i_2 \bullet}, \tilde{0}) - d(\overline{\tilde{X}}_{\bullet \bullet \bullet}, \tilde{0}).$$

Therefore, by decomposing the expression $\sum_{i_1=1}^{g_1} \sum_{i_2=1}^{g_2} \sum_{j=1}^{n} \left(d(\overline{\tilde{X}}_{i_1 i_2 \bullet}, \overline{\tilde{X}}_{\bullet \bullet \bullet})\right)^2$ and substituting the distance of $\overline{\tilde{X}}_{i_1 i_2 \bullet}$ to the fuzzy origin by $d(\overline{\tilde{X}}_{i_1 \bullet \bullet}, \tilde{0}) + d(\overline{\tilde{X}}_{\bullet i_2 \bullet}, \tilde{0}) - d(\overline{\tilde{X}}_{\bullet \bullet \bullet}, \tilde{0})$, we get the following:

$$\sum_{i_1=1}^{g_1} \sum_{i_2=1}^{g_2} \sum_{j=1}^{n} \left(d(\overline{\tilde{X}}_{i_1 i_2 \bullet}, \overline{\tilde{X}}_{\bullet \bullet \bullet})\right)^2 = \sum_{i_1=1}^{g_1} \sum_{i_2=1}^{g_2} n\left(d(\overline{\tilde{X}}_{i_1 i_2 \bullet}, \overline{\tilde{X}}_{\bullet \bullet \bullet})\right)^2$$

$$= \sum_{i_1=1}^{g_1} \sum_{i_2=1}^{g_2} n\left[d(\overline{\tilde{X}}_{i_1 i_2 \bullet}, \tilde{0}) - d(\overline{\tilde{X}}_{\bullet \bullet \bullet}, \tilde{0})\right]^2$$

$$= \sum_{i_1=1}^{g_1} \sum_{i_2=1}^{g_2} n\left[d(\overline{\tilde{X}}_{i_1 \bullet \bullet}, \tilde{0}) + d(\overline{\tilde{X}}_{\bullet i_2 \bullet}, \tilde{0}) - d(\overline{\tilde{X}}_{\bullet \bullet \bullet}, \tilde{0}) - d(\overline{\tilde{X}}_{\bullet \bullet \bullet}, \tilde{0})\right]^2$$

$$= \sum_{i_1=1}^{g_1} \sum_{i_2=1}^{g_2} n\left[d(\overline{\tilde{X}}_{i_1 \bullet \bullet}, \overline{\tilde{X}}_{\bullet \bullet \bullet}) + d(\overline{\tilde{X}}_{\bullet i_2 \bullet}, \overline{\tilde{X}}_{\bullet \bullet \bullet})\right]^2$$

$$= \sum_{i_1=1}^{g_1} \sum_{i_2=1}^{g_2} n\left[d(\overline{\tilde{X}}_{i_1 \bullet \bullet}, \overline{\tilde{X}}_{\bullet \bullet \bullet})\right]^2 + \sum_{i_1=1}^{g_1} \sum_{i_2=1}^{g_2} n\left[d(\overline{\tilde{X}}_{\bullet i_2 \bullet}, \overline{\tilde{X}}_{\bullet \bullet \bullet})\right]^2$$

$$+ \sum_{i_1=1}^{g_1} \sum_{i_2=1}^{g_2} n\left[2d(\overline{\tilde{X}}_{i_1 \bullet \bullet}, \overline{\tilde{X}}_{\bullet \bullet \bullet}) d(\overline{\tilde{X}}_{\bullet i_2 \bullet}, \overline{\tilde{X}}_{\bullet \bullet \bullet})\right]$$

$$= B_{1d} + B_{2d} + \sum_{i_1=1}^{g_1} \sum_{i_2=1}^{g_2} n\left[2d(\overline{\tilde{X}}_{i_1 \bullet \bullet}, \overline{\tilde{X}}_{\bullet \bullet \bullet}) d(\overline{\tilde{X}}_{\bullet i_2 \bullet}, \overline{\tilde{X}}_{\bullet \bullet \bullet})\right],$$

$$= B_{1d} + B_{2d} + 2n \sum_{i_1=1}^{g_1} d(\overline{\tilde{X}}_{i_1 \bullet \bullet}, \overline{\tilde{X}}_{\bullet \bullet \bullet}) \sum_{i_2=1}^{g_2} d(\overline{\tilde{X}}_{\bullet i_2 \bullet}, \overline{\tilde{X}}_{\bullet \bullet \bullet}).$$

Let us now prove that

$$2n \sum_{i_1=1}^{g_1} d\left(\overline{\tilde{X}}_{i_1\bullet\bullet}, \overline{\tilde{X}}_{\bullet\bullet\bullet}\right) \sum_{i_2=1}^{g_2} d\left(\overline{\tilde{X}}_{\bullet i_2\bullet}, \overline{\tilde{X}}_{\bullet\bullet\bullet}\right) = 0.$$

From Eqs. 8.23 and 8.24, we know that

$$\overline{\tilde{X}}_{i_1\bullet\bullet} = \frac{1}{g_2 n} \sum_{i_2=1}^{g_2} \sum_{j=1}^{n} \tilde{X}_{i_1 i_2 j} \quad \text{and} \quad \overline{\tilde{X}}_{\bullet i_2\bullet} = \frac{1}{g_1 n} \sum_{i_1=1}^{g_1} \sum_{j=1}^{n} \tilde{X}_{i_1 i_2 j},$$

which induces the following relations:

$$\frac{1}{g_1} \sum_{i_1=1}^{g_1} \overline{\tilde{X}}_{i_1\bullet\bullet} = \frac{1}{g_1} \sum_{i_1=1}^{g_1} \frac{1}{g_2 n} \sum_{i_2=1}^{g_2} \sum_{j=1}^{n} \tilde{X}_{i_1 i_2 j} = \overline{\tilde{X}}_{\bullet\bullet\bullet}$$

$$\text{and} \quad \frac{1}{g_2} \sum_{i_2=1}^{g_2} \overline{\tilde{X}}_{\bullet i_2\bullet} = \frac{1}{g_2} \sum_{i_2=1}^{g_2} \frac{1}{g_1 n} \sum_{i_1=1}^{g_1} \sum_{j=1}^{n} \tilde{X}_{i_1 i_2 j} = \overline{\tilde{X}}_{\bullet\bullet\bullet}.$$

Thus, it is clear to see that the distance of $\overline{\tilde{X}}_{\bullet\bullet\bullet}$ to the fuzzy origin can be written as:

$$d\left(\overline{\tilde{X}}_{\bullet\bullet\bullet}, \tilde{0}\right) = \frac{1}{g_1} \sum_{i_1=1}^{g_1} d\left(\overline{\tilde{X}}_{i_1\bullet\bullet}, \tilde{0}\right) = \frac{1}{g_2} \sum_{i_2=1}^{g_2} d\left(\overline{\tilde{X}}_{\bullet i_2\bullet}, \tilde{0}\right).$$

Then, we get the following:

$$\sum_{i_1=1}^{g_1} d\left(\overline{\tilde{X}}_{i_1\bullet\bullet}, \overline{\tilde{X}}_{\bullet\bullet\bullet}\right) = \sum_{i_1=1}^{g_1} \left[d\left(\overline{\tilde{X}}_{i_1\bullet\bullet}, \tilde{0}\right) - d\left(\overline{\tilde{X}}_{\bullet\bullet\bullet}, \tilde{0}\right) \right]$$

$$= \sum_{i_1=1}^{g_1} d\left(\overline{\tilde{X}}_{i_1\bullet\bullet}, \tilde{0}\right) - g_1 d\left(\overline{\tilde{X}}_{\bullet\bullet\bullet}, \tilde{0}\right) = 0.$$

It is the same for $\sum_{i_2=1}^{g_2} d\left(\overline{\tilde{X}}_{\bullet i_2\bullet}, \overline{\tilde{X}}_{\bullet\bullet\bullet}\right) = 0.$

We then deduce that

$$2n \sum_{i_1=1}^{g_1} d\left(\tilde{X}_{i_1\bullet\bullet}, \overline{\tilde{X}}_{\bullet\bullet\bullet}\right) \sum_{i_2=1}^{g_2} d\left(\tilde{X}_{\bullet i_2\bullet}, \overline{\tilde{X}}_{\bullet\bullet\bullet}\right) = 0.$$

2. For the proof of the Eq. 8.47, we can write that

$$\sum_{i_1=1}^{g_1}\sum_{i_2=1}^{g_2}\sum_{j=1}^{n} 2d\big(\tilde{X}_{i_1i_2j}, \overline{\tilde{X}}_{i_1i_2\bullet}\big)\, d\big(\overline{\tilde{X}}_{i_1i_2\bullet}, \overline{\tilde{X}}_{\bullet\bullet\bullet}\big)$$

$$= 2\sum_{i_1=1}^{g_1}\sum_{i_2=1}^{g_2} d\big(\overline{\tilde{X}}_{i_1i_2\bullet}, \overline{\tilde{X}}_{\bullet\bullet\bullet}\big) \sum_{j=1}^{n} d\big(\tilde{X}_{i_1i_2j}, \overline{\tilde{X}}_{i_1i_2\bullet}\big).$$

We know from Eq. 8.26 that $\overline{\tilde{X}}_{i_1i_2\bullet} = \frac{1}{n}\sum_{j=1}^{n}\tilde{X}_{i_1i_2j}$. By calculating the distance of $\overline{\tilde{X}}_{i_1i_2\bullet}$ to the fuzzy origin $\tilde{0}$, we get the following expression:

$$d\big(\overline{\tilde{X}}_{i_1i_2\bullet}, \tilde{0}\big) = \frac{1}{n}\sum_{j=1}^{n} d\big(\tilde{X}_{i_1i_2j}, \tilde{0}\big).$$

Therefore, Eq. 8.47 can be written in the following manner:

$$2\sum_{i_1=1}^{g_1}\sum_{i_2=1}^{g_2} d\big(\overline{\tilde{X}}_{i_1i_2\bullet}, \overline{\tilde{X}}_{\bullet\bullet\bullet}\big) \sum_{j=1}^{n} d\big(\tilde{X}_{i_1i_2j}, \overline{\tilde{X}}_{i_1i_2\bullet}\big)$$

$$= 2\sum_{i_1=1}^{g_1}\sum_{i_2=1}^{g_2} d\big(\overline{\tilde{X}}_{i_1i_2\bullet}, \overline{\tilde{X}}_{\bullet\bullet\bullet}\big) \sum_{j=1}^{n} \Big[d\big(\tilde{X}_{i_1i_2j}, \tilde{0}\big) - d\big(\overline{\tilde{X}}_{i_1i_2\bullet}, \tilde{0}\big)\Big]$$

$$= 2\sum_{i_1=1}^{g_1}\sum_{i_2=1}^{g_2} d\big(\overline{\tilde{X}}_{i_1i_2\bullet}, \overline{\tilde{X}}_{\bullet\bullet\bullet}\big) \Big[\sum_{j=1}^{n} d\big(\tilde{X}_{i_1i_2j}, \tilde{0}\big) - nd\big(\overline{\tilde{X}}_{i_1i_2\bullet}, \tilde{0}\big)\Big] = 0.$$

Consequently, the decomposition given in Eq. 8.45 is verified.   □

The produced sums of squares are now on crisp. We denote by T_{dist} the test statistic related to a given factor k. It can be written as follows:

$$T_{\text{dist}} = \frac{B_{kd}}{E_d}, \tag{8.48}$$

where B_{kd} is defined as the mean of the treatment sum of squares for the factor k and E_d is the mean of the error sum of squares.

Moreover, the crisp entities B_{kd} and E_d are measurable and independent. In the conventional approach, a typical statistic of the form of T_{dist} divided by the ratio of degrees of freedom is proved to be taken from the Fisher-Snedecor distribution with $df_{B_{kd}}$ and df_{E_d} degrees of freedom at the significance level δ. The derivation of the theoretical distribution is possible only if suitable assumptions are made in the modelling process. The classical assumption in such cases is the normality of the error terms of the constructed model. In our fuzzy model, we did not assume till now any particular distribution. Researchers have explored several methods to solve this problem. Many of them have used bootstrap techniques in similar contexts, such as Gonzalez-Rodriguez et al. (2012), Ramos-Guajardo and Lubiano (2012), and Gil et al. (2006) where the authors have proposed to construct bootstrap test statistics approximated by the Monte Carlo methods to estimate the distribution of the test statistic T_{dist}.

Jiryaei et al. (2013) proved that in the context of a one-way fuzzy analysis of variance using symmetrical triangular and exponential fuzzy numbers, the calculated test statistic is taken from a Fisher-Snedecor distribution under the null hypothesis, similarly to the classical approach. As for us, we will benefit from the constructed Algorithms 1 and 2 of the first method to perform a bootstrap analysis to quantify the distribution of the test statistic, where we denote by ζ its $1 - \delta$-quantile. We will after be interested in experimentally investigating the proposition of Jiryaei et al. (2013) but in the multi-ways situation.

Note that in this case, the calculation of the bootstrap p-value p^{boot} is exactly the same as Eq. 8.36. Accordingly, the decision rule of this hypotheses test can be written in terms of ζ as follows:

Decision Rule The decision rule related to this test can be depicted by comparing the test statistic T_{dist} to the quantile ζ of the corresponding bootstrap distribution at the significance level δ. The decision rule can be expressed by the following:

- If $T_{dist} > \zeta$, the null hypothesis H_0 is rejected;
- Otherwise, H_0 is not rejected.

Finally, we have to highlight that the coefficients of the concerned model are also fuzzy, and have to be estimated. Their calculations are made similarly to the procedure made in the classical context. In other terms, we suppose that these coefficients can be found by the ordinary least-squares method applied using the distances between the fuzzy response variable and the fuzzy parameters from one side, and to the fuzzy origin from another one.

8.3.1 Particular Cases

Practitioners often use triangular and trapezoidal fuzzy numbers because of their simplicity. Gaussian fuzzy numbers are used because of the smoothness of the shape. It is interesting to investigate the Mult-FANOVA using particular shapes of fuzzy numbers specifically when the signed distance is considered, and the fuzzy numbers are supposed to be symmetrical around the core. In the procedure discussed in Berkachy and Donzé (2018), we provided similar particular cases but in one-factor setups. Note that the interpretations of Berkachy and Donzé (2018) are exactly the same for multi-factors and one-factor cases.

Consider first symmetrical triangular and trapezoidal fuzzy numbers given, respectively, by the tuple (p, q, r) where $p \leq q \leq r$ and the quadruple (p, q, r, s)

with $p \le q \le r \le s$. Since the considered fuzzy numbers are symmetrical around their core, we can directly deduce that $p + r = 2q$ for the triangular case, and $p + s = q + r$ for the trapezoidal one. Let us also consider the distance d to be the SGD described in Sect. 4.6. For the multi-ways analysis of variance, we have to re-write the Eqs. 8.41–8.44 with respect to the SGD. We know that the SGD of a triangular fuzzy number $\tilde{X} = (p, q, r)$ is $d(\tilde{X}, \tilde{0}) = \frac{1}{4}(p + 2q + r)$, and the SGD of a trapezoidal fuzzy number $\tilde{X} = (p, q, r, s)$ is $d(\tilde{X}, \tilde{0}) = \frac{1}{4}(p + q + r + s)$. Then, for the case of triangular symmetrical fuzzy numbers, the sums of squares are the following:

$$T_{SGD} = \sum_{i_1=1}^{g_1} \sum_{i_2=1}^{g_2} \sum_{j=1}^{n} \left(q_{i_1 i_2 j} - q_{\bullet\bullet\bullet} \right)^2, \tag{8.49}$$

$$B_{1 SGD} = \sum_{i_1=1}^{g_1} n g_2 \left(q_{i_1 \bullet\bullet} - q_{\bullet\bullet\bullet} \right)^2, \tag{8.50}$$

$$B_{2 SGD} = \sum_{i_2=1}^{g_2} n g_1 \left(q_{\bullet i_2 \bullet} - q_{\bullet\bullet\bullet} \right)^2, \tag{8.51}$$

$$E_{SGD} = \sum_{i_1=1}^{g_1} \sum_{i_2=1}^{g_2} \sum_{j=1}^{n} \left(q_{i_1 i_2 j} - q_{i_1 i_2 \bullet} \right)^2, \tag{8.52}$$

where $q_{i_1 i_2 j}$, $q_{i_1 \bullet\bullet}$, $q_{\bullet i_2 \bullet}$, $q_{i_1 i_2 \bullet}$, and $q_{\bullet\bullet\bullet}$ are the respective cores of the fuzzy numbers $\tilde{X}_{i_1 i_2 j}$, $\tilde{X}_{i_1 \bullet\bullet}$, $\tilde{X}_{\bullet i_2 \bullet}$, $\tilde{X}_{i_1 i_2 \bullet}$, and $\tilde{X}_{\bullet\bullet\bullet}$, for given values of i_1, i_2 and j.
For the trapezoidal case, the sums are as follows:

$$T_{SGD} = \sum_{i_1=1}^{g_1} \sum_{i_2=1}^{g_2} \sum_{j=1}^{n} \left(\frac{q_{i_1 i_2 j} + r_{i_1 i_2 j}}{2} - \frac{q_{\bullet\bullet\bullet} + r_{\bullet\bullet\bullet}}{2} \right)^2, \tag{8.53}$$

$$B_{1 SGD} = \sum_{i_1=1}^{g_1} n g_2 \left(\frac{q_{i_1 \bullet\bullet} + r_{i_1 \bullet\bullet}}{2} - \frac{q_{\bullet\bullet\bullet} + r_{\bullet\bullet\bullet}}{2} \right)^2, \tag{8.54}$$

$$B_{2 SGD} = \sum_{i_2=1}^{g_2} n g_1 \left(\frac{q_{\bullet i_2 \bullet} + r_{\bullet i_2 \bullet}}{2} - \frac{q_{\bullet\bullet\bullet} + r_{\bullet\bullet\bullet}}{2} \right)^2, \tag{8.55}$$

$$E_{SGD} = \sum_{i_1=1}^{g_1} \sum_{i_2=1}^{g_2} \sum_{j=1}^{n} \left(\frac{q_{i_1 i_2 j} + r_{i_1 i_2 j}}{2} - \frac{q_{i_1 i_2 \bullet} + r_{i_1 i_2 \bullet}}{2} \right)^2, \tag{8.56}$$

where the intervals $[q_{i_1i_2j}; r_{i_1i_2j}]$, $[q_{i_1\bullet\bullet}; r_{i_1\bullet\bullet}]$, $[q_{\bullet i_2\bullet}; r_{\bullet i_2\bullet}]$, $[q_{i_1i_2\bullet}; r_{i_1i_2\bullet}]$,and $[q_{\bullet\bullet\bullet}; r_{\bullet\bullet\bullet}]$ are the respective core intervals of the fuzzy numbers $\tilde{X}_{i_1i_2j}$, $\overline{\tilde{X}}_{i_1\bullet\bullet}$, $\overline{\tilde{X}}_{\bullet i_2\bullet}$, $\overline{\tilde{X}}_{i_1i_2\bullet}$, and $\overline{\tilde{X}}_{\bullet\bullet\bullet}$, for particular values of i_1, i_2, and j.

We would like now to propose the case using gaussian fuzzy numbers. Therefore, we suppose that the fuzzy numbers modelling the response variable have gaussian shapes with their parameters η and σ. We know that the SGD of a gaussian fuzzy number $\tilde{X}$ with the couple of parameters (η, σ) is $d(\tilde{X}, \tilde{0}) = \eta$. We highlight that this distance is independent from σ. The sums of squares can then be given by

$$\text{T}_{SGD} = \sum_{i_1=1}^{g_1} \sum_{i_2=1}^{g_2} \sum_{j=1}^{n} \left(\eta_{i_1i_2j} - \eta_{\bullet\bullet\bullet}\right)^2, \tag{8.57}$$

$$\text{B}_{1SGD} = \sum_{i_1=1}^{g_1} ng_2 \left(\eta_{i_1\bullet\bullet} - \eta_{\bullet\bullet\bullet}\right)^2, \tag{8.58}$$

$$\text{B}_{2SGD} = \sum_{i_2=1}^{g_2} ng_1 \left(\eta_{\bullet i_2\bullet} - \eta_{\bullet\bullet\bullet}\right)^2, \tag{8.59}$$

$$\text{E}_{SGD} = \sum_{i_1=1}^{g_1} \sum_{i_2=1}^{g_2} \sum_{j=1}^{n} \left(\eta_{i_1i_2j} - \eta_{i_1i_2\bullet}\right)^2, \tag{8.60}$$

where for given values i_1, i_2, and j, $\eta_{i_1i_2j}$, $\eta_{i_1\bullet\bullet}$, $\eta_{\bullet i_2\bullet}$, $\eta_{i_1i_2\bullet}$, and $\eta_{\bullet\bullet\bullet}$ are the means of the gaussian fuzzy numbers $\tilde{X}_{i_1i_2j}$, $\overline{\tilde{X}}_{i_1\bullet\bullet}$, $\overline{\tilde{X}}_{\bullet i_2\bullet}$, $\overline{\tilde{X}}_{i_1i_2\bullet}$, and $\overline{\tilde{X}}_{\bullet\bullet\bullet}$.

Regardless of the value of the parameter σ of the modelling fuzzy numbers, it is easy to see that the approach with gaussian fuzzy numbers gives instinctively the same result as the classical approach. In a more general sense, we could deduce that the 3 cases (triangular, trapezoidal, and gaussian) gave the same conclusion. In the considered setups, these three calculations return exactly the result by the conventional approach. This leads to say that in such contexts, by choosing symmetrical shapes around the core of the modelling fuzzy numbers, we can return exactly the same results as in the classical approach.

We could finally remark that our defended fuzzy approach with the signed distance seems to be a reproduction of the classical approach, but with more flexibility on modelling the uncertainty of the response variable. However, choosing symmetrical shapes around the core of the considered fuzzy numbers should somehow be avoided while using the signed distance in order to benefit from the properties of the fuzzy approaches.

8.3.2   Ordering Effect and Sequential Sums

We proposed above a Mult-FANOVA model based on a balanced design. Such designs are often used in experimental studies. The reason for this use is that balanced designs make it easier to recognize the effect of a given level or factor and would eventually decrease the influence of the non-homogeneity in the sample. However, in survey studies, unbalanced designs are more frequent. In other terms, the number of observations in each set of treatments is not necessarily the same. On the other hand, we know that the purpose of introducing an additional independent variable to a model is to explain more variability of the dependent one, and by that, to decrease as much as possible the error of prediction of the model. The consequence of considering unbalanced designs in a multi-ways analysis of variance is that the order of introducing the predictors should no more be neglected. This order now on matters. As such, in the Mult-ANOVA models, due to the ordering effect of adding one or more factors, the sequential sums of squares are introduced. It could be defined as an increase in the amount of explained variability, or a decrease in the error sum of squares, i.e. the amount of reduced error. These sums of squares therefore depend on which factor is added before the other. A method of monitoring the influence of ordering the independent variables introduced in the Mult-ANOVA model on the sum of squares is then proposed.

Similarly to the conventional approach, we introduce the expressions of the sequential sums of squares in the fuzzy context using the considered distance d. Let us first define the different components of these sequential sums.

We denote by $E_d(F_1, \ldots, F_k, \ldots, F_p)$ and $B_d(F_1, \ldots, F_k, \ldots, F_p)$, respectively, the error sum of squares and the regression sum of squares in the model composed by the p factors $F_1, \ldots, F_k, \ldots, F_p$. As instance, if the factor F_1 is the only independent variable in the model, then the sums of squares will be written as $E_d(F_1)$ and $B_d(F_1)$. The entities E_d and B_d correspond to the expression of the error and hypotheses sum of squares using the metric d given in Eqs. 8.42–8.44.

In the same way, we denote by $B_d(F_k \mid F_1, \ldots, F_{k-1})$ the sequential sum of squares, defined as the sum of squares obtained by introducing the factor F_k to the model composed by the factors $F_1, \ldots, F_{k-1}$. The sequential sum of squares is then given by the following expression:

$$B_d(F_k \mid F_1, \ldots, F_{k-1}) = E_d(F_1, \ldots, F_{k-1}) - E_d(F_1, \ldots, F_{k-1}, F_k). \quad (8.61)$$

It is easy to see that this expression is the difference between the residuals of the model composed by the factors $F_1, \ldots, F_{k-1}$ and the residuals of the model with the factors $F_1, \ldots, F_{k-1}, F_k$.

The computations should then be done recursively by adding at once one factor only. We add that for these operations, the total sum of squares T_d remains the same through all the models. The variations happen only in the decomposition between

Table 8.2 The notations of the sums of squares affected by the ordering of the factors $F_1, \ldots, F_k$

Models	T_d	E_d	B_d
F_1	T	$E_d(F_1)$	$B_d(F_1)$
$F_1 + F_2$	T	$E_d(F_1, F_2)$	$B_d(F_2 \mid F_1)$
$\vdots$	$\vdots$	$\vdots$	$\vdots$
$F_1 + \ldots + F_{k-1}$	T	$E_d(F_1, \ldots, F_{k-1})$	$B_d(F_{k-1} \mid F_1, \ldots, F_{k-2})$
$F_1 + \ldots + F_k$	T	$E_d(F_1, \ldots, F_k)$	$B_d(F_k \mid F_1, \ldots, F_{k-1})$

the error and the treatment sums of squares. The detailed steps of the calculation of sums of squares are given in the following algorithm:

Algorithm:

1. Introduce the factor F_1. Calculate T_d, $B_d(F_1)$ and $E_d(F_1)$ for the model composed by F_1 only—Eqs. 8.41, 8.42, 8.43 and 8.44.
2. Introduce the factor F_2. Calculate $E_d(F_1, F_2)$ for the model composed by F_1 and F_2—Eq. 8.44.
3. Calculate $B_d(F_2 \mid F_1) = E_d(F_1) - E_d(F_1, F_2)$—Eq. 8.61.
4. Recursively do the same calculations for the cases $3, \ldots, k - 1$.
5. Introduce the factor F_k. Calculate $E_d(F_1, F_2, \ldots, F_k)$ for the model composed by $F_1, F_2, \ldots, F_k$—Eq. 8.44.
6. Calculate $B_d(F_k \mid F_1, \ldots, F_{k-1})$—Eq. 8.61.

The obtained sums of squares are summarized in Table 8.2. All the procedures described in this section are illustrated by the following two empirical applications.

8.4   Applications

We propose now two empirical applications of the multi-ways analysis of variance in the fuzzy context as introduced in Sect. 8.3. The fuzzy approach followed by the classical one is given. We close the section by a discussion and a comparison between both approaches.

As mentioned in the introduction, we remind that the main objective of this chapter is to show an application of such methodologies and how one could benefit from the idea of ranking fuzzy numbers by the signed distance and the generalized one. We use bootstrap techniques to empirically construct the distribution of the test statistic.

The objectives of the proposed applications are reduced to the following ones:

- To show how one can use the proposed approaches and consequently make a decision related to each factor of the model;
- To understand the influence of the shapes of the modelling fuzzy numbers;
- To understand the influence of the choice of a particular distance to be used in the process of estimation.

8.4.1 Application 1: Nestlé Data Set

For the first application, we use the experimental data set from Nestlé beverages described in Appendix D. The data set is composed of 160 observations resulting from 32 trials of an experiment on a given beverage. We are interested in the variable FILLING collected by experts judging the filling of the beverage. We consider as well the 11 factors $F_1, \ldots, F_{11}$ of the data set. These factors describe different aspects of the beverage. In Berkachy (2018), we exposed a study showing a wide range of Mult-FANOVA examples using this data set in order to understand the pros and cons of applying such tools on the considered experimental set of data.

At the significance level $\delta = 0.05$, we would like to test the hypothesis of equality of population means across all the factor levels. In addition, we suppose that the expert's perception of the filling of the beverage is subject to fuzziness. In the process of testing, this latter will then be taken into account using a convenient treatment approach. We mention that in this application, the design is balanced, i.e. the numbers of observations in all the factor levels are the same. Thus, the order of introduction of the factors does not matter. No effect of this ordering exists. Last but not least, we will perform the procedure of Mult-FANOVA modelled by the Aumann-type expectation.

The first step is to fuzzify the variable FILLING. We then have to choose a corresponding set of membership functions. This choice is usually done according to the experience of the practitioner in the field. Yet, in an ideal context, we would expect to have the membership functions chosen directly by the experts at the moment of the experiment.

For this application, we will use the set of membership functions where we consider non-symmetrical trapezoidal fuzzy numbers for modelling the uncertainty of the variable FILLING. This variable will then be denoted by $\tilde{Z}$. The corresponding α-cuts are seen in Table 8.3.

We propose a Mult-FANOVA model where we include the response variable, i.e. FILLING and the 11 factors existing in our data set. Each of the factors is divided into 2 categories (*Down* or *Up*). We remind that the ordering of the independent variables is not important for our design. In addition, we assume that there is no

Table 8.3 The α-cuts corresponding to the variable FILLING—Sect. 8.4.1

The fuzzy number $\{\tilde{Z}_t\}_{t=0,...,5}$	The left α-cut $(\tilde{Z}_t)_\alpha^L$	The right α-cut $(\tilde{Z}_t)_\alpha^R$
$\tilde{Z}_0 = (0, 0, 2.3)$	$(\tilde{Z}_0)_\alpha^L = 0$	$(\tilde{Z}_0)_\alpha^R = 2.3 - 2.3\,\alpha$
$\tilde{Z}_1 = (0, 0.8, 1.1, 3.25)$	$(\tilde{Z}_1)_\alpha^L = 0.8\alpha$	$(\tilde{Z}_1)_\alpha^R = 3.25 - 2.15\,\alpha$
$\tilde{Z}_2 = (0.9, 1.4, 2.15, 4)$	$(\tilde{Z}_2)_\alpha^L = 0.9 + 0.5\,\alpha$	$(\tilde{Z}_2)_\alpha^R = 4 - 1.85\,\alpha$
$\tilde{Z}_3 = (1.5, 2.5, 3.25, 4.5)$	$(\tilde{Z}_3)_\alpha^L = 1.5 + \alpha$	$(\tilde{Z}_3)_\alpha^R = 4.5 - 1.25\alpha$
$\tilde{Z}_4 = (1.25, 3.7, 4.1, 4.75)$	$(\tilde{Z}_4)_\alpha^L = 1.25 + 2.45\,\alpha$	$(\tilde{Z}_4)_\alpha^R = 4.75 - 0.65\alpha$
$\tilde{Z}_5 = (2.7, 5, 5, 5)$	$(\tilde{Z}_5)_\alpha^L = 2.7 + 2.3\,\alpha$	$(\tilde{Z}_5)_\alpha^R = 5$

interaction between the concerned factors.[3] The complete model explaining the variable FILLING is expressed as follows:

$$\tilde{Z} \sim F_1 + F_2 + F_3 + F_4 + F_5 + F_6 + F_7 + F_8 + F_9 + F_{10} + F_{11}, \qquad (8.62)$$

where $\tilde{Z}$ is the fuzzy response variable, and $F_1, \ldots, F_{11}$ are the 11 factors of the model, i.e. the independent variables.

Consider that the distance d is the generalized signed distance and that the corresponding parameter $\theta^\star$ representing the weight of the shape of the fuzzy numbers is equal to 1. The overall fuzzy mean $\overline{\tilde{Z}}$ of the variable FILLING given in Eq. 8.22 is as follows:

$$\overline{\tilde{Z}}_{\bullet\bullet\bullet} = (1.1981, 2.6513, 3.2053, 4.4031),$$

with its generalized signed distance measured from the fuzzy origin $d_{GSGD}(\overline{\tilde{Z}}_{\bullet\bullet\bullet}, \tilde{0}) = 3.0147$.

We have to calculate after the fuzzy means at each level related to a given factor, as seen in Eqs. 8.23 and 8.24. Since the modelling membership functions are trapezoidal, it is evident to expect that the concerned fuzzy means will be trapezoidal as well. Table 8.4 gives a summary of all the calculated means per factor level (Up and Down for each factor).

It is natural to see that whenever we take into consideration the fuzziness in our data set and we decide to model this uncertainty by fuzzy numbers, the resulting error, treatment and total sums of squares are consequently expected to be composed of fuzzy numbers. However, using the metric d, the different sums of squares E_d, B_d, and T_d calculated as seen in Eqs. 8.41–8.44 will be crisp. For our case, we used the GSGD to obtain crisp values of these elements. We calculated after the treatment mean sums of squares given by the treatment sum of squares of each factor divided by its degree of freedom. A summary of the different calculations related to the variable FILLING is given in Table 8.5. In addition, we estimated by the bootstrap

---

[3]The case with interactions can be similarly conceivable. Models which include interactions can be found in Berkachy (2018).

Table 8.4 Fuzzy means related to the levels *Up* and *Down* of each factor given by the trapezoidal number (p, q, r, s)—Sect. 8.4.1

Factor	Up				Down			
	p	q	r	s	p	q	r	s
F_1	1.32	2.77	3.30	4.43	1.08	2.54	3.11	4.37
F_2	1.22	2.66	3.21	4.40	1.17	2.64	3.20	4.40
F_3	1.44	3.80	4.17	4.77	0.95	1.50	2.24	4.04
F_4	1.14	2.58	3.16	4.39	1.25	2.72	3.25	4.42
F_5	1.31	2.78	3.31	4.44	1.08	2.52	3.10	4.37
F_6	1.21	2.67	3.23	4.42	1.19	2.63	3.18	4.39
F_7	1.15	2.61	3.18	4.39	1.25	2.69	3.23	4.41
F_8	1.15	2.58	3.15	4.38	1.24	2.72	3.26	4.42
F_9	1.14	2.58	3.16	4.39	1.25	2.72	3.25	4.42
F_{10}	1.21	2.66	3.21	4.41	1.18	2.64	3.20	4.40
F_{11}	1.22	2.64	3.20	4.40	1.18	2.66	3.21	4.40

Table 8.5 The summary of the Mult-FANOVA model for the fuzzy variable FILLING—Sect. 8.4.1

Factors	Df	T_d	E_d	B_d	B_d mean Sq.	T_{dist}	T_{dist}^{boot}	$Pr(> T_{dist}^{boot})$
F_1	1	93.45	92.04	1.41	1.41	14.37	11.05	0.00***
F_2	1	93.45	93.43	0.02	0.02	0.25	7.69	0.78
F_3	1	93.45	19.03	74.42	74.42	755.82	279.84	0.00***
F_4	1	93.45	93.05	0.40	0.40	4.06	3.66	0.05**
F_5	1	93.45	91.85	1.60	1.60	16.26	12.65	0.00***
F_6	1	93.45	93.41	0.04	0.04	0.45	2.00	0.29
F_7	1	93.45	93.26	0.19	0.19	1.98	2.54	0.11
F_8	1	93.45	93.08	0.37	0.37	3.74	3.32	0.05**
F_9	1	93.45	93.05	0.40	0.40	4.06	3.34	0.04**
F_{10}	1	93.45	93.44	0.01	0.01	0.12	1.28	0.53
F_{11}	1	93.45	93.45	0.00	0.00	0.02	2.00	0.80
Residuals	148				0.0985			

The significance levels as follows: ***p-value ≤ 0.01; **p-value ≤ 0.05; *p-value ≤ 0.10

technique the test statistic distributions, and the corresponding p-values related to each factor of the model where 1000 iterations are considered.

The hypotheses test of the Mult-ANOVA approach consists to test the equality of means of the response variable by level factor. For the factor F_1, this can be done by comparing the statistic T_{dist} to the $1 - \delta$-quantile of the bootstrap distribution T_{dist}^{boot} written as ζ. The decision related to this factor will then be the following:

Decision The test statistic $T_{dist} = 14.37$ is greater than the bootstrap quantile $\zeta = 11.05$ at the significance level 5%. The null hypothesis H_0 of equality of the group means is then rejected at the 0.05 significance level.

This decision can also be seen from the p-value calculated for each factor of the model. Thus, for the factor F_1, the corresponding p-value indicates a strong presumption against the null hypothesis.

Application with the Classical Approach

It is interesting to compare the results of the fuzzy and the classical approaches of Mult-ANOVA. For this reason, we perform the Mult-ANOVA method with the same setups but in the classical approach, i.e. with the procedure presented in Sect. 8.2. In this case, the assumption of fuzziness of the response variable is not considered. Thus, the variable FILLING is assumed to be precise. We briefly display the results of this analysis. We note that the overall mean of the variable FILLING is 3.0813. Table 8.6 gives the summary of the classical Mult-ANOVA model, the sums of squares with their respective degrees of freedom, the F-statistics and their p-values, etc. The F-statistic of the complete model on 11 and 148 degrees of freedom is 104.5373 with a p-value of 0.

For the factor F_1, the decision by the classical approach is as follows:

Decision Since the test statistic 9.92 is greater than the 95% fractile of the Fisher-Snedecor distribution with 1 and 148 degrees of freedom, i.e. $F_{(0.95; 1, 148)} = 3.9051$, then the decision is to reject the null hypothesis H_0 at the confidence level 95%.

For this factor, the decision is the same for the classical and the fuzzy approaches. Similar analyses should be done for the rest of the factors.

Table 8.6 The summary of the Mult-ANOVA model for the variable FILLING—Sect. 8.4.1

Factors	Df	T	E	B	B mean Sq.	F value	Pr($>$F)
F_1	1	183.94	182.54	1.41	1.41	9.92	0.00***
F_2	1	183.94	183.94	0.01	0.01	0.04	0.83
F_3	1	183.94	25.94	158.01	158.01	1114.90	0.00***
F_4	1	183.94	183.44	0.51	0.51	3.57	0.06*
F_5	1	183.94	182.14	1.81	1.81	12.74	0.00***
F_6	1	183.94	183.89	0.06	0.06	0.40	0.53
F_7	1	183.94	183.79	0.16	0.16	1.10	0.30
F_8	1	183.94	183.44	0.51	0.51	3.57	0.06*
F_9	1	183.94	183.44	0.51	0.51	3.57	0.06*
F_{10}	1	183.94	183.94	0.01	0.01	0.04	0.83
F_{11}	1	183.94	183.94	0.01	0.01	0.04	0.83
Residuals	148				0.1417		

The significance levels as follows: *** p-value ≤ 0.01; ** p-value ≤ 0.05; * p-value ≤ 0.10

Discussion

For the comparison between the defended fuzzy approach and the classical well-known approach concerning each factor of both models, we can clearly see from Tables 8.5 and 8.6 that the decomposition of the sums of squares per independent variable changes between both analyses. This behavior is stronger for the sums of squares and the test statistics than the p-values. The total, treatment and error sums of squares of the fuzzy analysis are relatively smaller than the classical one. In addition, we can easily see that the ratios Error/Total and Treatment/Total are different. We believe that the way these sums are behaving in the fuzzy approach is worth investigating so far, specifically depending on the shapes and spreads of modelling fuzzy numbers, or on the metric used in the process. Even though the test statistics on each factor of both analyses change, we can remark that both of them gave almost the same decision related to rejecting or not the null hypothesis H_0. However, we should highlight that by the fuzzy approach, our presumption against the null hypothesis is stronger than the classical approach. For example, for the factors F_8 and F_9, the null hypothesis is rejected at a significance level of 10% in the conventional Mult-ANOVA model, and at a 5% significance level in the fuzzy one.

Regarding the residuals of the model, the normality and the homogeneity of the variances between groups have been tested as a final step. For the normality of the residuals, we used two inference tests: the Shapiro–Wilks and the Anderson–Darling tests. For the homogeneity of the variances between groups, we used the Bartlett's test applied on each factor. The results are given in Table 8.7.

For the analyses on the residuals of our models related to the normality assumption, the p-values of the Shapiro–Wilks and the Anderson–Darling tests are less than 0.05 implying that the distributions of the residuals are significantly different from the normal distribution on a 5% significance level as seen in Table 8.7 for the fuzzy and the classical approaches. Normality of the residuals cannot be assumed. This result could be due to different facts such as the limited number of observations by category, the few categories, etc. For the skewness and the excess of kurtosis of the distributions of residuals, we can remark that the characteristics related to the fuzzy approach are closer to the ones of the normal distribution than the classical approach. As such, the skewness by the fuzzy approach is close to be 0. Thus, one could suspect that the distribution of the residuals by the fuzzy approach is overall a bit closer to the normal distribution than the one by the conventional approach.

For the assumption of the homogeneity of the variances between groups, we performed the Bartlett's tests for each factor of the model. The results are shown in Table 8.7. The statistics of these tests and their p-values are different between both analyses, implying that the set of variables assuming the homogeneity of the variances between groups is not the same. We remind that a p-value greater than 0.05 refers to the decision of not rejecting the hypothesis of equality of variances between groups. Accordingly, a conclusion is that by the classical approach, we tend to reject more often than the fuzzy approach the hypothesis of homogeneity of the

Table 8.7 The different tests for normality and homogeneity of variances of the residuals obtained from the Mult-FANOVA and the Mult-ANOVA models explaining the variable FILLING— Sect. 8.4.1

	Fuzzy approach		Classical approach	
	Statistic	p-value	Statistic	p-value
Shapiro–Wilks test	0.8168	$6.959e^{-13}$ ***	0.7837	$4.207e^{-14}$ ***
Anderson–Darling test	13.688	$2.2e^{-16}$ ***	8.2562	$2.2e^{-16}$ ***
Bartlett test				
F_1	3.9843	0.0459**	63.7065	$1.4441e^{-15}$ ***
F_2	0.6137	0.4334	15.3422	$8.9688e^{-5}$ ***
F_3	0.0336	0.8545	0.0938	0.7593
F_4	2.5988	0.1069	7.1774	0.0074***
F_5	4.1783	0.0409 **	49.3127	$2.1824e^{-12}$ ***
F_6	0.0645	0.7995	27.9212	$1.2635e^{-7}$ ***
F_7	0.3571	0.5501	4.2002	0.0404**
F_8	0.1884	0.6642	14.5413	0.0001***
F_9	2.4746	0.1157	5.0291	0.0249**
F_{10}	0.0112	0.9158	5.1291	0.0235**
F_{11}	0.0410	0.8395	1.2393	0.2656
Skewness	0.3308		1.9666	
Excess of kurtosis	−1.1906		15.5690	

The significance levels as follows: ***p-value ≤ 0.01; **p-value ≤ 0.05; *p-value ≤ 0.10

variances between the groups. This fact violates a basic assumption of application of the Mult-ANOVA models. Therefore, in terms of the homogeneity of variances, one can conclude that the classical Mult-ANOVA is more pessimistic than the fuzzy one.

In addition, in the same context of using the GSGD in the process of calculations of the sums of squares, we could perform a Tukey's HSD (honestly significant difference) post-hoc test[4] adapted to the fuzzy sets using this distance, in the purpose of completing our study. Well, based on these calculations, we got very close outputs in terms of interpretations to the calculated p-values, which give us indications about whether to reject or not the hypotheses of equality of means between the two groups *Up* and *Down* of the different factors.

8.4.2 Application 2: Finanzplatz Data Set

For the second empirical application, we perform a Mult-ANOVA related to the Finanzplatz data set described in Appendix B. We use the factors "branch" and "size" of the concerned firms. Consider that the variable "the expected state of

---

[4]For more information about this test, Tukey (1949) described the detailed procedure.

business in the next 12 months" contains uncertainty. We would like to know whether the means of this variable differ between the branch and size groups at the significance level 0.05.

The aim is to estimate a Mult-ANOVA model where the response variable is "the expected state of business in the next 12 months," and the independent variables are the size and branch of firms. Note that the variable "size" is composed by three groups (small, medium, big), and the variable "branch" is composed by four groups (banks, insurances, investment and financial service companies). For this data set, the design is unbalanced. The ordering of the integration of the covariates has then an influence in the model. In addition, we suppose that no interactions between the factors exist. We would like to take into consideration the uncertainty contained in the response variable. Let us denote the fuzzy version of this latter by $\tilde{S}$. The concerned model is then given as follows:

$$\tilde{S} \sim \text{size} + \text{branch}, \tag{8.63}$$

where $\tilde{S}$ is the fuzzy response variable, size and branch are the factors of the model.

For the fuzzification step, we propose to consider 2 cases in order to analyze the difference between them. For the first case, we chose arbitrarily to model each observation of the variable S by considering symmetrical triangles of spread 2 units. Since the idea is to use the procedure of Mult-FANOVA by the Aumann-type expectation, we propose to use the GSGD, for which the parameter $\theta^\star$ is equal to 1. We remind that using the considered set of membership functions, if the distance d is the signed distance, the method gives results exactly the same as the classical approach as seen in Sect. 8.3.1.

The overall fuzzy mean $\overline{\tilde{S}}$ of the response variable expressed by the Eq. 8.22 is given by

$$\overline{\tilde{S}}_{\bullet\bullet\bullet} = (2.5812, 3.5812, 4.5812).$$

Its generalized signed distance measured from the fuzzy origin is $d_{GSGD}(\overline{\tilde{S}}_{\bullet\bullet\bullet}, \tilde{0}) = 3.5928$. One has to calculate now the fuzzy means related to each level factor. These calculations will lead us to compute the total, the treatment and the error sums of squares. However, it is important to remark that the ordering is important in this case. Thus, we have to calculate the related sequential sums of squares as given in the algorithm of Sect. 8.3.2. Using the considered distance, the resulting sums of squares related to each factor with the corresponding degrees of freedom are shown in Table 8.8. For this application, we consider the bootstrap Algorithms 1 and 2 given in Sect. 8.3 with 1000 iterations. The test statistics T_{dist}, the bootstrap ones $T_{\text{dist}}^{\text{boot}}$, and the corresponding bootstrap p-values are presented in Table 8.9. In addition, we mention that the test statistic of the model is 1.8518, where the bootstrap p-value by the Algorithm 1 is 0.14 and the one by the Algorithm 2 is 0.13.

Table 8.8 The summary of the Mult-FANOVA model for the fuzzy variable "expected state of business in the next 12 months"—Case of symmetrical FN with the GSGD distance—Sect. 8.4.2

Factors	Df	T_S	E_S	B_S	B_S mean Sq.
Size	2	144.9573	$E_S(\text{size}) = 143.6574$	$B_S(\text{size}) = 1.2999$	0.6499
Branch	3	144.9573	$E_S(\text{branch, size}) = 142.6401$	$B_S(\text{branch} \mid \text{size}) = 1.0171$	0.3391
Residuals	228				0.6256

The significance levels as follows: *** p-value ≤ 0.01; ** p-value ≤ 0.05; * p-value ≤ 0.10

Table 8.9 The summary of the test statistics related to the Mult-FANOVA model for the fuzzy variable "expected state of business in the next 12 months" using the bootstrap Algorithms 1 and 2—Case of symmetrical FN with the GSGD distance—Sect. 8.4.2

Factors	T_{dist}	Algorithm 1		Algorithm 2	
		$T_{\text{dist}}^{\text{boot}}$	$\Pr\left(> T_{\text{dist}}^{\text{boot}}\right)$	$T_{\text{dist}}^{\text{boot}}$	$\Pr\left(> T_{\text{dist}}^{\text{boot}}\right)$
Size	1.0389	4.15	0.17	3.95	0.17
Branch	0.5419	5.25	0.34	5.37	0.33

The significance levels as follows: *** p-value ≤ 0.01; ** p-value ≤ 0.05; * p-value ≤ 0.10

By the results of Table 8.9, we could conclude that for both factors, we cannot reject the hypothesis of equality of means between groups, particularly by the fact that the bootstrap p-values are higher than the significance level 0.05. However, for the whole model, it is clear that the p-value is greater than the threshold of 5%. Thus, our model seems to be not statistically significant.

One could expect that by changing the chosen metric, we could get different values for the sums of squares, or different decisions related to the factors of the model. For our present case using symmetrical triangles, we replicated the same analysis with the different distances described in Chap. 4. The results obtained using all the distances and from the classical approach are very similar, almost all at the order of 10^{-3}. The present case with fuzziness seems to be a reproduction of the classical model. This conclusion is proved in Sect. 8.3.1 for the signed distance.

In addition, it is clear to see that the bootstrap techniques given in Algorithms 1 and 2 give very close results in terms of the p-values and the decisions related to the model and to each factor. Thus, the choice between both algorithms seems to be not influencing the decisions made.

From another side, we propose now to use non-symmetrical membership functions for the fuzzification step. Consider then the trapezoidal fuzzy terms modelling the different answers of the variable "expected state of business in the next 12 months" given as follows:

$$\tilde{S}_1 = (0, 0.5, 1.1, 3.7),$$

$$\tilde{S}_2 = (0.9, 1.8, 2.45, 4.5),$$

$$\tilde{S}_3 = (1.9, 2.6, 3.65, 4.9),$$

Table 8.10 The summary of the Mult-FANOVA model for the fuzzy variable "expected state of business in the next 12 months"—Case of non-symmetrical FN with the GSGD distance—Sect. 8.4.2

Factors	Df	T_S	E_S	B_S	B_S mean Sq.
size	2	54.2188	$E_S(\text{size}) = $ 53.4102	$B_S(\text{size}) = 0.8086$	0.4043
branch	3	54.2188	$E_S(\text{branch, size}) = $ 52.6736	$B_S(\text{branch} \mid \text{size}) = $ 0.7366	0.2455
Residuals	228				0.2310

The significance levels as follows: ***p-value ≤ 0.01; **p-value ≤ 0.05; *p-value ≤ 0.10

Table 8.11 The summary of the test statistics related to the Mult-FANOVA model for the fuzzy variable "expected state of business in the next 12 months" using the bootstrap Algorithms 1 and 2—Case of non-symmetrical FN with the GSGD distance—Sect. 8.4.2

Factors	T_{dist}	Algorithm 1		Algorithm 2	
		$T_{\text{dist}}^{\text{boot}}$	$\Pr\left(> T_{\text{dist}}^{\text{boot}}\right)$	$T_{\text{dist}}^{\text{boot}}$	$\Pr\left(> T_{\text{dist}}^{\text{boot}}\right)$
size	1.75	8.90	0.20	8.19	0.20
branch	1.06	5.88	0.36	6.08	0.34

The significance levels as follows: ***p-value ≤ 0.01; **p-value ≤ 0.05; *p-value ≤ 0.10

$$\tilde{S}_4 = (1.25, 3.7, 4.3, 4.85),$$

$$\tilde{S}_5 = (2.7, 5, 5, 5).$$

Table 8.10 provides a summary of the Mult-FANOVA table related to our analysis. For both factors, the related p-values lead us to not reject the equality of means between the groups. In addition, both algorithms gave p-values close to the case shown in Table 8.8. The test statistic of the model T_{dist} is 3.3442 where the bootstrap quantile by the Algorithm 1, for instance, is 16.33 with a p-value of 0.1498. This p-value indicates that our model does not appear to be statistically significant at the significance level 0.05.

The interpretations of this case using non-symmetrical fuzzy numbers is exactly the same as the case using symmetrical ones. In particular, by Table 8.11 we could clearly remark that the choice between the Algorithms 1 and 2 does not have any effect on the decisions made. Therefore, both algorithms could be useful in such situations. In further research, it would be interesting to investigate cases where the difference between both could eventually be glaring.

From another side, we know that in the conventional approach, a given test statistic in such analysis of variance is considered to be derived from a Fisher distribution. Since this latter cannot be insured in the fuzzy approach, the bootstrap technique has been used. However, one could suspect that a particular relationship between the bootstrap distributions of test statistics and the Fisher theoretical distribution could exist. For this reason, we propose the qq-plots of the bootstrap distributions with the corresponding theoretical Fisher-Snedecor distribution as

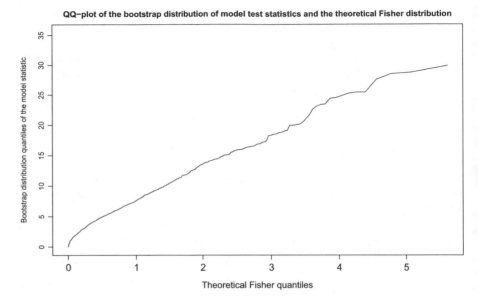

Fig. 8.1 QQ-plot of the bootstrap distribution of the model test statistic and the theoretical Fisher distribution—Sect. 8.4.2

shown in Figs. 8.1 and 8.2. From these latter, it seems that effectively a possible relation between the proposed distributions could be remarked.

Let us now perform a Mult-FANOVA where we calculate numerically the fuzzy sums of squares. It would then be interesting to proceed the fuzzy approach by the heuristic decision rule presented in Sect. 8.3. The decisions will be made according to each variable of the model. Consider the variable "branch" for this test, and consider also the fuzzy membership functions to be symmetrical triangular as described above. Note that for these calculations the difference between fuzzy numbers is approximated. Following these setups, we calculated the Mult-FANOVA model and we got the fuzzy numbers defining, respectively, the fuzzy treatment and the fuzzy weighted residuals sums of squares, shown in Fig. 8.3.

In order to make a decision regarding the equality of means between the different branches of the firms, we have to calculate either the surfaces under the membership functions of the fuzzy numbers, or either the signed distances of these latter, as given in Eqs. 8.33–8.38. Let us use the signed distance measured from $\tilde{0}$. By the Algorithm 1 of the bootstrap technique presented in Sect. 8.3, let us consider κ to be the 95%-quantile of the bootstrap distribution of the test statistic for the factor branch. At 1000 iterations, we found that $\kappa = 1.4315$. We get the following results:

$$d_{SGD}(\overline{\tilde{E}}^{*}, \tilde{0}) = d_{SGD}(\overline{\tilde{E}} \times \kappa, \tilde{0}) = 2.013;$$

$$d_{SGD}(\overline{\tilde{B}}, \tilde{0}) = 0.3077,$$

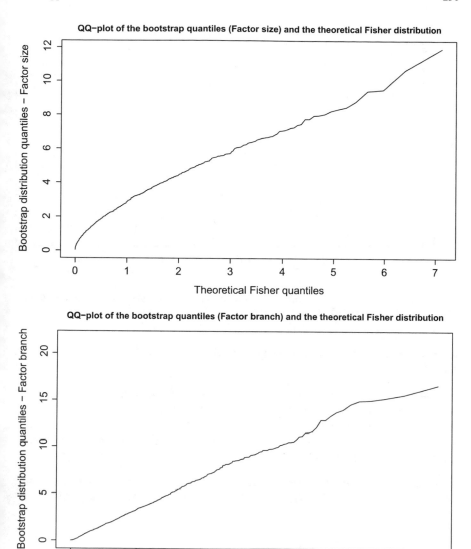

Fig. 8.2 QQ-plots of the bootstrap distributions of the test statistics related to the factors size and branch, respectively, with the theoretical Fisher distribution—Sect. 8.4.2

where the trapezoidal fuzzy numbers $\overline{\tilde{E}} = (1.1908, 1.2055, 1.6069, 1.6216)$ and $\overline{\tilde{B}} = (0.2436, \quad 0.2758, 0.3391, 0.3722)$ are, respectively, the fuzzy residuals mean sum of squares and the fuzzy treatment mean sum of squares related to the factor branch.

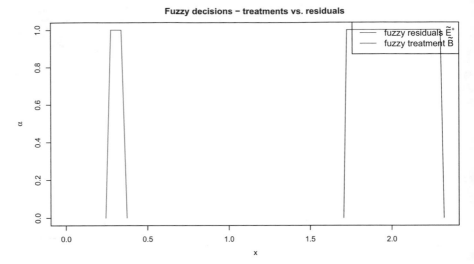

Fig. 8.3 Fuzzy means of the treatment and the residuals sums of squares related to the variable "branch" of the Mult-FANOVA model—Sect. 8.4.2

According to these distances, we are now able to make a decision. It is then given as follows:

Decision Rule Since $d_{SGD}(\overline{\tilde{E}}^*, \tilde{0}) > d_{SGD}(\overline{\tilde{B}}, \tilde{0})$, then we tend to not reject the null hypothesis of equality of means between groups of the variable branch at the significance level $\delta = 0.05$.

Thus, we deduce that the different economic branches might have close visions related to the next 12 months with respect to their means.

We could also calculate the contribution of both sums of squares to the total sums of squares of the Mult-FANOVA model as given in Eqs. 8.39 and 8.40, and we get

$$C(\overline{\tilde{E}}^*) = \frac{\mathrm{Surf}(\overline{\tilde{E}}^*)}{\mathrm{Surf}(\overline{\tilde{B}}) + \mathrm{Surf}(\overline{\tilde{E}}^*)} = 0.867,$$

$$C(\overline{\tilde{B}}) = \frac{\mathrm{Surf}(\overline{\tilde{B}})}{\mathrm{Surf}(\overline{\tilde{B}}) + \mathrm{Surf}(\overline{\tilde{E}}^*)} = 0.133.$$

The previous decision rule can also be viewed by the big gap between the contributions of both fuzzy sums. On the other hand and in terms of the bootstrap distribution of the test statistic of this analysis, we could see from the qq-plots shown in Fig. 8.4 that the estimated distribution seems to be somehow lacking from being drawn from a Fisher-Snedecor distribution. This could be due, for instance, to the fuzzy approximations.

QQ-plot of the bootstrap quantiles (Factor size) and the theoretical Fisher distribution

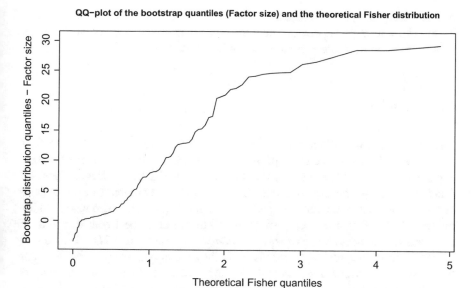

QQ-plot of the bootstrap quantiles (Factor branch) and the theoretical Fisher distribution

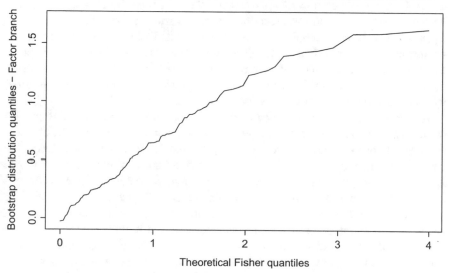

Fig. 8.4 QQ-plots of the bootstrap distributions of the test statistics related to the factors size and branch, respectively, with the theoretical Fisher distribution—Sect. 8.4.2

As a conclusion, we could say that the Mult-FANOVA approach using a pre-defined distance, and the one preserving the fuzzy nature of the sums of squares gave exactly the same decision for the variable "branch." In further situations, the decision relying on the fuzzy type sums might be difficult to make since overlapping between these fuzzy sums of squares related to the treatment and the residuals could be somehow strongly expected. Thus, in such situations, a heuristic decision rule, similar to the proposed one, could be a prudent idea.

8.5   Conclusion

The aim of this chapter is to present an approach of the Mult-ANOVA in the fuzzy environment, based on the concept of fuzzy regression where the signed distance and the generalized one are used as ranking methods. For this model composed by a range of factors, we assumed that the uncertainty is contained in the response variable, and in the model parameters. Two methods were introduced for this purpose. In the first procedure, we proposed to preserve the fuzzy nature of the numerically calculated sums of squares, and we exposed a heuristic decision rule related to the corresponding test statistic. In the second one, we used the Aumann-type expectation where a convenient distance is needed. As a particular case, we used the signed distance with a set of membership function having the particularity of being symmetrical around the core of the fuzzy number. The method related to the Aumann expectation returned exactly the results of the classical approach. Furthermore, we showed the procedure of calculations of the sequential sums of squares when the considered design is not balanced. To illustrate our approaches, we displayed two empirical applications where the objectives were to understand the influence of the choice of symmetry in the shapes of the modelling fuzzy numbers, and the use of the distances on the Mult-FANOVA models. For this latter, we completed our study by the classical approaches.

An interesting finding of this chapter is that the more we get closer to precise modelling fuzzy numbers, the more we tend to obtain results similar to the classical approach. In other terms, the more we suppose that the fuzziness contained in the data is high, the more we should expect to get different results than the conventional approach. Thus, the fuzzy approach seems to be a reproduction of the classical one in terms of decisions made, but with more flexibility on the treatment of uncertainty of the response variable. One could say that the fuzzy approach is in some sense a generalization of the classical one. In the cases exposed in this chapter, we overall got similar decisions related to the null hypothesis, by the classical and the fuzzy approaches. Although the decisions made using both approaches are the same, one has to clearly see that the decomposition of the sums of squares is different. An investigation of this result would be interesting in order to reveal the analytical characteristics of these sums.

To confirm our hypotheses, we performed in previous researches the same Mult-FANOVA model using different symmetrical and non-symmetrical shapes of modelling fuzzy numbers. The conclusions of the tested cases are comparable. The interpretations of the parameter vector and their p-values, the residuals and the predicted values in both approaches differ sometimes. For the fuzzy case, regarding the assumption of the equality of variance, we see that whenever we get farther from symmetrical shapes, one could expect to get slightly different decisions according to the hypothesis postulated for the model. However, using the classical approach, we tend more strongly to reject the normality of the residuals of the Mult-ANOVA model than the fuzzy approach. This latter can be seen as a disadvantage of the classical Mult-ANOVA compared to the fuzzy one. A deeper investigation of the behavior of the residuals is needed.

Finally, we have to highlight that the main purpose of adopting such fuzzy approaches is the ability to efficiently take into account the uncertainty and fuzziness occurring in data sets. Hence, additional tests have to be performed in order to diagnose the influence of different variations of the choice of the distance, the shape, and the spread of the fuzzy numbers, the number of observations per category and the number of categories, the algorithm chosen for bootstrap, etc. in the purpose of being able to stress the use of each one of these latter.

Appendix

This Appendix is devoted to the description of the Nestlé data set as follows:

D   Description of the Nestlé data base

This data base is collected by the beverage section of the company Nestlé. An expert is asked to rate different aspects of a given confidential beverage. The collected data set is composed by 160 observations resulting from 32 trials of an experiment on the beverage. The full data base comprised of 13 variables among them 11 factors having 2 levels each given by "Up" and "Down," and the variables FLOW and FILLING. A description of the variables constituting this data set is given in Table 8.12.

Table 8.12 The variables of the Nestlé data set—Sect. D

Variables	Possible answers	Description
F_1	Factor with two levels: Down (1) and Up (2)	Aspect 1 of the beverage
F_2	Factor with two levels: Down (1) and Up (2)	Aspect 2 of the beverage
F_3	Factor with two levels: Down (1) and Up (2)	Aspect 3 of the beverage
F_4	Factor with two levels: Down (1) and Up (2)	Aspect 4 of the beverage
F_5	Factor with two levels: Down (1) and Up (2)	Aspect 5 of the beverage
F_6	Factor with two levels: Down (1) and Up (2)	Aspect 6 of the beverage
F_7	Factor with two levels: Down (1) and Up (2)	Aspect 7 of the beverage
F_8	Factor with two levels: Down (1) and Up (2)	Aspect 8 of the beverage
F_9	Factor with two levels: Down (1) and Up (2)	Aspect 9 of the beverage
F_{10}	Factor with two levels: Down (1) and Up (2)	Aspect 10 of the beverage
F_{11}	Factor with two levels: Down (1) and Up (2)	Aspect 11 of the beverage
FLOW	Four linguistic answers: 0 to 3.	Flow of the liquid
FILLING	Six linguistic answers: 0 to 5.	Filling of the liquid in the container

References

Berkachy, R. (2018). *Fuzzy Multivariate Analysis of Variance using the Signed Distance; Application on a Data Set, Nestlé Beverages*, Technical Report Nestlé beverages Orbe.

Berkachy, R., & Donzé, L. (2018). Fuzzy one-way ANOVA using the signed distance method to approximate the fuzzy product. In Collectif LFA (Eds.), *LFA 2018 - Rencontres francophones sur la Logique Floue et ses Applications Cépaduès* (pp. 253–264).

Bonferroni, C. E. (1935). *Il calcolo delle assicurazioni su gruppi di teste*. Tipografia del Senato. https://books.google.ch/books?id=xnQ4cgAACAAJ

Buckley, J. J., & Jowers, L. J. (2008). Fuzzy linear regression I. In *Monte Carlo Methods in Fuzzy Optimization* (pp. 117–125). Berlin: Springer. ISBN: 978-3-540-76290-4. https://doi.org/10.1007/978-3-540-76290-4_11

Celmins, A. (1987). Least squares model fitting to fuzzy vector data. *Fuzzy Sets and Systems, 22*(3), 245–269. ISSN: 0165-0114. https://doi.org/10.1016/0165-0114(87)90070-4. http://www.sciencedirect.com/science/article/pii/0165011487900704

Diamond, P. (1988). Fuzzy least squares. *Information Sciences, 46*(3), 141–157. ISSN: 0020-0255. https://doi.org/10.1016/0020-0255(88)90047-3. http://www.sciencedirect.com/science/article/pii/0020025588900473

Gil, M. A., Montenegro, M., González-Rodríguez, G., Colubi, A., & Casals, M. R. (2006). Bootstrap approach to the multi-sample test of means with imprecise data. *Computational Statistics & Data Analysis, 51*(1), 148–162. ISSN: 0167-9473. https://doi.org/10.1016/j.csda.2006.04.018. http://www.sciencedirect.com/science/article/pii/S0167947306001174

Gonzalez-Rodriguez, G., Colubi, A., & Gil, M. A. (2012). Fuzzy data treated as functional data: A one-way ANOVA test approach. *Computational Statistics & Data Analysis, 56*(4), 943–955.

Gonzalez-Rodriguez, G., Colubi, A., Gil, M. A., & D'Urso, P. (2006). An asymptotic two dependent samples test of equality of means of fuzzy random variables. In: *Proceedings of the 17th IASC-ERS*.

Guo, P., & Tanaka, H. (2006). Dual models for possibilistic regression analysis. *Computational Statistics & Data Analysis, 51*(1), 253–266. ISSN: 0167-9473. https://doi.org/10.1016/j.csda.2006.04.005. http://www.sciencedirect.com/science/article/pii/S0167947306000910

Hotelling, H. (1951). A generalized T test and measure of multivariate dispersion. In: *Proceedings of the Second Berkeley Symposium on Mathematical Statistics and Probability (Univ of Calif Press, 1951)* (pp. 23–41).

Jiryaei, A., Parchami, A., & Mashinchi, M. (2013). One-way Anova and least squares method based on fuzzy random variables. *TJFS: Turkish Journal of Fuzzy Systems, 4*(1), 18–33.

Kang, M.-K., & Han, S.-I. (2004). Multivariate analysis of variance for fuzzy data. *International Journal of Fuzzy Logic and Intelligent Systems, 4*(1), 97–100.

Lawley, D. N. (1938). A generalization of Fisher's Z test. *Biometrika, 30*(1–2), 180–187. ISSN: 0006-3444. https://doi.org/10.1093/biomet/30.1-2.180. http://oup.prod.sis.lan/biomet/article-pdf/3/1-2/180/422459/30-1-2-180.pdf

Montenegro, M., Gonzalez-Rodriguez, G., Gil, M. A., Colubi, A., & Casals, M. R. (2004). Introduction to ANOVA with fuzzy random variables. In *Soft methodology and random information systems* (pp. 487–494). Berlin: Springer.

Nather, W. (2006). Regression with fuzzy random data. *Computational Statistics & Data Analysis, 51*(1), 235–252. ISSN: 0167-9473. https://doi.org/10.1016/j.csda.2006.02.021

Parchami, A., Nourbakhsh, M., & Mashinchi, M. (2017). Analysis of variance in uncertain environments. *Complex & Intelligent Systems, 3*(3), 189–196

Pillai, K. C. S. (1965). On the distribution of the largest characteristic root of a matrix in multivariate analysis. *Biometrika, 52*, 405–414. https://doi.org/10.1093/biomet/52.34.405

Ramos-Guajardo, A. B., & Lubiano, M. A. (2012). K-sample tests for equality of variances of random fuzzy sets. *Computational Statistics & Data Analysis, 56*(4), 956–966. ISSN: 0167-9473. https://doi.org/10.1016/j.csda2010.11025. http://www.sciencedirect.com/science/article/pii/S0167947310004536

Tanaka, H., Uejima, S., & Asai, K. (1982). Linear regression analysis with fuzzy model. *IEEE Transactions on Systems Man, and Cybernetics, 12*(6), 903–907. ISSN: 0018-9472. https://doi.org/10.1109/TSMC.1982.4308925

Tukey, J. W. (1949). Comparing individual means in the analysis of variance. *Biometrics, 5*(2), 99–114. ISSN: 0006-341X, 1541-0420. http://www.jstor.org/stable/3001913.

Wilks, S. S. (1932). Certain generalizations in the analysis of variance. *Biometrika, 24*, 471–494. https://doi.org/10.1093/biomet/24.3-4.471

Wilks, S. S. (1946). Sample criteria for testing equality of means equality of variances and equality of covariances in a normal multivariate distribution. *The Annals of Mathematical Statistics, 17*(3), 257–281. ISSN: 00034851. http://www.jstor.org/stable/2236125

Wu, H.-C. (2007). Analysis of variance for fuzzy data. *International Journal of Systems Science, 38*(3), 235–246.

Part III
An R Package for Fuzzy Statistical Analysis: A Detailed Description

Chapter 9
`FuzzySTs`: Fuzzy Statistical Tools: A Detailed Description

Despite the fact that the approaches presented in the first two parts are appealing from a theoretical point of view, being able to apply them in statistical analysis with real data sets seems to be another level of challenge. These analyses are always computed by functions describing the corresponding procedures in the R statistical software found in R Core Team (2015). Several researchers have already developed R functions used in similar contexts. We briefly review some of them. For basic introduction of fuzzy numbers, the package `FuzzyNumbers` described in Gagolewski (2014) is very known in the field. It provides methods to deal with basic fuzzy numbers. In addition, based on the extension principle, this package gives the required tools to perform arithmetic operations on fuzzy numbers. This task can be accomplished by an approximation or by exact fuzzy values, often based on trapezoidal or piecewise continuous fuzzy numbers. These operations are described theoretically in Chap. 2.

Several packages have treated the subject of linguistic terms by fuzzy tools. Many of them considered fuzzy systems as presented in Chap. 3. We note, for example, the package `FuzzyToolkitUoN` described in Knott et al. (2013), which consists of a framework of fuzzy inference systems based on different inference and aggregation rules, and using some known defuzzification operators. In the same way, the package `lfl` ("Linguistic Fuzzy Logic") presented in Burda (2018) and Burda (2015) provides algorithms of computations of fuzzy rule-based systems.

From another side, from a statistical programming point of view, the approach by Féron–Puri and Ralescu of Sect. 5.2.2 is more prevalent than the Kwakernaak–Kruse and Meyer approach of Sect. 5.2.1. Researchers tended more often to implement the statistical measures by the first approach for multiple reasons, such that its practical and light computational burden, its crisp nature, and the related facility to be implemented and interpreted, etc. However, for the Kwakernaak–Kruse and Meyer approach, the computations are known to be more tedious, since one has to deal with fuzzy type sets all along the calculations. This fact increases the computational burden of such tasks.

© The Author(s), under exclusive license to Springer Nature Switzerland AG 2021
R. Berkachy, *The Signed Distance Measure in Fuzzy Statistical Analysis*,
Fuzzy Management Methods, https://doi.org/10.1007/978-3-030-76916-1_9

We know that the Féron–Puri and Ralescu methodology needs a distance as discussed in Chap. 5. Some distances described in Chap. 4 are implemented in available packages. Trutschnig and Lubiano (2018), the authors of the package SAFD, i.e. "Statistical Analysis of Fuzzy Data," provide an implementation of the Bertoluzza distance described in Sect. 4.2. The authors also expose univariate statistical tools where they consider fuzzy data having the shape of a polygon. Using this package, basic operations on fuzzy numbers can be computed. We note, for example, the addition, the scalar multiplication, the mean, the median etc. The sample variance is also implemented as proposed in Sect. 5.4, where the Bertoluzza distance is used. The authors finally propose one and multi-sample bootstrap tests for the equality of means as given in Gil et al. (2006). For the package FuzzyStatTra given in Lubiano and de la Rosa de Saa (2017), the authors exposed some fundamental functions of statistical summaries using trapezoidal fuzzy numbers. As an instance, they proposed ways to calculate the mean, the variance, the median, and some other scale measures based on the mid/spr and the ϕ-wabl/ldev/rdev metric. These ideas were particularly developed by Blanco-Fernandez et al. (2013) and Sinova et al. (2013). This package also introduced the ρ_1 distance, in addition to the implemented versions of the so-called "M-estimators of scale with loss function" and the "Gini-Simpson diversity index," presented in Sinova et al. (2016).

From another side, the maximum likelihood method was often involved in R functions. For example, the estimations by this method were used extensively in the context of the EM-algorithm with incomplete data. For the fuzzy environment, the package EM.Fuzzy described in Parchami (2018b) has the aim of providing a tool to apply these algorithms based on the maximum likelihood method using triangular and trapezoidal fuzzy numbers. The unknown parameter, assumed to be a fuzzy number, is then estimated by the corresponding functions as proposed in Denoeux (2011).

For the topic of statistical inference described in Chap. 6, multiple fuzzy programming packages have been introduced. Parchami (2017) proposed a package called Fuzzy.p.value based on the papers Filzmoser and Viertl (2004) and Parchami et al. (2010), where the main goal is to supply the membership function of a fuzzy type p-value, based either on fuzzy hypotheses or on data. Nevertheless, the authors of this package asserted that if the required p-value is fuzzy, then it would be convenient to assume that the significance level is fuzzy as well. For this purpose, they proposed to compare the obtained fuzzy p-value to the significance level by a corresponding evaluation method.

Some packages were proposed to compute hypotheses tests related to the equality of means as mentioned in Chap. 8. Beside the package SAFD, which introduces bootstrap tests for the equality of means, Parchami (2018a) proposed a package, the so-called ANOVA.TFNs, for univariate ANOVA implementations, as described theoretically in Parchami et al. (2017). This package, mainly reduced to the function FANOVA for the fuzzy one-way ANOVA procedure, gives the test statistic with its corresponding p-value summarized in an ANOVA table. A generalization to the multi-factor case as proposed in Chap. 8 is an asset.

The complex and expensive computations of fuzzy type numbers occurring in statistical methods are two big disadvantages of applying fuzzy logic in such contexts. Convenient computation methods taking into account this burden are needed to be able to overcome the problem from one side. From another one, a package resuming, upgrading, and generalizing all these functions on a more compact programming interface is required.

We created an R package with numerous functions. Our main aims are to

- implement all the theoretical tools developed previously,
- have a coherent programming environment,
- have a user-friendly analytical procedures,
- propose a panoply of calculation methods, and
- construct a complete self-sustaining package.

We propose then a package of R functions called FuzzySTs described in Berkachy and Donzé (2020), which aims to support the different theoretical approaches. To complement the package FuzzyNumbers as an instance, we introduce Gaussian and Gaussian bell fuzzy numbers to the class of fuzzy numbers. In terms of fuzzy arithmetic operations with respect to the extension principle given in Chap. 2, we propose a numerical procedure to compute the fuzzy difference and the fuzzy square, where the polynomial forms of their α-cuts are implemented. According to our package, when the data are considered to be fuzzy rather than crisp, two ways of fuzzification given by two R functions are proposed. The choice of functions is related to the shape of the modelling fuzzy numbers, such that a function based on trapezoidal fuzzy numbers can be defined by their quadruples, and another one can be given by their numerical α-level sets. Note that our functions have the great advantage of being able to consider any shape of fuzzy numbers.

Furthermore, since by the Féron–Puri and Ralescu approach, a distance is needed in the process, and then we provide a function performing the different possibilities of distances illustrated in Chap. 4. Statistical measures such as the mean, the variance, the moment, etc. are consequently implemented. The contribution of this package is that, in compliance with the concepts given in Chap. 5, a numerical method of computation of the fuzzy variance by the Kwakernaak–Kruse and Meyer approach is introduced, in addition to the Féron–Puri and Ralescu one. We propose, after, a function that enables to test fuzzy hypotheses and/or fuzzy data by both approaches using convenient functions. These functions cover all the notions of Chap. 6. In other words, for the Kwakernaak–Kruse and Meyer approach, our functions allow to draw the fuzzy decisions based on a hypotheses test by fuzzy confidence intervals as given in Sect. 6.3. All the situations of intersection between the fuzzy confidence intervals and the fuzzy hypotheses are taken into account. In addition, it is up to the user to choose the method of calculation of the confidence interval, i.e. using the traditional expression or the maximum likelihood method. In the same way, for the Féron–Puri and Ralescu approach, the fuzzy p-value can be easily computed and defuzzified. Related decisions are easily listed.

For the idea of the global and individual evaluations expressed in Chap. 7, no previous implementations in similar sense were found in the R repository. For this reason, we propose functions accomplishing this task, where our computations can be used with all the abovementioned distances. We have to note that the indicators of information rate are also exposed. In addition, the weight factor and the missingness in the data can be taken into account by corresponding weight readjustment tools as shown in Sect. 7.2. Finally, for the concepts of Chap. 8, we introduce numerical estimations of multi-ways analysis of variance models in the fuzzy context. The unbalance in the considered designs is also foreseen.

To sum up, we can obviously say that the defended package provides a complete framework of the tools described in the theoretical and applied sections, including a big panoply of possibilities in terms of computations. This chapter intends then to show the documentation of the R functions implemented to follow up the theoretical and application parts. For each function, we give a detailed description followed by the corresponding syntax, its arguments, and the expected output. Note that the functions described in this chapter are designed and developed from scratch in compliance with the needed concepts of the theoretical part. However, these functions depend only on the package FuzzyNumbers presenting the basic notions of introduction of fuzzy numbers, on the package polynom for computing polynoms, and on the package EM.Fuzzy for estimating MLE estimators by the EM-algorithm in addition to the basic functions of R. Thereafter, they have been validated on multiple data bases from different sources. For our current case, simple examples illustrate each proposed function. In order to be more user-friendly, note that eventual optimization and enhancement are needed from a technical point of view.

9.1   Description of the Package

The objective of the package FuzzySTs is to provide an implementation of the theoretical approaches presented in the first two parts. Despite the fact that our functions are all coherent and constitute a whole, in order to present our package, we describe each function in the same order as their content appears in the previous chapters. For each function, we expose a detailed description, its arguments, the obtained outputs, and the different function dependencies. They are finally illustrated by simple examples. We have to mention that the numerical integrations are mainly made by the Simpson method.

9.1.1   Dependencies

FuzzyNumbers, polynom, EM.Fuzzy

9.1.2 List of Functions

9.1.3   Description of the Functions

| is.alphacuts | *Verifies if a matrix is a set of left and right α-cuts* | 1 |

Description

This function verifies if a given matrix can be a matrix of numerical α-cuts. It is then verified if the considered matrix has the following requirements: it should be composed of 2 vectors in columns with no missing values, for which the lengths of both vectors are equal. The first vector should be increasing and named "L" as left, and the second one should be decreasing and named "U" as upper. The last value of the first vector should be less than the first value of the second vector.

This function intentionally requires a form of the set of numerical α-cuts similar to the output of the function `alphacut` of a fuzzy number by the package `FuzzyNumbers` described in Gagolewski (2014).

Usage

```
is.alphacuts (matrix)
```

Arguments

matrix   a matrix of 2 equal length columns with no NA.

Value

Returns a value TRUE if the concerned object can be a set of numerical left and right α-cuts, FALSE otherwise.

See Also

```
is.fuzzification 7, is.trfuzzification 8, nbreakpoints 2,
tr.gfuzz 9.
```

Examples

```
mat <- matrix(c(1,2,3,7,6,5), ncol = 2)
is.alphacuts(mat)
```

---

nbreakpoints   *Calculates the number of breakpoints of a numerical*
matrix of α-cuts   2

Description

For a given fuzzy number object expressed by its numerical α-cuts such that
is.alphacuts = TRUE, the function nbreakpoints calculates the number
of breakpoints chosen to numerically construct these α-cuts. In other terms, it is the
number of sample points on which the membership functions are evaluated.

Usage

nbreakpoints (object)

Arguments

object   a matrix of numerical α-cuts or a 3-dimensional array. No NA is allowed.

Value

Returns a numerical positive integer.

See Also

is.alphacuts 1, is.fuzzification 7, is.trfuzzification 8,
tr.gfuzz 9.

Examples

```
X <- TrapezoidalFuzzyNumber(1,2,3,4)
alpha.X <- alphacut(X, seq(0,1,0.01))
nbreakpoints(alpha.X)
```

GaussianFuzzyNumber	*Creates a Gaussian fuzzy number*	3

Description

This function creates a Gaussian fuzzy number from the class of Gaussian fuzzy numbers as presented in Definition 2.2.3. This fuzzy number is given by its numerical α-cuts such that is.alphacuts = TRUE. The smoothness of the constructed curve is defined by the step, the margin, and the precision fixed by default to 0.01, [3*sigma;3*sigma], and 4.

Usage

```
GaussianFuzzyNumber   (mean, sigma, alphacuts = FALSE, margin =
                      c(3*sigma,                3*sigma), step
                      = 0.01, breakpoints = 100, precision=4,
                      plot=FALSE)
```

Arguments

mean	a numerical value of the parameter μ of the Gaussian curve.
sigma	a numerical value of the parameter σ of the Gaussian curve.
alphacuts	fixed by default to "FALSE". No α-cuts are printed in this case.
margin	an optional numerical couple of values representing the range of calculations of the Gaussian curve written as [mean - 3*sigma; mean + 3*sigma] by default.
step	a numerical value fixing the step between two knots dividing the interval [mean - 3*sigma; mean + 3*sigma].
breakpoints	a positive arbitrary integer representing the number of breaks chosen to build the numerical α-cuts. It is fixed to 100 by default.
precision	an integer specifying the number of decimals for which the calculations are made. These latter are set by default to be at the order of $\frac{1}{10^4}$.
plot	fixed by default to "FALSE". plot="TRUE" if a plot of the fuzzy number is required.

Value

If the parameter `alphacuts`="TRUE", the function returns a matrix composed of 2 vectors representing the left and right α-cuts. For this output, `is.alphacuts` = TRUE. If the parameter `alphacuts`="FALSE", the function returns a list composed of the class, the mean, the sigma, and the vectors of the left and right α-cuts.

See Also

`is.alphacuts` 1, `nbreakpoints` 2, `GaussianBellFuzzyNumber` 4.

Examples

```
GFN <- GaussianFuzzyNumber(mean = 0, sigma = 1, alphacuts
= TRUE, plot=TRUE)
is.alphacuts(GFN)
```

`GaussianBellFuzzyNumber`   *Creates a Gaussian two-sided bell fuzzy*
number   4

Description

This function creates a two-sided Gaussian bell fuzzy number from the class of Gaussian bell fuzzy numbers as seen in Definition 2.2.4. The constructed curve is a combination of two Gaussian curves. The smoothness of the constructed curve is defined by the `step`, the `margin`, and the `precision` fixed by default to 0.01, `[3*left.sigma;3*right.sigma]`, and 4.

Usage

```
GaussianBellFuzzyNumber    (left.mean, left.sigma, right.mean,
                           right.sigma, alphacuts = FALSE, margin
                           = c(3*left.sigma,3*right.sigma), step
                           = 0.01, breakpoints = 100, precision=4,
                           plot=FALSE)
```

Arguments

`left.mean`	a numerical value of the parameter μ of the left Gaussian curve.
`right.mean`	a numerical value of the parameter μ of the right Gaussian curve.
`left.sigma`	a numerical value of the parameter σ of the left Gaussian curve.
`right.sigma`	a numerical value of the parameter σ of the right Gaussian curve.
`alphacuts`	fixed by default to "FALSE". No α-cuts are printed in this case.
`margin`	an optional numerical couple of values representing the range of calculations of the Gaussian curve written as [`left.mean` - 3*`left.sigma`; `right.mean` + 3*`right.sigma`] by default.
`step`	numerical value fixing the step between two knots dividing the interval [`left.mean` - 3*`left.sigma`; `right.mean` + 3*`right.sigma`].
`breakpoints`	a positive arbitrary integer representing the number of breaks chosen to build the α-cuts. `breakpoints` is fixed to 100 by default.
`precision`	an integer specifying the number of decimals for which the calculations are made. These latter are set by default to be at the order of $\frac{1}{10^4}$.
`plot`	fixed by default to "FALSE". `plot="TRUE"` if a plot of the fuzzy number is required.

Value

If the parameter `alphacuts="TRUE"`, the function returns a matrix composed of 2 vectors representing the left and right α-cuts. For this output, `is.alphacuts = TRUE`. If the parameter `alphacuts="FALSE"`, the function returns a list composed of the class, the mean, the sigma, and the vectors of the left and right α-cuts.

See Also

is.alphacuts 1, nbreakpoints 2, GaussianFuzzyNumber 3.

Examples

```
GBFN <- GaussianBellFuzzyNumber(left.mean = -1, left.sigma
= 1, right.mean = 2,
right.sigma = 1, alphacuts = TRUE, plot=TRUE)
is.alphacuts(GBFN)
```

Fuzzy.Difference   *Calculates the difference between two fuzzy num-*
bers   5

Description

This function calculates the difference between two trapezoidal fuzzy numbers
with respect to the extension principle given in Definition 2.6.5. In case the fuzzy
numbers are not trapezoidal and written by their numerical α-cuts, the function
constructs a trapezoidal approximation composed of the minimum and maximum
values of the left and right α-cuts of the fuzzy number.

Usage

```
Fuzzy.Difference(X, Y, alphacuts=FALSE, breakpoints =
100)
```

Arguments

X	a fuzzy number of any type.
Y	a fuzzy number of any type.
alphacuts	fixed by default to "FALSE". No α-cuts are printed in this case.
breakpoints	a positive arbitrary integer representing the number of breaks chosen to build the α-cuts. breakpoints is fixed to 100 by default.

Value

If the parameter alphacuts="TRUE", the function returns a matrix composed of 2 vectors representing the left and right α-cuts. For this output, is.alphacuts = TRUE. If the parameter alphacuts="FALSE", the function returns a trapezoidal fuzzy number given by the quadruple (p,q,r,s), such that $p \leq q \leq r \leq s$.

See Also

is.alphacuts 1,   nbreakpoints 2,   GaussianFuzzyNumber 3,
GaussianBellFuzzyNumber 4,
Fuzzy.Square 6.

Examples

```
X <- TrapezoidalFuzzyNumber(5,6,7,8)
Y <- TrapezoidalFuzzyNumber(1,2,3,4)
Fuzzy.Difference(X,Y)
```

Fuzzy.Square   *Calculates numerically the square of a fuzzy number*   6

Description

This function calculates numerically the square of a fuzzy number with respect to the extension principle. The computations are made according to Definition 2.6.2 of the fuzzy product. The considered input fuzzy number is supposed to be trapezoidal or triangular. However, if a fuzzy number of some other types is introduced by its numerical α-cuts, a trapezoidal approximative fuzzy number is constructed by considering the minimum and maximum values of the left and right α-cuts of the fuzzy number. The calculations are based on resolving second-order polynoms.

Usage

```
Fuzzy.Square(X, breakpoints=100, plot = FALSE)
```

Arguments

X   a fuzzy number.

breakpoints   a positive arbitrary integer representing the number of breaks chosen to build the α-cuts. breakpoints is fixed to 100 by default.

plot   fixed by default to "FALSE". plot="TRUE" if a plot of the fuzzy number is required.

Value

Returns a matrix composed of 2 vectors representing the numerical left and right α-cuts. For this output, is.alphacuts = TRUE.

If the polynomial expressions of the left and right α-level sets are needed, the functions

```
Fuzzy.Square.poly.left (X, breakpoints=100)
```

and `Fuzzy.Square.poly.right (X, breakpoints=100)`

can be, respectively, used to print the related polynoms at the corresponding definition domains.

See Also

is.alphacuts 1, nbreakpoints 2, GaussianFuzzyNumber 3, GaussianBellFuzzyNumber 4, Fuzzy.Difference 5.

Examples

```
X <- TrapezoidalFuzzyNumber(1,2,3,4)
Fuzzy.Square(X, plot=TRUE)
```

---

is.fuzzification   *Verifies if a matrix is a fuzzification matrix*   7

Description

This function checks if an array can be a fuzzification matrix of a given variable. If it is verified, the considered array should fulfill the following requirements: it should be an array composed of 3 dimensions (m, n, p), where the number of lines m is the number of observations in the data set, the number of columns n is the number of chosen breakpoints, and the third dimension should be fixed to the value $p=2$ (related to the left and right α-cuts of a fuzzy number). The numbers of lines and columns should be the same for dimensions $p=1$, i.e. left, and $p=2$, i.e. right.

Usage

```
is.fuzzification (object)
```

Arguments

object   an array of 3 dimensions c(m, n, 2), with m lines and n columns. No NA is allowed.

Value

Returns a value TRUE if the concerned object is a numerical fuzzification matrix, FALSE otherwise.

See Also

is.alphacuts 1, nbreakpoints 2, is.trfuzzification 8, tr.gfuzz
9.

Examples

```
data <- array(c(1,1,2,2,3,3,5,5,6,6,7,7),dim=c(2,3,2))
is.fuzzification(data)
```

---

is.trfuzzification   *Verifies if a matrix is a fuzzification matrix of*
trapezoidal fuzzy numbers   8

Description

This function checks if a considered matrix can be a matrix of fuzzification
of a variable modelled by trapezoidal fuzzy numbers written by the quadruple
(p, q, r, s). The matrix should be composed of 4 columns (p, q, r, s), such
that $p \leq q \leq r \leq s$, where the number of lines corresponds to the number of
observations of the data set. Note that for triangular fuzzy numbers, the column q is
equal to the column r.

Usage

```
is.trfuzzification (object)
```

Arguments

object   a matrix of 4 columns (p, q, r, s), where $p \leq q \leq r \leq s$. No NA is allowed.

Value

Returns a value TRUE if the concerned object is a trapezoidal or triangular fuzzification matrix, FALSE otherwise.

See Also

is.alphacuts 1, nbreakpoints 2, is.fuzzification 7, tr.gfuzz 9.

Examples

```
data <- matrix(c(1,1,2,2,3,3,4,4),ncol=4)
is.trfuzzification(data)
```

---

tr.gfuzz   *Transforms a trapezoidal fuzzification matrix into a numerical*
one   9

Description

This function transforms a trapezoidal fuzzification matrix such that is.trfuzzi fication = TRUE, to a 3-dimensional numerical fuzzification matrix for which is.fuzzification = TRUE.

Usage

```
is.trfuzzification (object, breakpoints = 100)
```

Arguments

object	a matrix of 4 columns (p, q, r, s), where $p \leq q \leq r \leq s$. No NA is allowed.
breakpoints	a positive arbitrary integer representing the number of breaks chosen to build the numerical α-cuts. breakpoints is fixed to 100 by default.

Value

Returns a 3-dimensional array with dimensions $(m, n, 2)$, i.e. m lines, n columns, with no NA.

See Also

is.alphacuts 1, nbreakpoints 2, is.fuzzification 7, is.trfuzzi fication 8.

Examples

```
data <- matrix(c(1,1,2,2,3,3,4,4),ncol=4)
data.tr <- tr.gfuzz(data); is.fuzzification(data.tr)
```

FUZZ	*Fuzzifies a variable modelled by trapezoidal or triangular fuzzy*
numbers	10

Description

This function fuzzifies a variable modelled by trapezoidal or triangular fuzzy numbers as described in Sect. 3.1. The output is such that is.trfuzzification = TRUE. The membership functions are required. They should be introduced in the following manner: they should be called "MF" attached to the index of the main-item followed by one of the sub-items, i.e. the variable, and finally the index of the linguistic term. Note that the decomposition in main-items and sub-items is

Table 9.1   Decomposition 1

Main-item (mi=1)						Main-item 2 (mi=2)			
Sub-item 1 (si=1)		Sub-item 2 (si=2)		Sub-item 3 (si=3)		Sub-item 1 (si=1)		Sub-item 2 (si=2)	
$\tilde{L}_{111}$	MF111	$\tilde{L}_{121}$	MF121	$\tilde{L}_{131}$	MF131	$\tilde{L}_{211}$	MF211	$\tilde{L}_{221}$	MF221
$\tilde{L}_{112}$	MF112	$\tilde{L}_{122}$	MF122	$\tilde{L}_{132}$	MF132	$\tilde{L}_{212}$	MF212	$\tilde{L}_{222}$	MF222
$\tilde{L}_{113}$	MF113	$\tilde{L}_{123}$	MF123	$\tilde{L}_{133}$	MF133	$\tilde{L}_{213}$	MF213		
$\tilde{L}_{114}$	MF114			$\tilde{L}_{134}$	MF134	$\tilde{L}_{214}$	MF214		

Table 9.2   Decomposition 2

Main-item 1 (mi=1)									
Sub-item 1 (si=1)		Sub-item 2 (si=2)		Sub-item 3 (si=3)		Sub-item 4 (si=4)		Sub-item 5 (si=5)	
$\tilde{L}_{111}$	MF111	$\tilde{L}_{121}$	MF121	$\tilde{L}_{131}$	MF131	$\tilde{L}_{141}$	MF141	$\tilde{L}_{151}$	MF151
$\tilde{L}_{112}$	MF112	$\tilde{L}_{122}$	MF122	$\tilde{L}_{132}$	MF132	$\tilde{L}_{142}$	MF142	$\tilde{L}_{152}$	MF152
$\tilde{L}_{113}$	MF113	$\tilde{L}_{123}$	MF123	$\tilde{L}_{133}$	MF133	$\tilde{L}_{143}$	MF143		
$\tilde{L}_{114}$	MF114			$\tilde{L}_{134}$	MF134	$\tilde{L}_{144}$	MF144		

described in Sect. 7.1. This syntax is the same as the one used in the fuzzification process of the package FuzzyToolkitUoN given in Knott et al. (2013).

To illustrate the programming language related to the syntax of the membership functions of the modelling fuzzy numbers, we propose the following example.

Consider a data set composed of 5 linguistic variables. The aim is to fuzzify the fourth variable of a given data base. For this example, we suppose that the decomposition in main and sub-items is needed. We propose to divide the data set into 2 main-items as given in Table 9.1. For each linguistic term of the considered variable, one has to choose a corresponding modelling fuzzy number. Suppose we want to model the linguistic L_{213}, i.e. the third linguistic term of the first variable found in the second main-item, by a fuzzy number written as $\tilde{L}_{213}$. Its membership function should then be expressed by MF213. This case is seen in Table 9.1.

If the decomposition in main and sub-items is not required, it is then recommended to fix the index of the main-item to 1. Consequently, the index of the sub-item will be nothing but the column position of the variable in the architecture of the data set. For the same example previously described where the fourth variable has to be fuzzified, suppose that no decomposition is desired, then the architecture becomes as seen in Table 9.2. Therefore, the considered linguistic term L_{213} of the previous example is now on called L_{143} with its corresponding fuzzy number $\tilde{L}_{143}$. Its membership function will afterward be denoted by MF143 as seen in Table 9.2.

Usage

```
FUZZ (object,mi,si,PA)
```

Arguments

object	a data set.
mi	the index of the main-item containing the concerned variable.
si	the index of the sub-item of a given main-item mi.
PA	a vector of the linguistic terms of the considered variable.

Value

Returns a fuzzification matrix composed of 4 columns $c(p,q,r,s)$ and m lines, i.e. the number of observations. No NA is allowed.

See Also

is.alphacuts 1, is.trfuzzification 8, tr.gfuzz 9, GFUZZ 11.

Examples

```
# Fuzzification of the first sub-item of a data set -
No decomposition is required
data <- matrix(c(1,2,3,2,2,1,1,3,1,2),ncol=1)
MF111 <- TrapezoidalFuzzyNumber(0,1,1,2)
MF112 <- TrapezoidalFuzzyNumber(1,2,2,3)
MF113 <- TrapezoidalFuzzyNumber(2,3,3,3)
PA11 <- c(1,2,3)
data.fuzzified <- FUZZ(data,mi=1,si=1,PA=PA11)
is.trfuzzification(data.fuzzified)
```

GFUZZ	*Fuzzifies a variable modelled by any type of fuzzy numbers*	11

Description

This function fuzzifies a variable modelled by any type of fuzzy numbers as described in Sect. 3.1. The output is such that is.fuzzification = TRUE. The list of possible types of fuzzy numbers is the one by the FuzzyNumbers including the functions GaussianFuzzyNumber 3 and GaussianBellFuzzyNumber 4.

For the construction of the fuzzification matrix, the membership functions are required. One has the possibility to consider identical membership functions. In other terms, we mean that every linguistic will be modelled by different or identical fuzzy number across all the observations of the concerned variable. The membership functions should be called "MF" attached to the index of the main-item followed by one of the sub-items, i.e. the variable, and finally the index of the linguistic term, as described in the decomposition shown in Sect. 7.1. A detailed example of the syntax of these membership functions can be found in the description of the function FUZZ 10.

Usage

```
GFUZZ (object,mi,si,PA,spec="Identical", breakpoints =
100)
```

Arguments

object	a data set.
mi	the index of the main-item containing the concerned variable.
si	the index of the sub-item of a given main-item mi.
PA	a vector of the linguistic terms of the considered variable.
spec	specification of the fuzzification matrix. The possible values are "Identical" and "Not Identical".
breakpoints	a positive arbitrary integer representing the number of breaks chosen to build the numerical α-cuts. breakpoints is fixed to 100 by default.

Value

Returns a numerical fuzzification array of 3 dimensions $(m, n, 2)$, with m lines, n columns, and no NA.

See Also

is.alphacuts 1, is.fuzzification 7, tr.gfuzz 9, FUZZ 10.

Examples

```
# Fuzzification of the first sub-item of a data set -
No decomposition
# is required
data <- matrix(c(1,2,3,2,2,1,1,3,1,2),ncol=1)
MF111 <- TrapezoidalFuzzyNumber(0,1,1,2)
MF112 <- TrapezoidalFuzzyNumber(1,2,2,3)
MF113 <- TrapezoidalFuzzyNumber(2,3,3,3)
PA11 <- c(1,2,3)
data.fuzzified <- GFUZZ(data,mi=1,si=1,PA=PA11)
is.fuzzification(data.fuzzified)
```

---

distance	*Calculates the distance between fuzzy numbers*	12

Description

This function calculates the distance between two fuzzy numbers with respect to a distance chosen from the family of distances presented in Chap. 4. The family of distances is composed by the following ones: ρ_1, ρ_2, $d_{Bertoluzza}$, $\rho_p^\star$, $\delta_{p,q}^\star$, $d_{mid/spr}$, $d_{\phi-wabl/ldev/rdev}$, d_{SGD}, $d_{SGD}^{\theta^\star}$, and the d_{GSGD}. Note that the numerical integrations are made by the Simpson method.

Usage

```
distance    (X,Y, type, i=1, j=1, theta = 1/3, thetas = 1, p=2,
            q=0.5, breakpoints=100)
```

Arguments

X	a fuzzy number.
Y	a fuzzy number.
type	type of distance chosen from the family of distances. The different choices are given by "Rho1", "Rho2", "Bertoluzza", "Rhop", "Delta.pq", "Mid/Spr", "wabl", "DSGD", "DSGD.G", and "GSGD".
i and j	parameters of the density function of the Beta distribution, fixed by default to $i = 1$ and $j = 1$.
theta	a numerical value between 0 and 1, representing a weighting parameter. By default, theta is fixed to $\frac{1}{3}$ referring to the Lebesgue space. This measure is used in the calculations of the following distances: $d_{Bertoluzza}$, $d_{mid/spr}$ and $d_{\phi-wabl/ldev/rdev}$.
thetas	a decimal value between 0 and 1, representing the weight given to the shape of the fuzzy number. By default, thetas is fixed to 1. This parameter is used in the calculations of the $d_{SGD}^{\theta^*}$ and the d_{GSGD} distances.
p	a positive integer such that $1 \leq p \leq \infty$, referring to the parameter of the metric ρ_p^* or $\delta_{p,q}^*$.
q	a decimal value between 0 and 1, referring to the parameter of the metric $\delta_{p,q}^*$.
breakpoints	a positive arbitrary integer representing the number of breaks chosen to build the numerical α-cuts. It is fixed to 100 by default.

Value

Returns a numerical value.

See Also

is.alphacuts 1, GaussianFuzzyNumber 3, GaussianBellFuzzy Number 4, Fuzzy.Difference 5, GFUZZ 9.

Examples

```
X <- TrapezoidalFuzzyNumber(1,2,3,4)
Y <- TrapezoidalFuzzyNumber(4,5,6,7)
distance(X, Y, type = "DSGD.G")
distance(X, Y, type = "GSGD")
```

Fuzzy.sample.mean   *Calculates the fuzzy sample mean*   13

Description

This function calculates the fuzzy sample mean of a given fuzzy variable as shown in Definition 5.5.2. If the variable is encoded by trapezoidal fuzzy numbers written by their quadruple, this function can return a fuzzy number of the same form. The function can return as well a fuzzy number given by its numerical α-cuts.

Usage

```
Fuzzy.sample.mean(object, breakpoints = 100, alphacuts=
FALSE)
```

Arguments

object	a fuzzification matrix constructed by a call to the function FUZZ or the function GFUZZ, or a similar matrix. No NA is allowed.
breakpoints	a positive arbitrary integer representing the number of breaks chosen to build the numerical α-cuts. It is fixed to 100 by default.
alphacuts	fixed by default to "FALSE". No α-cuts are printed in this case.

Value

If the parameter alphacuts="TRUE", the function returns a matrix composed of 2 vectors representing the numerical left and right α-cuts. For this output, is.alphacuts = TRUE. If the parameter alphacuts="FALSE", the function returns a trapezoidal fuzzy number given by the quadruple (p,q,r,s).

See Also

is.alphacuts 1, nbreakpoints 2, GaussianFuzzyNumber 3, GaussianBellFuzzyNumber 3, FUZZ 10, GFUZZ 11, Weighted.fuzzy. mean 14.

Examples

```
# Simple example
mat <- matrix(c(1,2,2,3,3,4,4,5), ncol =4)
Fuzzy.sample.mean(mat)
```

Weighted.fuzzy.mean   *Calculates the weighted fuzzy sample mean* 14

Description

This function calculates the weighted fuzzy sample mean of a given fuzzy variable. If the variable is encoded by trapezoidal fuzzy numbers written by their quadruple, this function can return a fuzzy number of the same form. The function can return as well a fuzzy number given by its numerical α-cuts.

Usage

```
Weighted.fuzzy.mean(object, weight, breakpoints = 100,
alphacuts=FALSE)
```

Arguments

object   a fuzzification matrix constructed by a call to the function FUZZ or the
                 function GFUZZ, or a similar matrix. No NA is allowed.

weight   a weighting vector of the same length of the fuzzification matrix. No NA
                 allowed.

breakpoints   a positive arbitrary integer representing the number of breaks chosen to build
                 the numerical α-cuts. It is fixed to 100 by default.

alphacuts   fixed by default to "FALSE". No α-cuts are printed in this case.

Value

If the parameter alphacuts="TRUE", the function returns a matrix composed
of 2 vectors representing the numerical left and right α-cuts. For this output,
is.alphacuts = TRUE. If the parameter alphacuts="FALSE", the function
returns a trapezoidal fuzzy number given by the quadruple (p,q,r,s).

See Also

is.alphacuts 1, nbreakpoints 2, GaussianFuzzyNumber 3,
GaussianBellFuzzyNumber 4, FUZZ 10, GFUZZ 11, Fuzzy.sample.
mean 13.

Examples

```
# Simple example
mat <- matrix(c(1,2,2,3,3,4,4,5), ncol =4)
w <- c(1,3)
Weighted.fuzzy.mean(mat, w)
```

`Moment`	*Calculates a central sample moment of a random fuzzy variable* 15

Description

This function calculates the k-th classical central sample moment of a random fuzzy variable by the Féron–Puri and Ralescu approach, as given in Definition 5.5.3. This moment can be calculated using a distance chosen from the family of distances presented in Chap. 4. By this function, one can easily compute the skewness and the kurtosis of the random fuzzy variable.

Usage

```
Moment    (object, k, dist.type, i=1, j=1, theta = 1/3, thetas =
          1, p=2, q=0.5, breakpoints=100)
```

Arguments

`object`	a fuzzification matrix constructed by a call to the function `FUZZ` or the function `GFUZZ`, or a similar matrix. No `NA` is allowed.
`k`	the order of the moment.
`dist.type`	type of distance chosen from the family of distances. The different choices are given by "Rho1", "Rho2", "Bertoluzza", "Rhop", "Delta.pq", "Mid/Spr", "wabl", "DSGD", "DSGD.G", and "GSGD".
`i and j`	parameters of the density function of the Beta distribution, fixed by default to $i = 1$ and $j = 1$.
`theta`	a numerical value between 0 and 1, representing a weighting parameter. By default, `theta` is fixed to $\frac{1}{3}$ referring to the Lebesgue space. This measure is used in the calculations of the following distances: $d_{Bertoluzza}$, $d_{mid/spr}$, and $d_{\phi-wabl/ldev/rdev}$.
`thetas`	a decimal value between 0 and 1, representing the weight given to the shape of the fuzzy number. By default, `thetas` is fixed to 1. This parameter is used in the calculations of the $d_{SGD}^{\theta^\star}$ and the d_{GSGD} distances.
`p`	a positive integer such that $1 \leq p \leq \infty$, referring to the parameter of the metric $\rho_p^\star$ or $\delta_{p,q}^\star$.
`q`	a decimal value between 0 and 1, referring to the parameter of the metric $\delta_{p,q}^\star$.
`breakpoints`	a positive arbitrary integer representing the number of breaks chosen to build the numerical α-cuts. It is fixed to 100 by default.

Value

Returns a numerical value.

See Also

is.alphacuts 1, FUZZ 10, GFUZZ 11, Skewness 16, Kurtosis 17.

Examples

```
# Simple example
mat <- matrix(c(1,2,2,3,3,4,4,5), ncol =4)
Moment(mat, k=4, dist.type = "GSGD")
```

Skewness   *Calculates the skewness of a random fuzzy variable*   16

Description

This function calculates the skewness of a random fuzzy variable based on the expression of the central sample moments shown in Definition 5.5.3. The calculation is made using the function Moment 15. For a random fuzzy variable $\tilde{X}$, the skewness is given by the following ratio:

$$\text{Skewness}(\tilde{X}) = \frac{v_3(\tilde{X})}{(v_2(\tilde{X}))^{3/2}}, \tag{9.1}$$

where $v_3(\tilde{X})$ is the third central sample moment of the variable $\tilde{X}$, and $v_2(\tilde{X})$ is its second central sample moment.

Usage

```
Skewness    (object, dist.type, i=1, j=1, theta = 1/3, thetas = 1,
            p=2, q=0.5, breakpoints=100)
```

Arguments

object	a fuzzification matrix constructed by a call to the function FUZZ or the function GFUZZ, or a similar matrix. No NA is allowed.
dist.type	type of distance chosen from the family of distances. The different choices are given by "Rho1", "Rho2", "Bertoluzza", "Rhop", "Delta.pq", "Mid/Spr", "wabl", "DSGD", "DSGD.G", and "GSGD".
i and j	parameters of the density function of the Beta distribution, fixed by default to $i = 1$ and $j = 1$.
theta	a numerical value between 0 and 1, representing a weighting parameter. By default, theta is fixed to $\frac{1}{3}$ referring to the Lebesgue space. This measure is used in the calculations of the following distances: $d_{Bertoluzza}$, $d_{mid/spr}$, and $d_{\phi-wabl/ldev/rdev}$.
thetas	a decimal value between 0 and 1, representing the weight given to the shape of the fuzzy number. By default, thetas is fixed to 1. This parameter is used in the calculations of the $d_{SGD}^{\theta^\star}$ and the d_{GSGD} distances.
p	a positive integer such that $1 \leq p \leq \infty$, referring to the parameter of the metric $\rho_p^\star$ or $\delta_{p,q}^\star$.
q	a decimal value between 0 and 1, referring to the parameter of the metric $\delta_{p,q}^\star$.
breakpoints	a positive arbitrary integer representing the number of breaks chosen to build the numerical α-cuts. It is fixed to 100 by default.

Value

Returns a numerical value.

See Also

is.alphacuts 1, FUZZ 10, GFUZZ 11, Moment 15, Kurtosis 17.

Examples

```
# Simple example
mat <- matrix(c(1,2,0.25,1.8,2,2.6,0.5,3,3,2.6,3.8,4,4,
4.2,3.9,5), ncol =4)
Skewness(mat, dist.type = "GSGD")
```

Kurtosis   *Calculates the excess of kurtosis of a random fuzzy variable* 17

Description

This function calculates the excess of kurtosis of a random fuzzy variable based on the expression of the central sample moments shown in Definition 5.5.3. The calculation is made using the function Moment 15. For a random fuzzy variable $\tilde{X}$, the excess of kurtosis, denoted by Kurtosis($\tilde{X}$), is given by the following ratio:

$$\text{Kurtosis}(\tilde{X}) = \frac{v_4(\tilde{X})}{(v_2(\tilde{X}))^2} - 3, \tag{9.2}$$

where $v_4(\tilde{X})$ is the fourth central sample moment of the variable $\tilde{X}$, and $v_2(\tilde{X})$ is its second central sample moment.

Usage

```
Kurtosis  (object, dist.type, i=1, j=1, theta = 1/3, thetas =
          1, p=2, q=0.5, breakpoints=100)
```

Arguments

object
: a fuzzification matrix constructed by a call to the function FUZZ or the function GFUZZ, or a similar matrix. No NA is allowed.

dist.type
: type of distance chosen from the family of distances. The different choices are given by "Rho1", "Rho2", "Bertoluzza", "Rhop", "Delta.pq", "Mid/Spr", "wabl", "DSGD", "DSGD.G", and "GSGD".

i and j
: parameters of the density function of the Beta distribution, fixed by default to $i = 1$ and $j = 1$.

theta
: a numerical value between 0 and 1, representing a weighting parameter. By default, theta is fixed to $\frac{1}{3}$ referring to the Lebesgue space. This measure is used in the calculations of the following distances: $d_{Bertoluzza}$, $d_{mid/spr}$, and $d_{\phi-wabl/ldev/rdev}$.

thetas
: a decimal value between 0 and 1, representing the weight given to the shape of the fuzzy number. By default, thetas is fixed to 1. This parameter is used in the calculations of the $d_{SGD}^{\theta^\star}$ and the d_{GSGD} distances.

p
: a positive integer such that $1 \le p \le \infty$, referring to the parameter of the metric $\rho_p^\star$ or $\delta_{p,q}^\star$.

q
: a decimal value between 0 and 1, referring to the parameter of the metric $\delta_{p,q}^\star$.

breakpoints
: a positive arbitrary integer representing the number of breaks chosen to build the numerical α-cuts. It is fixed to 100 by default.

Value

Returns a numerical value.

See Also

is.alphacuts 1, FUZZ 10, GFUZZ 11, Moment 15, Skewness 16.

Examples

```
# Simple example
mat <- matrix(c(1,2,0.25,1.8,2,2.6,0.5,3,3,2.6,3.8,4,4,
4.2,3.9,5), ncol =4)
Kurtosis(mat, dist.type = "GSGD")
```

`Fuzzy.variance`	*Calculates the variance of a fuzzy variable*	18

Description

This function calculates the variance of a fuzzy variable. By this function, one could compute the following types of variances:

- the Fréchet variance of a random fuzzy variable as defined in Definition 5.4.5.
- a numerical "point to point" fuzzy type variance for trapezoidal and triangular fuzzy numbers using the functions `Fuzzy.Difference` 5 and `Fuzzy.Square` 6.

  We have to mention that for the calculations, any type of fuzzy numbers can be used. However, these numbers will be treated as trapezoidal fuzzy numbers. A condition on the monotony of the left of the right α-level sets of the produced fuzzy number is postulated. Thus, the left side should always be ascending, and the right one should always be descending.

  For the fuzzy variance by this method, we propose two additional functions defined as

  `Fuzzy.exact.variance.poly.left (object, breakpoints = 100)`

  and `Fuzzy.exact.variance.poly.right (object, breakpoints = 100)`,

  which produce the polynomial forms of the numerical α-cuts obtained using the function

  `Fuzzy.variance(object, method = "exact", ...).`

  The output is a table composed of the coefficients of the second-order equations of the left and the right sides, given at the corresponding definition domains.
- a cheap fuzzy type variance basically for trapezoidal and triangular fuzzy numbers by 5 different approximations of the fuzzy square. Note that for these computations, the function `Fuzzy.Difference` 5 is used. We add that the fuzzification matrix should be obtained by the function FUZZ for trapezoidal and triangular fuzzy numbers. The different approximations can be written as follows:

  1. Approximation 1: For a trapezoidal fuzzy number $\tilde{X} = (p, q, r, s)$, $p \leq q \leq r \leq s$, the fuzzy square of $\tilde{X}$ can be approximated by

  $$\tilde{X} \otimes \tilde{X} = (p, q, r, s) \otimes (p, q, r, s) = \left(q^2 - 2qp, q^2, r^2, 2rs - r^2\right).$$

2. Approximation 2: For a trapezoidal fuzzy number $\tilde{X} = (p, q, r, s)$, $p \leq q \leq r \leq s$, the fuzzy square of $\tilde{X}$ can be approximated by

$$\tilde{X} \otimes \tilde{X} = (p, q, r, s) \otimes (p, q, r, s) = \left(\min(p^2, ps), q^2, r^2, \max(ps, s^2) \right).$$

3. Approximation 3: Consider a trapezoidal fuzzy number $\tilde{X} = (p, q, r, s)$, $p \leq q \leq r \leq s$. Let $T(\tilde{X}) = (p_r, q_r, r_r, s_r)$ be the trapezoidal fuzzy number resulting from the approximation of $\tilde{Y} = \tilde{X} \otimes \tilde{X}$. The quadruple representing $T(\tilde{X})$ is given by the following:

$$p_r = \max \left\{ y \in \tilde{Y}_L^\alpha \mid \mu_{\tilde{X} \otimes \tilde{X}}(y) = 0 \right\},$$

$$q_r = \min \left\{ y \in \tilde{Y}_L^\alpha \mid \mu_{\tilde{X} \otimes \tilde{X}}(y) = 1 \right\},$$

$$r_r = \max \left\{ y \in \tilde{Y}_R^\alpha \mid \mu_{\tilde{X} \otimes \tilde{X}}(y) = 1 \right\},$$

$$s_r = \min \left\{ y \in \tilde{Y}_R^\alpha \mid \mu_{\tilde{X} \otimes \tilde{X}}(y) = 0 \right\}.$$

4. Approximation 4: For a fuzzy number defined as a piecewise one, we approximate its fuzzy square using the product operation $*$ defined by the FuzzyNumbers package.
5. Approximation 5: For a trapezoidal fuzzy number $\tilde{X} = (p, q, r, s)$, $p \leq q \leq r \leq s$, the fuzzy square of $\tilde{X}$ can be approximated by

$$\tilde{X} \otimes \tilde{X} = (p, q, r, s) \otimes (p, q, r, s)$$

$$= \left(\min(p^2, ps, s^2), \min(q^2, qr, r^2), \max(q^2, qr, r^2), \max(p^2, ps, s^2) \right).$$

Using almost all of these approximations, a computational complexity induced by the approximation operation is expected to occur. It is related to the ordering of the obtained non-positive elements of the quadruples defining the fuzzy numbers. This fact violates the principles of the direction of the left and right α-cuts of an LR fuzzy number. Therefore, we proposed to solve the problem using the shifting technique, also known as the translation technique.

Usage

```
Fuzzy.variance    (object, method, dist.type = "DSGD", i=1,
                   j=1, theta = 1/3, thetas = 1, p=2, q=0.5,
                   breakpoints=100, plot=FALSE)
```

Arguments

object	a fuzzification matrix constructed by a call to the function FUZZ or the function GFUZZ, or a similar matrix. No NA is allowed.
method	the choices are the following: "distance", "exact", "approximation1", "approximation2", "approximation3", "approximation4", "approximation5".
dist.type	type of distance chosen from the family of distances. The different choices are given by "Rho1", "Rho2", "Bertoluzza", "Rhop", "Delta.pq", "Mid/Spr", "wabl", "DSGD", "DSGD.G", and "GSGD".
i and j	parameters of the density function of the Beta distribution, fixed by default to $i = 1$ and $j = 1$.
theta	a numerical value between 0 and 1, representing a weighting parameter. By default, theta is fixed to $\frac{1}{3}$ referring to the Lebesgue space. This measure is used in the calculations of the following distances: $d_{Bertoluzza}$, $d_{mid/spr}$, and $d_{\phi-wabl/ldev/rdev}$.
thetas	a decimal value between 0 and 1, representing the weight given to the shape of the fuzzy number. By default, thetas is fixed to 1. This parameter is used in the calculations of the $d_{SGD}^{\theta^\star}$ and the d_{GSGD} distances.
p	a positive integer such that $1 \le p \le \infty$, referring to the parameter of the metric $\rho_p^\star$ or $\delta_{p,q}^\star$.
q	a decimal value between 0 and 1, referring to the parameter of the metric $\delta_{p,q}^\star$.
breakpoints	a positive arbitrary integer representing the number of breaks chosen to build the numerical α-cuts. It is fixed to 100 by default.
plot	fixed by default to "FALSE". plot="TRUE" if a plot of the fuzzy number is required.

Value

If the parameter method = "distance", returns a numerical value, else returns the numerical α-cuts of the estimated fuzzy variance.

See Also

Fuzzy.Difference 5, Fuzzy.Square 6, is.fuzzification 7, is.trfuzzification 8, FUZZ 10, GFUZZ 11, distance 12.

Examples

```
data <- matrix(c(1,2,3,2,2,1,1,3,1,2),ncol=1)
MF111 <- TrapezoidalFuzzyNumber(0,1,1,2)
MF112 <- TrapezoidalFuzzyNumber(1,2,2,3)
MF113 <- TrapezoidalFuzzyNumber(2,3,3,3)
PA11 <- c(1,2,3)

data.fuzzified <- FUZZ(data,mi=1,si=1,PA=PA11)
Fuzzy.variance(data.fuzzified, method = "approximation5",
plot=TRUE)
Fuzzy.variance(data.fuzzified, method = "exact", plot=TRUE)
Fuzzy.variance(data.fuzzified, method = "distance")

data.fuzzified2 <- GFUZZ(data,mi=1,si=1,PA=PA11)
Fuzzy.variance(data.fuzzified2, method = "exact", plot=TRUE)
Fuzzy.variance(data.fuzzified2, method = "distance")
```

---

boot.ml	*Estimates the bootstrap distribution of the likelihood ratio (LR)*
19	

Description

This function estimates the empirical distribution of the likelihood ratio (LR) by the bootstrap technique as exposed in Sect. 6.3.2. It produces a vector of replications of LR for several random drawings from a primary data set as presented in Algorithms 1 and 2 of Sect. 6.3.2. The coefficient η proposed in Eq. 6.37 is then nothing but the $1 - \delta$-quantile of this distribution. This function can till now be used to the following distributions: the normal, the Poisson, and the Student distributions. The related density functions are known and their likelihood functions can be accordingly computed. In addition, this function computes internally the MLE estimator by the EM-algorithm using the function EM.Trapezoidal by the EM.Fuzzy package proposed in Parchami (2018b). A fuzzy number modelling the crisp estimator can be added. The default spread of this number is 2.

The number of replications, the smoothness, and the margins of calculations of the obtained distributions are defined by the nsim, step, and the margin fixed by default to 100, 0.05, and $c(5, 5)$, respectively.

Usage

```
boot.ml  (object, algorithm, distribution, sig, ct=c(1,1),
          nsim=100, mu=NA, sigma=NA, step = 0.05, margin =
          c(5,5), breakpoints = 100, plot=TRUE)
```

Arguments

object	a fuzzification matrix constructed by a call to the function FUZZ or the function GFUZZ, or a similar matrix. No NA is allowed.
algorithm	an algorithm chosen between "algo1" and "algo2".
distribution	a distribution chosen between "normal", "poisson", and "Student".
sig	a numerical value representing the significance level of the interval.
ct	an optional numerical couple of values fixed to [1;1], the difference between the lower bounds of the core set and the support set of the fuzzy number modelling the MLE estimator from one side, and the difference between the upper bounds of both sets from another one.
nsim	an integer giving the number of replications needed in the bootstrap procedure. It is set to 100 by default.
mu	if the mean of the normal distribution is known, mu should be a numerical value; otherwise, the argument mu is fixed to NA.
sigma	if the standard deviation of the normal distribution is known, sigma should be a numerical value; otherwise, the argument sigma is fixed to NA.
margin	an optional numerical couple of values fixed to $c(5, 5)$, representing the range of calculations around the parameter, i.e. [parameter-5; parameter+5].
step	a numerical value fixed to 0.05, defining the step of iterations on the interval [parameter-5; parameter+5].
breakpoints	a positive arbitrary integer representing the number of breaks chosen to build the numerical α-cuts. It is fixed to 100 by default.
plot	fixed by default to "FALSE". plot="TRUE" if a plot of the fuzzy number is required.

Value

Returns a vector of decimals representing the bootstrap distribution of LR.

See Also

`is.alphacuts` 1, `is.fuzzification` 7, `FUZZ` 10, `GFUZZ` 11, `Fuzzy.sample.mean` 13, `fci.ml` 20.

Examples

```
data <- matrix(c(1,2,3,2,2,1,1,3,1,2),ncol=1)
MF111 <- TrapezoidalFuzzyNumber(0,1,1,2)
MF112 <- TrapezoidalFuzzyNumber(1,2,2,3)
MF113 <- TrapezoidalFuzzyNumber(2,3,3,3)
PA11 <- c(1,2,3)
data.fuzzified <- FUZZ(data,mi=1,si=1,PA=PA11)

emp.dist <- boot.ml(data.fuzzified, algorithm = "algo1",
distribution = "normal", sig = 0.05, sigma = 0.62)
eta.boot <- quantile(emp.dist, probs = 95/100)
```

`fci.ml`	*Estimates a fuzzy confidence interval by the likelihood method*	20

Description

This function estimates the fuzzy confidence interval by the likelihood method as described in Sect. 6.3.2. It produces the left and right α-cuts as shown, respectively, in Eqs. 6.59 and 6.60. The proposed method can be used to compute the interval without any specification of the distribution to estimate a given parameter. However, for our current situation, we restrict ourselves to distributions drawn from the normal, the Poisson, and the Student distribution since the related density functions are known and their likelihood functions can be easily computed. An eventual upgrade to this function is welcomed in order to be able to introduce empirical density functions as an instance. The smoothness and the margins of calculations of the constructed interval are defined by the `step` and the `margin` fixed by default to 0.05 and $c(5, 5)$.

Usage

```
fci.ml  (object, t, distribution, sig, coef.boot, mu=NA,
        sigma=NA, step = 0.05, margin = c(5,5), breakpoints =
        100, plot=TRUE)
```

Arguments

object	a fuzzification matrix constructed by a call to the function FUZZ or the function GFUZZ, or a similar matrix. No NA is allowed.
t	a given numerical or fuzzy type parameter of the distribution.
distribution	a distribution chosen between "normal", "poisson", and "Student".
sig	a numerical value representing the significance level of the interval.
coef.boot	a decimal representing the 1-sig-quantile of the bootstrap distribution of LR.
mu	if the mean of the normal distribution is known, mu should be a numerical value; otherwise, the argument mu is fixed to NA.
sigma	if the standard deviation of the normal distribution is known, sigma should be a numerical value; otherwise, the argument sigma is fixed to NA.
margin	an optional numerical couple of values fixed to $c(5, 5)$, representing the range of calculations around the parameter t.
step	a numerical value fixed to 0.05, defining the step of iterations on the interval [t-5; t+5].
breakpoints	a positive arbitrary integer representing the number of breaks chosen to build the numerical α-cuts. It is fixed to 100 by default.
plot	fixed by default to "FALSE". plot="TRUE" if a plot of the fuzzy number is required.

Value

Returns a matrix composed of 2 vectors representing the numerical left and right α-cuts. For this output, is.alphacuts = TRUE.

See Also

is.alphacuts 1, is.fuzzification 7, FUZZ 10, GFUZZ 11,
Fuzzy.sample.mean 13,
Fuzzy.variance 18, boot.ml 19.

Examples

```
data <- matrix(c(1,2,3,2,2,1,1,3,1,2),ncol=1)
MF111 <- TrapezoidalFuzzyNumber(0,1,1,2)
MF112 <- TrapezoidalFuzzyNumber(1,2,2,3)
MF113 <- TrapezoidalFuzzyNumber(2,3,3,3)
PA11 <- c(1,2,3)
data.fuzzified <- FUZZ(data,mi=1,si=1,PA=PA11)
Fmean <- Fuzzy.sample.mean(data.fuzzified)
fci.ml(data.fuzzified, t = Fmean, distribution = "normal",
sig= 0.05,
coef.boot = 1.8225, sigma = 0.62)
```

Fuzzy.decisions   *Computes the fuzzy decisions of a fuzzy inference test*
by the traditional fuzzy confidence intervals   21

Description

This function calculates the fuzzy decisions obtained from a fuzzy inference test based on the fuzzy confidence intervals defined in Eqs. 6.24 and 6.25. The corresponding construction of these decisions is given in Sect. 6.4.1. We have to mention that for this function, all the cases of intersection between the fuzzy confidence intervals and the fuzzy null hypothesis are taken into account. The different possible positions for the two-sided and the one-sided cases are given, respectively, in Tables 9.5, 9.3 and 9.4. In addition, by this function, one could get the defuzzification of the obtained fuzzy decisions. This task can be made using the distance of these fuzzy numbers to the fuzzy origin. The different distances are shown in Chap. 4. Note that for the likelihood method, an analog function is called

```
Fuzzy.decisions.ML (data.fuzzified, H0, H1, t,

            coef.boot, mu=NA, sigma=NA, etc).
```

Usage

```
Fuzzy.decisions   (type, H0, H1, t, s.d, n, sig, distribution,
                  distance.type="DSGD", i=1, j=1, theta = 1/3,
                  thetas=1, p=2, q=0.5, breakpoints=100)
```

Arguments

`type`	a category between "0," "1," and "2." The category "0" refers to a bilateral test, the category "1" for a lower unilateral one, and "2" for an upper unilateral test.
`H0`	a trapezoidal or a triangular fuzzy number representing the fuzzy null hypothesis, written by its numerical α-cuts.
`H1`	a trapezoidal or a triangular fuzzy number representing the fuzzy alternative hypothesis, written by its numerical α-cuts.
`t`	a given numerical or fuzzy type value representing the parameter of the distribution.
`s.d`	a numerical value for the standard deviation of the distribution.
`n`	the total number of observations of the data set.
`sig`	a numerical value representing the significance level of the test.
`distribution`	a distribution chosen between "normal", "poisson", and "Student".
`distance.type`	type of distance chosen from the family of distances, set by default to the signed distance. The different choices are given by "Rho1", "Rho2", "Bertoluzza", "Rhop", "Delta.pq", "Mid/Spr", "wabl", "DSGD", "DSGD.G", and "GSGD".
`i and j`	parameters of the density function of the Beta distribution, fixed by default to $i = 1$ and $j = 1$.
`theta`	a numerical value between 0 and 1, representing a weighting parameter. By default, `theta` is fixed to $\frac{1}{3}$ referring to the Lebesgue space. This measure is used in the calculations of the following distances: $d_{Bertoluzza}$, $d_{mid/spr}$, and $d_{\phi-wabl/ldev/rdev}$.
`thetas`	a decimal value between 0 and 1, representing the weight given to the shape of the fuzzy number. By default, `thetas` is fixed to 1. This parameter is used in the calculations of the $d_{SGD}^{\theta^\star}$ and the d_{GSGD} distances.
`p`	a positive integer such that $1 \leq p \leq \infty$, referring to the parameter of the metric $\rho_p^\star$ or $\delta_{p,q}^\star$.
`q`	a decimal value between 0 and 1, referring to the parameter of the metric $\delta_{p,q}^\star$.
`breakpoints`	a positive arbitrary integer representing the number of breaks chosen to build the numerical α-cuts. It is fixed to 100 by default.

Value

Returns a list composed of the arguments, the fuzzy confidence intervals and their complements, the fuzzy decisions, and the defuzzified values.

See Also

Examples

```
H0 <- alphacut (TriangularFuzzyNumber (2.9,3,3.1),
seq(0,1, 0.01))
H1 <- alphacut (TriangularFuzzyNumber (3,3,5),
seq(0,1,0.01))
t <- alphacut (TriangularFuzzyNumber (0.8,1.80,2.80),
seq(0,1,0.01))
res <- Fuzzy.decisions(type = 0, H0, H1, t = t,
s.d = 0.79, n = 10, sig = 0.05,
distribution = "normal", distance.type = "GSGD")
```

Fuzzy.CI.test confidence intervals	*Computes a fuzzy inference test by the traditional fuzzy* 22

Description

This function tests a fuzzy null hypothesis against a fuzzy alternative one, based on traditional fuzzy confidence intervals as proposed in Sect. 6.4. This test is computed using the function `Fuzzy.decisions` 21, which provides the fuzzy decisions related to the test. As seen in Sect. 6.4.2, these decisions are afterward defuzzified by calculating their distance to the fuzzy origin, with respect to the different distances given in Chap. 4. This operation is described in Eq. 6.99.

Note that this function is made for the case of the mean only. This function is designed to be used with the normal, the Poisson, and the Student distributions.

Usage

Fuzzy.CI.test   (type, H0, H1, t, s.d, n, sig, distribution,
                 distance.type="DSGD", i=1, j=1, theta = 1/3,
                 thetas=1, p=2, q=0.5, breakpoints=100)

Arguments

type	a category between "0," "1," and "2." The category "0" refers to a bilateral test, the category "1" for a lower unilateral one, and "2" for an upper unilateral test.
H0	a trapezoidal or a triangular fuzzy number representing the fuzzy null hypothesis.
H1	a trapezoidal or a triangular fuzzy number representing the fuzzy alternative hypothesis.
t	a given numerical or fuzzy type value representing the parameter of the distribution.
s.d	a numerical value for the standard deviation of the distribution.
n	the total number of observations of the data set.
sig	a numerical value representing the significance level of the test.
distribution	a distribution chosen between "normal", "poisson", and "Student".
distance.type	type of distance chosen from the family of distances, set by default to the signed distance. The different choices are given by "Rho1", "Rho2", "Bertoluzza", "Rhop", "Delta.pq", "Mid/Spr", "wabl", "DSGD", "DSGD.G", and "GSGD".
i and j	parameters of the density function of the Beta distribution, fixed by default to $i = 1$ and $j = 1$.
theta	a numerical value between 0 and 1, representing a weighting parameter. By default, theta is fixed to $\frac{1}{3}$ referring to the Lebesgue space. This measure is used in the calculations of the following distances: $d_{Bertoluzza}$, $d_{mid/spr}$, and $d_{\phi-wabl/ldev/rdev}$.
thetas	a decimal value between 0 and 1, representing the weight given to the shape of the fuzzy number. By default, thetas is fixed to 1. This parameter is used in the calculations of the $d_{SGD}^{\theta^\star}$ and the d_{GSGD} distances.
p	a positive integer such that $1 \leq p \leq \infty$, referring to the parameter of the metric $\rho_p^\star$ or $\delta_{p,q}^\star$.
q	a decimal value between 0 and 1, referring to the parameter of the metric $\delta_{p,q}^\star$.
breakpoints	a positive arbitrary integer representing the number of breaks chosen to build the numerical α-cuts. It is fixed to 100 by default.

Value

Returns a list composed of the arguments, the fuzzy confidence intervals, the fuzzy decisions, the defuzzified values, and the decision made.

See Also

`is.alphacuts` 1, `Fuzzy.sample.mean` 13, `Fuzzy.variance` 18, `boot.ml` 19, `fci.ml` 20, `Fuzzy.decisions` 21, `Fuzzy.decisions.ML` 21, `Fuzzy.CI.ML.test` 23.

Examples

```
H0 <- TriangularFuzzyNumber(2.9,3,3.1)
H1 <- TriangularFuzzyNumber(3,3,5)
res <- Fuzzy.CI.test(type = 0, H0, H1,
t = TriangularFuzzyNumber(0.8,1.80,2.80), s.d = 0.79,
n = 10, sig = 0.05, distribution = "normal",
distance.type="GSGD")
```

Fuzzy.CI.ML.test	*Computes a fuzzy inference test by the fuzzy confidence intervals method calculated by the likelihood method* 23

Description

This function tests a fuzzy null hypothesis against a fuzzy alternative one, based on the fuzzy confidence interval constructed using the likelihood method as shown in Sect. 6.4. This interval is computed using the function `fci.ml` 20 for the computation of the confidence interval and `Fuzzy.decisions.ML` 21 for the computation of the fuzzy decisions. These latter are then defuzzified by a distance chosen from the family of distances using the function `distance` 12. The use of the function `Fuzzy.CI.ML.test` is restricted to the distributions drawn from the normal, the Poisson, and the Student distributions. An eventual improvement of these functions is to consider the empirical distributions or any other known distribution.

Usage

```
Fuzzy.CI.ML.test    (object, H0, H1, t, mu=NA, sigma=NA, sig,
                     distribution, coef.boot, distance.type="DSGD",
                     i=1, j=1, theta = 1/3, thetas=1, p=2, q=0.5,
                     breakpoints=100, step = 0.05, margin = c(5,5),
                     plot=TRUE)
```

Arguments

object	a fuzzification matrix constructed by a call to the function FUZZ or the function GFUZZ, or a similar matrix. No NA is allowed.
H0	a trapezoidal or a triangular fuzzy number representing the fuzzy null hypothesis.
H1	a trapezoidal or a triangular fuzzy number representing the fuzzy alternative hypothesis.
t	a given numerical or fuzzy type parameter of the distribution.
mu	if the mean of the normal distribution is known, mu should be a numerical value; otherwise, the argument mu is fixed to NA.
sigma	if the standard deviation of the normal distribution is known, sigma should be a numerical value; otherwise, the argument sigma is fixed to NA.
sig	a numerical value representing the significance level of the test.
distribution	a distribution chosen between "normal", "poisson", and "Student".
coef.boot	a decimal representing the 1-sig-quantile of the bootstrap distribution of LR.
distance.type	type of distance chosen from the family of distances, set by default to the signed distance. The different choices are given by "Rho1", "Rho2", "Bertoluzza", "Rhop", "Delta.pq", "Mid/Spr", "wabl", "DSGD", "DSGD.G", and "GSGD".
i and j	parameters of the density function of the Beta distribution, fixed by default to $i = 1$ and $j = 1$.
theta	a numerical value between 0 and 1, representing a weighting parameter. By default, theta is fixed to $\frac{1}{3}$ referring to the Lebesgue space. This measure is used in the calculations of the following distances: $d_{Bertoluzza}$, $d_{mid/spr}$, and $d_{\phi-wabl/ldev/rdev}$.
thetas	a decimal value between 0 and 1, representing the weight given to the shape of the fuzzy number. By default, thetas is fixed to 1. This parameter is used in the calculations of the $d_{SGD}^{\theta^*}$ and the d_{GSGD} distances.

p	a positive integer such that $1 \leq p \leq \infty$, referring to the parameter of the metric $\rho_p^\star$ or $\delta_{p,q}^\star$.
q	a decimal value between 0 and 1, referring to the parameter of the metric $\delta_{p,q}^\star$.
breakpoints	a positive arbitrary integer representing the number of breaks chosen to build the numerical α-cuts. It is fixed to 100 by default.
step	a numerical value fixed to 0.05, defining the step of iterations on the interval [t-5; t+5].
margin	an optional numerical couple of values fixed to [5; 5], representing the range of calculations around the parameter t.
plot	fixed by default to "FALSE". plot="FALSE" if a plot of the fuzzy number is not required.

Value

Returns a list composed of the arguments, the fuzzy confidence intervals, the fuzzy decisions, the defuzzified values, and the decision made.

See Also

is.alphacuts 1, distance 12, Fuzzy.sample.mean 13,
Fuzzy.variance 18, boot.ml 19,
fci.ml 20, Fuzzy.decisions 21, Fuzzy.decisions.ML 21,
Fuzzy.CI.test 22.

Examples

```
data <- matrix(c(1,2,3,2,2,1,1,3,1,2),ncol=1)
MF111 <- TrapezoidalFuzzyNumber(0,1,1,2)
MF112 <- TrapezoidalFuzzyNumber(1,2,2,3)
MF113 <- TrapezoidalFuzzyNumber(2,3,3,3)
PA11 <- c(1,2,3)
data.fuzzified <- FUZZ(data,mi=1,si=1,PA=PA11)

Fmean <- Fuzzy.sample.mean(data.fuzzified)
H0 <- TriangularFuzzyNumber(2.2,2.5,3)
H1 <- TriangularFuzzyNumber(2.5,2.5,5)

emp.dist <- boot.ml(data.fuzzified, algorithm = "algo1",
distribution
```

```
= "normal", sig= 0.05, sigma = 0.7888)
eta.boot <- quantile(emp.dist, probs = 95/100)


(res <- Fuzzy.CI.ML.test(data.fuzzified, H0, H1, t =
Fmean, sigma=0.7888,
sig=0.05, coef.boot=eta.boot, distribution="normal",
distance.type="GSGD"))
res$decision
```

Fuzzy.p.value	*Computes the fuzzy p-value of a given fuzzy hypothesis*
test	24

Description

This function calculates the fuzzy *p*-value of a given hypothesis test as presented in Proposition 6.4.2. For this fuzzy *p*-value, the null and the alternative hypotheses have to be defined, as well as the corresponding parameter (the mean in this case), and the considered distribution. The normal, the Poisson, and the Student distributions can be used. We add that a defuzzification of the obtained fuzzy number is also proposed, as seen in Sect. 6.4.4. For this task, one can choose a distance from the family shown in Chap. 4.

Usage

```
Fuzzy.p.value   (type, H0, H1, t, s.d, n, sig, distribution,
                distance.type="DSGD", i=1, j=1, theta = 1/3,
                thetas = 1, p=2, q=0.5, breakpoints=100)
```

Arguments

`type`	a category between "0," "1," and "2." The category "0" refers to a bilateral test, the category "1" for a lower unilateral one, and "2" for an upper unilateral test.
`H0`	a trapezoidal or a triangular fuzzy number representing the fuzzy null hypothesis.
`H1`	a trapezoidal or a triangular fuzzy number representing the fuzzy alternative hypothesis.
`t`	a given numerical or fuzzy type parameter of the distribution.
`s.d`	a numerical value for the standard deviation of the distribution.
`n`	the total number of observations of the data set.
`sig`	a numerical value representing the significance level of the test.
`distribution`	a distribution chosen between "normal", "poisson", and "Student".
`distance.type`	type of distance chosen from the family of distances, set by default to the signed distance. The different choices are given by "Rho1", "Rho2", "Bertoluzza", "Rhop", "Delta.pq", "Mid/Spr", "wabl", "DSGD", "DSGD.G", and "GSGD".
`i and j`	parameters of the density function of the Beta distribution, fixed by default to $i = 1$ and $j = 1$.
`theta`	a numerical value between 0 and 1, representing a weighting parameter. By default, `theta` is fixed to $\frac{1}{3}$ referring to the Lebesgue space. This measure is used in the calculations of the following distances: $d_{Bertoluzza}$, $d_{mid/spr}$, and $d_{\phi-wabl/ldev/rdev}$.
`thetas`	a decimal value between 0 and 1, representing the weight given to the shape of the fuzzy number. By default, `thetas` is fixed to 1. This parameter is used in the calculations of the $d_{SGD}^{\theta^\star}$ and the d_{GSGD} distances.
`p`	a positive integer such that $1 \leq p \leq \infty$, referring to the parameter of the metric $\rho_p^\star$ or $\delta_{p,q}^\star$.
`q`	a decimal value between 0 and 1, referring to the parameter of the metric $\delta_{p,q}^\star$.
`breakpoints`	a positive arbitrary integer representing the number of breaks chosen to build the numerical α-cuts. It is fixed to 100 by default.

Value

Returns the defuzzified p-value, the decision made, and the plot of the fuzzy p-value with the corresponding significance level.

See Also

is.alphacuts 1, distance 12, Fuzzy.sample.mean 13,
Fuzzy.variance 18, fci.ml 20,

Fuzzy.decisions 21, Fuzzy.decisions.ML 21, Fuzzy.CI.test 22,
Fuzzy.CI.ML.test 23.

Examples

```
H0 <- TriangularFuzzyNumber(2.2,2.5,3)
H1 <- TriangularFuzzyNumber(2.5,2.5,5)
Fuzzy.p.value(type=1, H0, H1,
t=TriangularFuzzyNumber(0.8,1.8,2.8),
s.d=0.7888, n=10, sig=0.05, distribution="normal",
distance.type="GSGD")
```

---

adjusted.weight.SI   *Calculates the adjusted weight for a given sub-*
item of a linguistic questionnaire   25

Description

For a given observation, this function calculates the adjusted weight of a sub-item
of a linguistic questionnaire when non-response is present. This function, expressed
in Eq. 7.6, redistributes the weights on the non-missing answers. Counting the
answers in a given sub-item can be done based on the functions delta_jki and
Delta_jki defined, respectively, in Eqs. 7.1 and 7.5.

Usage

```
adjusted.weight.SI    (data, i, k, b_jk)
```

Arguments

data   the data set to evaluate.
i   an observation index.
k   a sub-item index
b_jk   an array referring to the initial weights given to each sub-item of the considered main-
        item. This array will be afterward recalculated.

Value

Returns a numerical value giving the readjusted weight of the sub-item k of the considered main-item for the observation i.

See Also

`adjusted.weight.MI` 26, `IND.EVAL` 27, `GLOB.EVAL` 28, R 29, Ri 30.

Examples

`adjusted.weight.SI(data, 17, 1, c(0.5,0.5))`

`adjusted.weight.MI` *item of a linguistic questionnaire*	*Calculates the adjusted weight for a given main-* 26

Description

When non-response is present, this function calculates the adjusted weight given to a main-item of a linguistic questionnaire related to a particular observation, as expressed in Eq. 7.7. This function redistributes the weights based on the missing answers occurring in a given main-item. The calculation of the readjusted weight in a sub-item is done using the function `adjusted.weight.SI` 25.

Usage

`adjusted.weight.MI    (data, i, j, b_j, b_jk, SI)`

Arguments

data   the data set to evaluate.

i   an observation index.

j   a main-item index

b_j   an array referring to the initial weights given to each main-item of the considered main-item. This array will be afterward recalculated.

b_jk   a matrix of length(b_j) rows and max(SI) columns expressing the initial weights of each sub-item of a given main-item. By analogy to the hypothesis of the method described in Section 7.1, this matrix should be given in the following form:

$$b_{jk} = \begin{pmatrix} b_{11} & \cdots & b_{1k} & \cdots & b_{1m_j} \\ \vdots & & \vdots & & \vdots \\ b_{j1} & \cdots & b_{jk} & \cdots & b_{jm_j} \\ \vdots & & \vdots & & \vdots \\ b_{r1} & \cdots & b_{rk} & \cdots & b_{rm_j} \end{pmatrix}.$$

Note that if the main-items of the questionnaire do not have the same number of sub-items, the weights of the excluded sub-items are null. For example, if the first main-item has 5 sub-items and the second one is composed of 3 only, then the matrix b_{jk} will be the following:

$$b_{jk} = \begin{pmatrix} b_{11} & b_{12} & b_{13} & b_{14} & b_{15} \\ b_{21} & b_{22} & b_{23} & 0 & 0 \end{pmatrix}.$$

SI   an array representing the total number of sub-items per main-item.

Value

Returns a numerical value giving the readjusted weight of the main-item j for the observation i.

See Also

adjusted.weight.SI 25, IND.EVAL 27, GLOB.EVAL 28, R 29, Ri 30.

Examples

```
b_j <- c(1/3,1/3,1/3)
b_jk <- matrix(c(0.5,0.125,0.2,0.5,0.125,0.2,0,0.125,
0.2,0,0.125,0.2,0,0.125,
0.2,0,0.125,0,0,0.125,0,0,0.125,0),nrow=MI)
SI <- c(2,8,5)
adjusted.weight.MI(data, 17, 1, b_j, b_jk, SI)
```

IND.EVAL	*Calculates the individual evaluations of a linguistic question-*	
naire		27

Description

This function calculates the individual evaluations expressed in Eq. 7.11, following the procedure described in Sect. 7.3. The user has to define the data set to be evaluated and the decomposition of the linguistic questionnaire by main- and sub-items. The related initial weights have to be defined as well. In addition, a distance has to be chosen from the family shown in Chap. 4. Finally, the fuzzy numbers modelling the different linguistic terms should be given. The computations can be made using all types of fuzzy numbers. We add that the argument spec = "Identical" refers to the fact that every linguistic L_q will be modelled by the identical fuzzy number across all the observations of the concerned variable. Each fuzzy linguistic should be called "MF" attached to the index of its main-item, followed by the index of its sub-item, and the one of the linguistic term, as explained in the description of the function FUZZ 10. We have to mention that adding up the answers of a given linguistic expressed by Eqs. 7.1 and 7.5 is implemented as the functions delta_jki and Delta_jki, respectively.

If missingness occurs in the data set, we proposed in Sect. 7.2 to readjust the weights of the main- and sub-items. Practically, this task will be performed using the functions adjusted. weight.MI and adjusted.weight.SI as developed in Eqs. 7.7 and 7.6.

Usage

```
IND.EVAL   (data, MI, bmi, SI, b_jkt, range, distance.type,
            i=1, j=1, theta = 1/3, thetas=1, p=2, q=0.5,
            breakpoints=100, spec="Identical")
```

Arguments

data   the data set to evaluate.

MI   a numerical value representing the total number of main-items dividing the linguistic questionnaire.

bmi   an array referring to the initial weights of the main-items.

SI   an array representing the total number of sub-items per main-item.

b_jkt   a matrix of MI rows and max(SI) columns expressing the initial weights of each sub-item of a given main-item. By analogy to the hypothesis of the method described in Sect. 7.1, this matrix should be given in the following form:

$$b_{jkt} = \begin{pmatrix} b_{11t} & \dots & b_{1kt} & \dots & b_{1m_{jt}} \\ \vdots & & \vdots & & \vdots \\ b_{j1t} & \dots & b_{jkt} & \dots & b_{jm_{jt}} \\ \vdots & & \vdots & & \vdots \\ b_{r1t} & \dots & b_{rkt} & \dots & b_{rm_{jt}} \end{pmatrix}.$$

range   a vector of 2 elements giving the range of definition of the produced individual evaluations. The range is usually chosen in the interval between 0 and the maximum of the support set of all the membership functions modelling the data set.

distance.type   type of distance chosen from the family of distances, set by default to the signed distance. The different choices are given by "Rho1", "Rho2", "Bertoluzza", "Rhop", "Delta.pq", "Mid/Spr", "wabl", "DSGD", "DSGD.G", and "GSGD".

i and j   parameters of the density function of the Beta distribution, fixed by default to $i = 1$ and $j = 1$.

theta   a numerical value between 0 and 1, representing a weighting parameter. By default, theta is fixed to $\frac{1}{3}$ referring to the Lebesgue space. This measure is used in the calculations of following distances: $d_{Bertoluzza}$, $d_{mid/spr}$, and $d_{\phi-wabl/ldev/rdev}$.

thetas   a decimal value between 0 and 1, representing the weight given to the shape of the fuzzy number. By default, thetas is fixed to 1. This parameter is used in the calculations of the $d_{SGD}^{\theta^\star}$ and the d_{GSGD} distances.

p   a positive integer such that $1 \leq p \leq \infty$, referring to the parameter of the metric $\rho_p^\star$ or $\delta_{p,q}^\star$.

q   a decimal value between 0 and 1, referring to the parameter of the metric $\delta_{p,q}^\star$.

breakpoints   a positive arbitrary integer representing the number of breaks chosen to build the numerical α-cuts. It is fixed to 100 by default.

spec   specification of the fuzzification matrix. The possible values are "Identical" and "Not Identical".

Value

Returns the data set of individual evaluations, for which the number of observations is exactly the same as the initial data set.

See Also

distance 12, adjusted.weight.SI 25, adjusted.weight.MI 26, GLOB.EVAL 28, R 29, Ri 30.

Examples

```
MI <- 3
bmi <- c(1/3,1/3,1/3)
b_jkt <- matrix(c(0.5,0.125,0.2,0.5,0.125,0.2,0,0.125,
0.2,0,0.125,0.2,0,0.125,
0.2,0,0.125,0,0,0.125,0,0,0.125,0),nrow=MI)
SI <- c(2,8,5)
range <- matrix(c(rep(0,15), rep(28,15)), ncol=2)
IND.EVAL(data, MI, bmi, SI, b_jkt, range = range,
distance.type ="DSGD.G")
```

GLOB.EVAL   *Calculates the global evaluation of a linguistic questionnaire*
28

Description

This function calculates the global evaluation of a linguistic questionnaire as expressed in Eq. 7.11. The arguments of this function are very similar to the ones of the function IND.EVAL 27.

We have to remind that under the assumption of non-missingness, the global evaluation is the weighted mean of the set of individual evaluations resulting from the function IND.EVAL 27, as seen in Proposition 7.3.2. Thus, one can get the global evaluation using the basic function Weighted.mean. Nevertheless, we introduce a function GLOB.EVAL.mean (ind.eval, p_ind) applied on the vector of individual evaluations called ind.eval, which executes this task using the vector of weights denoted by p_ind. We add that if no sampling weights are applied, the argument p_ind is set to be p_ind=rep(1, length(ind.eval)).

Usage

```
GLOB.EVAL   (data, MI, bmi, SI, b_jkt, p_ind =
             rep(1/nrow(data)), distance.type, i=1, j=1, theta
             = 1/3, thetas=1, p=2, q=0.5, breakpoints=100)
```

Arguments

data	the data set to evaluate.
MI	a numerical value representing the total number of main-items dividing the linguistic questionnaire.
bmi	an array referring to the initial weights of the main-items.
SI	an array representing the total number of sub-items per main-item.
b_jkt	a matrix of MI rows and max(SI) columns expressing the initial weights of each sub-item of a given main-item. By analogy to the hypothesis of the method described in Sect. 7.1, this matrix should be given in the following form:

$$
b_{jkt} = \begin{pmatrix} b_{11t} & \dots & b_{1kt} & \dots & b_{1m_jt} \\ \vdots & & \vdots & & \vdots \\ b_{j1t} & \dots & b_{jkt} & \dots & b_{jm_jt} \\ \vdots & & \vdots & & \vdots \\ b_{r1t} & \dots & b_{rkt} & \dots & b_{rm_jt} \end{pmatrix}.
$$

p_ind	a vector of the relative sampling weights of the units, for which length(p_ind) = nrow(data). If the weights are not relative, the following expression should be applied to the vector:

$$
\frac{p\_ind}{\sum_{i=1}^{n} p\_ind}.
$$

If no sampling weights are used, the vector of weights is reduced to a vector of values 1, i.e. rep(1, nrow(data)).

distance.type	type of distance chosen from the family of distances, set by default to the signed distance. The different choices are given by "Rho1", "Rho2", "Bertoluzza", "Rhop", "Delta.pq", "Mid/Spr", "wabl", "DSGD", "DSGD.G", and "GSGD".

i and j	parameters of the density function of the Beta distribution, fixed by default to $i = 1$ and $j = 1$.
theta	a numerical value between 0 and 1, representing a weighting parameter. By default, theta is fixed to $\frac{1}{3}$ referring to the Lebesgue space. This measure is used in the calculations of the following distances: $d_{Bertoluzza}$, $d_{mid/spr}$, and $d_{\phi-wabl/ldev/rdev}$.
thetas	a decimal value between 0 and 1, representing the weight given to the shape of the fuzzy number. By default, thetas is fixed to 1. This parameter is used in the calculations of the $d_{SGD}^{\theta^\star}$ and the d_{GSGD} distances.
p	a positive integer such that $1 \le p \le \infty$, referring to the parameter of the metric $\rho_p^\star$ or $\delta_{p,q}^\star$.
q	a decimal value between 0 and 1, referring to the parameter of the metric $\delta_{p,q}^\star$.
breakpoints	a positive arbitrary integer representing the number of breaks chosen to build the numerical α-cuts. It is fixed to 100 by default.

Value

Returns a numerical value representing the global evaluation of the linguistic questionnaire.

See Also

distance 12, adjusted.weight.SI 25, adjusted.weight.MI 26, IND.EVAL 27, R 29, Ri 30.

Examples

```
MI <- 3
bmi <- c(1/3,1/3,1/3)
b_jkt <- matrix(c(0.5,0.125,0.2,0.5,0.125,
0.2,0,0.125,0.2,0,0.125,0.2,0,0.125,
0.2,0,0.125,0,0,0.125,0,0,0.125,0),nrow=MI)
SI <- c(2,8,5)
range <- matrix(c(rep(0,15), rep(28,15)), ncol=2)
ind.eval <- IND.EVAL(data, MI, bmi, SI, b_jkt,
range=range, distance.type
="GSGD")
GLOB.mean <- GLOB.EVAL.mean(ind.eval)
GLOB <- GLOB.EVAL(data, MI, bmi, SI, b_jkt,
distance.type ="GSGD")
```

R	*Calculates the indicator of information rate of the data base*	29

Description

This function calculates the indicator of information rate of the complete data base as given in Eq. 7.13. This computation uses the functions adjusted.weight.SI 25 and Delta.jki.

Usage

R   (data, p_ind, b_jk, SI)

Arguments

data   the data set to evaluate.

p_ind   a vector of the relative sampling weights of the units. Thus, if the weights are not relative, the following expression should be applied to the vector:

$$\frac{\text{p\_ind}}{\sum_{i=1}^{n}\text{p\_ind}}.$$

b_jk   a matrix of length(b_j) rows and max(SI) columns expressing the initial weights of each sub-item of a given main-item. By analogy to the hypothesis of the method described in Sect. 7.1, this matrix should be given in the following form:

$$b_{jk} = \begin{pmatrix} b_{11} & \cdots & b_{1k} & \cdots & b_{1m_j} \\ \vdots & & \vdots & & \vdots \\ b_{j1} & \cdots & b_{jk} & \cdots & b_{jm_j} \\ \vdots & & \vdots & & \vdots \\ b_{r1} & \cdots & b_{rk} & \cdots & b_{rm_j} \end{pmatrix}.$$

SI   an array representing the total number of sub-items per main-item.

Value

Returns a numerical value giving the indicator of information rate of the complete linguistic questionnaire. Note that the obtained value is interpreted as the more it tends to the value 1, the less the complete questionnaire contains missing values.

See Also

adjusted.weight.SI 25, adjusted.weight.MI 26, IND.EVAL 27, GLOB.EVAL 28, Ri 30.

Examples

```
p_ind <- weight/sum(weight)
b_jk <- matrix(c(0.5,0.125,0.2,0.5,0.125,0.2,0,0.125,
0.2,0,0.125,0.2,0,0.125,
0.2,0,0.125,0,0,0.125,0,0,0.125,0),nrow=MI)
SI <- c(2,8,5)
R (data, p_ind, b_jk, SI)
```

Ri	*Calculates the indicator of information rate of the data base for a given*
unit	*30*

Description

This function calculates the indicator of information rate of the complete data base for a given unit as given in Eq. 7.14. This computation uses the functions adjusted.weight.SI 25 and Delta.jki.

Usage

```
Ri   (data, i, b_jk, SI)
```

Arguments

data   the data set to evaluate.

i   the observation index.

b_jk   a matrix of `length(b_j)` rows and `max(SI)` columns expressing the initial weights of each sub-item of a given main-item. By analogy to the hypothesis of the method described in Sect. 7.1, this matrix should be given in the following form:

$$b_{jk} = \begin{pmatrix} b_{11} & \cdots & b_{1k} & \cdots & b_{1m_j} \\ \vdots & & \vdots & & \vdots \\ b_{j1} & \cdots & b_{jk} & \cdots & b_{jm_j} \\ \vdots & & \vdots & & \vdots \\ b_{r1} & \cdots & b_{rk} & \cdots & b_{rm_j} \end{pmatrix}.$$

SI   an array representing the total number of sub-items per main-item.

Value

Returns a numerical value giving the indicator of information rate of the complete linguistic questionnaire for a particular observation. Note that the obtained value is interpreted as the more it tends to the value 1, the less the observation i contains missing values.

See Also

`adjusted.weight.SI` 25, `adjusted.weight.MI` 26, `IND.EVAL` 27, `GLOB.EVAL` 28, R 29.

Examples

```
b_jk <- matrix(c(0.5,0.125,0.2,0.5,0.125,0.2,0,0.125,
0.2,0,0.125,0.2,0,0.125,
0.2,0,0.125,0,0,0.125,0,0,0.125,0),nrow=MI)
SI <- c(2,8,5)
Ri (data, i = 4, b_jk, SI)
```

Description

This function estimates a Mult-FANOVA model based on the construction of linear regression models. For this model, multiple factors can be introduced. The detailed procedure is described in Sect. 8.3. This function can be computed by 3 different methods (`distance`, `exact`, and `approximation`) as given in the description of the function `Fuzzy.variance` 18. The descriptions of the three procedures are given as follows:

- `distance`: By respect to a conveniently chosen distance, the calculations can be done as proposed in the first part of Sect. 8.3 and in Berkachy and Donzé (2018) for the univariate one-way case using the signed distance. This method can be applied using the function `distance` 12. The computed distances used in the calculations of the sums of squares are seen as a defuzzification of the obtained fuzzy numbers. The decision rule is done by comparing the obtained test statistics with respect to the distances to the quantiles of the Fisher distribution.

- `exact`: For the method denoted by `exact`, the functions `Fuzzy.Difference` 5 and `Fuzzy. Square` 6 as numerical calculations of the difference and the square of two fuzzy numbers are used. In this case, we practically preserve the fuzzy nature of the sums of square. Thus, no defuzzification of these sums is directly performed. Fuzzy numbers referring to the mean of the fuzzy treatment sums of squares $\tilde{B}$ and the weighted mean of the fuzzy residuals sums of squares denoted by $\tilde{E}^\star$ are constructed in order to make a decision. However, since overlapping between these fuzzy numbers is often expected, then, a defuzzification as a last step is required. The decision rule for this test is based on the surfaces of both fuzzy numbers $\tilde{B}$ and $\tilde{E}^\star$. The contribution of each fuzzy number to the total variation of fuzzy sums of squares is afterward given.

- `approximation`: For the third case, i.e. `approximation`, the function is defined exactly in the same manner as the function by the `exact` method. The function `Fuzzy.Difference` 5 is used for the difference operation basically between trapezoidal or triangular fuzzy numbers. However, the calculation of the fuzzy square is based on an approximation for this case. The approximation used is the one given in Approximation 4 of the function `Fuzzy.variance` 18. In other terms, for the fuzzy product, we use the operator proposed by the package `FuzzyNumbers`. The decision rule is the same as the one for the `exact` procedure.

For the cases `exact` and `approximation`, we have to highlight that the outcome related to each factor could be printed at a time. No view of the overall set of factors can be exposed. Thus, the index of the factor in the model should be entered by the user.

From another side, we note that for the univariate case, a similar function is constructed. It is denoted by FANOVA and could be applied using exactly the same three methods previously described. For the case with the distance method, the procedure is described in Berkachy and Donzé (2018). For the cases with the exact and approximation methods, the function FANOVA returns fuzzy type decisions. Since the defuzzification is needed in these cases, a function called Defuzz.FANOVA is proposed. The distance to be used in this case is set by default to the signed distance. Yet, several metrics can be used for this calculation. The output of the function Defuzz.FANOVA is the same as the FANOVA one but with the defuzzified results. We add that the bootstrap technique is used in such procedures to estimate the distributions of the corresponding statistics. A final remark is that for this function, the data set should be attached.

Usage

```
FMANOVA   (formula, dataset, data.fuzzified, sig = 0.05, method,
          distance.type = "DSGD", index.var = NA, nsim=100,
          i=1, j=1, theta = 1/3, thetas = 1, p=2, q=0.5,
          breakpoints=100, int.method = "int.simpson", plot =
          TRUE)
```

Arguments

formula	a description of the model to be fitted.
dataset	the data frame containing all the variables of the model.
data.fuzzified	the fuzzified data set constructed by a call to the function FUZZ or the function GFUZZ, or a similar matrix.
sig	the significance level of the test. It is set by default to 0.05.
method	the choices are the following: "distance", "exact", and "approximation".
index.var	the column index of the considered variable for which the output will be printed. It is an argument of the Mult-FANOVA models by the exact and the approximation methods only.

nsim	an integer giving the number of replications needed in the bootstrap procedure. It is set to 100 by default.
distance.type	type of distance chosen from the family of distances. The different choices are given by "Rho1", "Rho2", "Bertoluzza", "Rhop", "Delta.pq", "Mid/Spr", "wabl", "DSGD", "DSGD.G", and "GSGD". It is set by default to "DSGD".
i and j	parameters of the density function of the Beta distribution, fixed by default to $i = 1$ and $j = 1$.
theta	a numerical value between 0 and 1, representing a weighting parameter. By default, theta is fixed to $\frac{1}{3}$ referring to the Lebesgue space. This measure is used in the calculations of the following distances: $d_{Bertoluzza}$, $d_{mid/spr}$, and $d_{\phi-wabl/ldev/rdev}$.
thetas	a decimal value between 0 and 1, representing the weight given to the shape of the fuzzy number. By default, thetas is fixed to 1. This parameter is used in the calculations of the $d^{\theta^*}_{SGD}$ and the d_{GSGD} distances.
p	a positive integer such that $1 \leq p \leq \infty$, referring to the parameter of the metric ρ^*_p or $\delta^*_{p,q}$.
q	a decimal value between 0 and 1, referring to the parameter of the metric $\delta^*_{p,q}$.
breakpoints	a positive arbitrary integer representing the number of breaks chosen to build the numerical α-cuts. It is fixed to 100 by default.
int.method	the method of numerical integration. It is set by default to the Simpson method, i.e. int.method="int.simpson".
plot	fixed by default to "TRUE". plot="FALSE" if a plot of the fuzzy number is not required.

Value

Returns a list of all the arguments of the function, the total, treatment and residuals sums of squares, the coefficients of the model, the test statistics with the corresponding p-values, and the decision made.

See Also

summary.FMANOVA 32, is.balanced 33, SEQ.ORDERING 26, FTukeyHSD 35, Ftests 36, distance 12, Fuzzy.Difference 5, Fuzzy.Square 6.

Examples

```
YExVar <- FUZZ(data,1,15,c(1,2,3,4,5))
attach(data)
formula <- ExVar ~ F1 + F2 + F1*F2
res <- FMANOVA(formula, data, YExVar,
method = "distance", distance.type = "wabl")
res1 <- FMANOVA(formula, data, YExVar, method = "exact",
index.var = 1)
```

summary.FMANOVA	*Prints the summary of the estimation of a Mult-*
FANOVA model	32

Description

In the case of the Mult-FANOVA model computed using a given distance, this function prints the summary of the estimation of the corresponding Mult-FANOVA model, resulting from the function FMANOVA 31. If the considered model includes interaction terms, then the function summary.interactions.FMANOVA (res) can be used to print the summary statistics related to these terms. We note that the obtained output is very similar to the one given by the known aov and lm functions of R. Thus, the elements of the result of a call of the function FMANOVA are compatible with the class of lm functions, as an instance with the functions terms, fitted.values, residuals, df.residuals, etc.

For the one-way case, an analog function denoted by summary.FANOVA is introduced as well, in order to be compatible with the function FANOVA.

Usage

```
summary.FMANOVA   (res)
```

Arguments

res   a result of a call of FMANOVA 31, where method = "distance".

Value

Returns a list of summary statistics of the estimated model given in res, shown in an FMANOVA table. In addition, the F-statistics with their *p*-values and the decision are given.

See Also

FMANOVA 31, is.balanced 33, SEQ.ORDERING 34, FTukeyHSD 35, Ftests 36.

Examples

```
YExVar <- FUZZ(data,1,15,c(1,2,3,4,5))
attach(data)
formula <- ExVar ~ F1 + F2 + F1*F2
res <- FMANOVA(formula, data, YExVar,
method = "distance", distance.type = "wabl")
summary.FMANOVA(res)
summary.interaction.FMANOVA(res)
```

---

is.balanced   *Verifies if a design is balanced*   33

Description

This function is used to verify if a considered fitting model is balanced, i.e. if the number of observations by factor levels is the same.

Usage

```
is.balanced    (nvar)
```

Arguments

nvar   a line array given by the contingency table related to the considered variable, often written as a result of a call of the function table.

Value

Returns a logical decision TRUE or FALSE, to indicate if a given design is, respectively, balanced or not.

See Also

FMANOVA 31, summary.FMANOVA 32, SEQ.ORDERING 34, FTukeyHSD 35, Ftests 36, table.

Examples

```
data <- matrix(c(1,2,3,2,2,1,1,3,1,2),ncol=1)
ni <- t(table(data))
is.balanced(ni)
```

SEQ.ORDERING   *Calculates the sequential sums of squares*   34

Description

If the design of the model is not balanced, such that is.balanced = FALSE, the ordering of the variables affects the model. This function recalculates then the fitting model but by taking into account the sequential ordering of the factors. It calculates as well the coefficients of the model, the predicted values, and the residuals according to the new model. We add that the coefficients of the model are calculated by compliance to the least-squares method. Finally note that 3 versions of this function, related to the 3 methods (`distance`, `exact`, and `approximation`), are proposed separately. These versions are, respectively, called `SEQ.ORDERING`, `SEQ.ORDERING.EXACT`, and `SEQ.ORDERING.APPROXIMATION`.

Usage

```
SEQ.ORDERING    (scope, data, f.response)
```

Arguments

`scope`	a description of the complete fitting model.
`data`	the data frame containing all the variables of the model.
`f.response`	the vector of distances of the fuzzy response variable to the fuzzy origin.

Value

Returns a list of the new sets of sums of squares, as well as the coefficients, the residuals, and the `fitted.values`.

See Also

`FMANOVA` 31, `summary.FMANOVA` 32, `is.balanced` 33, `FTukeyHSD` 35, `Ftests` 36, `distance` 12.

Examples

```
attach(data)
formula <- ExVar ~ F1 + F2 + F1*F2
f.response <- distance(ExVar,
TriangularFuzzyNumber(0,0,0), "GSGD")
```

```
SEQ.ORDERING (scope = formula, data = data, f.response
= f.response)
```

FTukeyHSD   *Calculates the Tukey HSD test corresponding to the fuzzy response*
variable   35

Description

In the case of the Mult-FANOVA model performed by the distance method, this function calculates the Tukey HSD test applied on the mean of the fuzzy response variable related to the different factor levels. We have to remind that this test is done by variable and not for the complete model.

Usage

```
FTukeyHSD    (test, variable, cont=c(1,-1), conf.level=0.95)
```

Arguments

test	a result of a call of FMANOVA (formula, method="distance", ...) 31.
variable	the name of a variable in the data set.
cont	the contrasts of the model. It is set by default to $[1; -1]$.
conf.level	the confidence level of the test. It is set by default to 95%.

Value

Returns a table of comparisons of means of the different levels of a given factor, two by two. The table contains the means of populations, the lower and upper bounds of the confidence intervals, and their p-values.

See Also

FMANOVA 31, summary.FMANOVA 32, is.balanced 33, SEQ.ORDERING 34, Ftests 36, distance 12.

Examples

```
YExVar <- FUZZ(data,1,15,c(1,2,3,4,5))
attach(data)
formula <- ExVar ~ F1 + F2 + F1*F2
res <- FMANOVA(formula, data, YExVar,
method = "distance", distance.type =
"wabl")
FTukeyHSD(res, "F1")[[1]]
```

Ftests	*Calculates multiple tests corresponding to the fuzzy response*
variable	36

Description

In the case of the Mult-FANOVA model performed by the distance method, this function calculates multiple indicators of the comparison between the means of the different level factors. We draw the attention that these indicators are constructed on the sums of squares related to the complete model. Thus, no particular factors are specifically involved.

Usage

```
Ftests   (test)
```

Arguments

`test`   a result of a call of `FMANOVA(formula, method="distance", ...)` 31.

Value

Returns a table of the following different indicators: "Wilks", "F-Wilks", "Hotelling-Lawley trace", and "Pillai Trace".

See Also

`FMANOVA` 31, `summary.FMANOVA` 32, `is.balanced` 33, `SEQ.ORDERING` 34, `FTukeyHSD` 35, `distance` 12.

Examples

```
YExVar <- FUZZ(data,1,15,c(1,2,3,4,5))
attach(data)
formula <- ExVar ~ F1 + F2 + F1*F2
res <- FMANOVA(formula, data, YExVar,
method = "distance", distance.type = "wabl")
Ftests(res)
```

Appendix

E   Intersection Situations Between the Fuzzy Confidence Intervals and the Fuzzy Hypothesis

We express the possible situations of intersection between a given fuzzy confidence interval and its complements from one side and the fuzzy null hypothesis from another one. The different FCIs are calculated by the traditional expression. The proposed situations correspond to the one-sided and the two-sided tests. As such, Table 9.3 displays the cases of a left one-sided test, Table 9.4 displays the cases of a right one-sided test, and Table 9.5 displays the ones of a two-sided test. For all these cases, we note that the fuzzy null hypothesis $\tilde{H}_0$ (the red curve), its core θ_0 (the green dotted line), the fuzzy confidence interval $\tilde{\Pi}$ (the blue curve), and its complement $\neg\tilde{\Pi}$ (the black curve) are proposed.

Table 9.3 The different situations of intersection between the fuzzy null hypothesis $\tilde{H}_0$ and the confidence intervals: case of left one-sided test

Left one-sided test	
Situations	Tuples of fuzzy intersection numbers
	$\tilde{D}_0 = (1, 1, 1) \mid \tilde{D}_0 = (0, 0, 0)$ $\tilde{D}_1 = (0, 0, 0) \mid \tilde{D}_1 = (1, 1, 1)$ H_0 is not rejected $\mid H_0$ is rejected
	$\tilde{D}_0 = (\mu_{\tilde{\pi}}(A_R), \mu_{\tilde{\pi}}(\theta_0), \mu_{\tilde{\pi}}(A_L))$ $\tilde{D}_1 = (0, \mu_{\neg\tilde{\pi}}(\theta_0), \mu_{\neg\tilde{\pi}}(R_R))$
	$\tilde{D}_0 = (1, 1, 1)$ $\tilde{D}_1 = (0, 0, \mu_{\neg\tilde{\pi}}(R_R))$

(continued)

Table 9.3 (continued)

Left one-sided test

Situations		Tuples of fuzzy intersection numbers
		$\tilde{D}_0 = (0, \mu_{\tilde{\pi}}(\theta_0), \mu_{\tilde{\pi}}(A_L))$ $\tilde{D}_1 = (\mu_{\neg\tilde{\Pi}}(R_L), \mu_{\neg\tilde{\Pi}}(\theta_0), \mu_{\neg\tilde{\Pi}}(R_R))$
		$\tilde{D}_0 = (0, 0, \mu_{\tilde{\pi}}(A_L))$ $\tilde{D}_1 = (1, 1, 1)$
		$\tilde{D}_0 = (\mu_{\tilde{\pi}}(A_R), \mu_{\tilde{\pi}}(\theta_0), \mu_{\tilde{\pi}}(A_L))$ $\tilde{D}_1 = (\mu_{\neg\tilde{\Pi}}(R_L), \mu_{\neg\tilde{\Pi}}(\theta_0), \mu_{\neg\tilde{\Pi}}(R_R))$
Else		The fuzziness is high

Table 9.4 The different situations of intersection between the fuzzy null hypothesis $\tilde{H}_0$ and the confidence intervals: case of right one-sided test

Right one-sided test

Situations	Tuples of fuzzy intersection numbers
	$\tilde{D}_0 = (0, 0, 0) \mid \tilde{D}_0 = (1, 1, 1)$ $\tilde{D}_1 = (1, 1, 1) \mid \tilde{D}_1 = (0, 0, 0)$ H_0 is rejected $\mid H_0$ is not rejected
	$\tilde{D}_0 = (0, \mu_{\tilde{\pi}}(\theta_0), \mu_{\tilde{\pi}}(A_R))$ $\tilde{D}_1 = (\mu_{\neg\tilde{\pi}}(R_R), \mu_{\neg\tilde{\pi}}(\theta_0), \mu_{\neg\tilde{\pi}}(R_L))$
	$\tilde{D}_0 = (0, 0, \mu_{\tilde{\pi}}(A_R))$ $\tilde{D}_1 = (1, 1, 1)$

(continued)

Table 9.4 (continued)

Right one-sided test

Situations	Tuples of fuzzy intersection numbers
	$\tilde{D}_0 = (\mu_{\tilde{\pi}}(A_L), \mu_{\tilde{\pi}}(\theta_0), \mu_{\tilde{\pi}}(A_R))$ $\tilde{D}_1 = (0, \mu_{\neg\tilde{\pi}}(\theta_0), \mu_{\neg\tilde{\pi}}(R_L))$
	$\tilde{D}_0 = (1, 1, 1)$ $\tilde{D}_1 = (0, 0, \mu_{\neg\tilde{\pi}}(R_L))$
	$\tilde{D}_0 = (\mu_{\tilde{\pi}}(A_L), \mu_{\tilde{\pi}}(\theta_0), \mu_{\tilde{\pi}}(A_R))$ $\tilde{D}_1 = (\mu_{\neg\tilde{\pi}}(R_R), \mu_{\neg\tilde{\pi}}(\theta_0), \mu_{\neg\tilde{\pi}}(R_L))$
Else	The fuzziness is high

Table 9.5 The different situations of intersection between the fuzzy null hypothesis $\tilde{H}_0$ and the confidence intervals: case of two-sided test

Two-sided test	
Situations	Tuples of fuzzy intersection numbers
	$\tilde{D}_0 = (0, 0, 0)$ $\tilde{D}_1 = (1, 1, 1)$
	$\tilde{D}_0 = (1, 1, 1)$ $\tilde{D}_1 = (0, 0, 0)$
	$\tilde{D}_0 = (0, 0, \mu_{\tilde{H}}(A_R))$ $\tilde{D}_1 = (\mu_{\neg\tilde{H}}(R_R), 1, 1)$

(continued)

Table 9.5 (continued)

Two-sided test	
Situations	Tuples of fuzzy intersection numbers
	$\tilde{D}_0 = (0, \mu_{\tilde{H}}(\theta_0), \mu_{\tilde{H}}(A_R))$ $\tilde{D}_1 = (0, \mu_{\neg\tilde{\pi}}(\theta_0), \mu_{\neg\tilde{\pi}}(R_L))$
	$\tilde{D}_0 = (0, \mu_{\tilde{H}}(\theta_0), \mu_{\tilde{H}}(A_R))$ $\tilde{D}_1 = (\mu_{\neg\tilde{\pi}}(R_R), \mu_{\neg\tilde{\pi}}(\theta_0), \mu_{\neg\tilde{\pi}}(R_L))$
	$\tilde{D}_0 = (0, 0, \mu_{\tilde{H}}(A_L))$ $\tilde{D}_1 = (\mu_{\neg\tilde{\pi}}(R_L), 1, 1)$

$$\tilde{D}_0 = (0, \mu_{\tilde{n}}(\theta_0), \mu_{\tilde{n}}(A_L))$$
$$\tilde{D}_1 = (0, \mu_{\neg\tilde{n}}(\theta_0), \mu_{\neg\tilde{n}}(R_R))$$

$$\tilde{D}_0 = (0, \mu_{\tilde{n}}(\theta_0), \mu_{\tilde{n}}(A_L))$$
$$\tilde{D}_1 = (\mu_{\neg\tilde{n}}(R_L), \mu_{\neg\tilde{n}}(\theta_0), \mu_{\neg\tilde{n}}(R_R))$$

$$\tilde{D}_0 = (\mu_{\tilde{n}}(A_L), \mu_{\tilde{n}}(\theta_0), \mu_{\tilde{n}}(A_R))$$
$$\tilde{D}_1 = (\mu_{\neg\tilde{n}}(R_R), \mu_{\neg\tilde{n}}(\theta_0), \mu_{\neg\tilde{n}}(R_L))$$

(continued)

Table 9.5 (continued)

Two-sided test

Situations	Tuples of fuzzy intersection numbers
	$\tilde{D}_0 = (\mu_{\tilde{\pi}}(A_R), \mu_{\tilde{\pi}}(\theta_0), \mu_{\tilde{\pi}}(A_L))$ $\tilde{D}_1 = (\mu_{\neg\tilde{\pi}}(R_L), \mu_{\neg\tilde{\pi}}(\theta_0), \mu_{\neg\tilde{\pi}}(R_R))$
	$\tilde{D}_0 = (1, 1, 1)$ $\tilde{D}_1 = (0, 0, \mu_{\neg\tilde{\pi}}(R_L))$
	$\tilde{D}_0 = (\mu_{\tilde{\pi}}(A_L), \mu_{\tilde{\pi}}(\theta_0), \mu_{\tilde{\pi}}(A_R))$ $\tilde{D}_1 = (\mu_{\neg\tilde{\pi}}(R_R), \mu_{\neg\tilde{\pi}}(\theta_0), \mu_{\neg\tilde{\pi}}(R_L))$

$$\tilde{D}_0 = (1, 1, 1)$$
$$\tilde{D}_1 = (0, 0, \mu_{\neg\tilde{\Pi}}(R_R))$$

$$\tilde{D}_0 = (\mu_{\tilde{\pi}}(A_R), \mu_{\tilde{\pi}}(\theta_0), \mu_{\tilde{\pi}}(A_L))$$
$$\tilde{D}_1 = (\mu_{\neg\tilde{\Pi}}(R_L), \mu_{\neg\tilde{\Pi}}(\theta_0), \mu_{\neg\tilde{\Pi}}(R_R))$$

Else | The fuzziness is high

F   Possible Positions of the Areas A_l and A_r Under the MF of $\tilde{t}$ on the Left and Right Sides of the Median $\tilde{M}$

See Figs. 9.1 and 9.2.

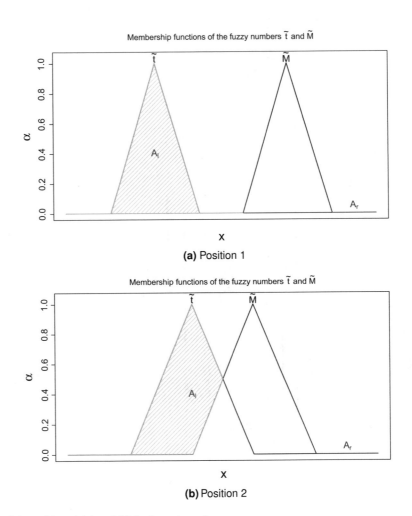

Fig. 9.1 Positions (**a**) 1 and (**b**) 2 where $A_l > A_r$

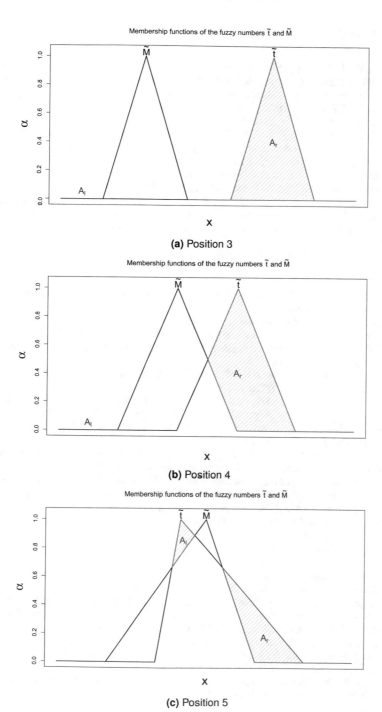

(a) Position 3

(b) Position 4

(c) Position 5

Fig. 9.2 Positions (**a**) 3, (**b**) 4, and (**c**) 5 where $A_l \leq A_r$

References

Berkachy, R., & Donzé, L. (2018). Fuzzy one-way ANOVA using the signed distance method to approximate the fuzzy product. In L.F.A. Collectif (Ed.), *LFA 2018 Rencontres francophones sur la Logique Floue et ses Applications*, Cépaduès (pp. 253–264).

Berkachy, R., & Donzé, L. (2020). *FuzzySTs: Fuzzy Statistical Tools R package.* https://CRAN.R-project.org/package=FuzzySTs.

Blanco-Fernandez, A., et al. (2013). Random fuzzy sets: A mathematical tool to develop statistical fuzzy data analysis. *Iranian Journal of Fuzzy Systems, 10*(2), 1–28. ISSN: 1735-0654. https://doi.org/10.22111/ijfs2013609. http://ijfs.usb.ac.ir/article_609html.

Burda, M. (2015). Linguistic fuzzy logic in R. In *Proc. IEEE International Conference On Fuzzy Systems (FUZZ-IEEE)* (pp. 1–7). Istanbul: IEEE. ISBN: 978-1-4673-7428-6.

Burda, M. (2018). *lfl: Linguistic Fuzzy Logic R Package.* https://CRAN.R-project.org/package=lfl.

Denoeux, T. (2011). Maximum likelihood estimation from fuzzy data using the EM algorithm. *Fuzzy Sets and Systems, 183*(1), 72–91. ISSN: 0165-0114. https://doi.org/1.1016/j.fss.2011.05.022.

Filzmoser, P., & Viertl, R. (2004). Testing hypotheses with fuzzy data: The fuzzy *p*-value. *Metrika* (vol. 59, pp. 21–29). Springer-Verlag. ISSN: 0026-1335. https://doi.org/10.1007/s001840300269.

Gagolewski, M. (2014). *FuzzyNumbers Package: Tools to Deal with Fuzzy Numbers in R.* http://FuzzyNumbers.rexamine.com/.

Gil, M. A., et al. (2006). Bootstrap approach to the multi-sample test of means with imprecise data. *Computational Statistics & Data Analysis, 51*(1), 148–162. ISSN: 0167-9473. https://doi.org/10.1016/j.csda.2006.04.018. http://www.sciencedirect.com/science/article/pii/S0167947306001174.

Knott, C., et al. (2013). *FuzzyToolkitUoN: Type 1 Fuzzy Logic Toolkit, R Package.* https://CRAN.Rproject.org/package=FuzzyToolkitUoN.

Lubiano, A., & de la Rosa de Saa, S. (2017). *FuzzyStatTra: Statistical Methods for Trapezoidal Fuzzy Numbers R Package.* https://CRAN.R-project.org/package=FuzzyStatTra.

Parchami, A. (2017). *Fuzzy.p.value: Computing Fuzzy p-Value R Package.* https://CRAN.R-project.org/package=Fuzzy.p.value.

Parchami, A. (2018a). *ANOVA.TFNs: One-Way Analysis of Variance Based on Triangular Fuzzy Numbers R Package.* https://CRAN.R-project.org/package=ANOVA.TFNs.

Parchami, A. (2018b). *EM.Fuzzy: EM Algorithm for Maximum Likelihood Estimation by Non-Precise Information, R Package.* https://CRAN.R-project.org/package=EM.Fuzzy.

Parchami, A., Taheri, S.M., & Mashinchi, M. (2010). Fuzzy *p*-value in testing fuzzy hypotheses with crisp data. *Statistical Papers, 51*(1), 209–226. ISSN: 0932-5026. https://doi.org/10.1007/s00362-008-0133-4.

Parchami, A., Nourbakhsh, M., & Mashinchi, M. (2017). Analysis of variance in uncertain environments. *Complex & Intelligent Systems, 3*(3), 189–196.

R Core Team. (2015). *R: A Language and Environment for Statistical Computing.* R Foundation for Statistical Computing. Vienna, Austria. https://www.R-project.org/.

Sinova, B., de la Rosa de Saa, S., & Gil, M.A. (2013). A generalized L1-type metric between fuzzy numbers for an approach to central tendency of fuzzy data. *Information Sciences, 242*, 22–34. ISSN: 0020-0255. https://doi.org/10.1016/j.ins.2013.03.063.

Sinova, B., Gil, M. A., & Van Aelst, S. (2016). M-estimates of location for the robust central tendency of fuzzy data. *IEEE Transactions on Fuzzy Systems, 24*(4), 945–956. ISSN: 1063-6706. https://doi.org/10.1109/TFUZZ.2015.2489245.

Trutschnig, W., & Lubiano, A. (2018). *SAFD: Statistical Analysis of Fuzzy Data, R Package.* https://CRAN.Rproject.org/package=SAFD.

Conclusion

The core objective of this three-part thesis has been to introduce improved versions of the signed distance measure and to provide a range of fuzzy statistical analyses based on these distances, from a theoretical, an empirical, and a programming points of view.

We proposed in the first part the foundations of the theory of fuzzy sets and known concepts of the Mamdani–Sugeno fuzzy process. We then exposed a review of some metrics used to measure the distance between two fuzzy numbers. Even though the signed distance measure appears to have nice properties, this distance presents some drawbacks. For this reason, we presented improved versions of it given by two L_2 distances called the $d_{SGD}^{\theta^\star}$ and the generalized signed distance d_{GSGD}. The d_{GSGD} distance is a directional one which has comparable properties to the signed distance, while the $d_{SGD}^{\theta^\star}$ distance is a non-directional one. The key strength of these distances is that they take into consideration the shape and the spreads of the involved fuzzy numbers. This fact makes their use more efficient and realistic than the original signed distance. In addition, for the d_{GSGD}, the directionality is conserved. Thus, the sign of this measure describes the direction of travel between the concerned fuzzy numbers. Our purpose was to investigate the use of these distances in various statistical methods.

We exposed after two approaches defining the concept of random variables in a fuzzy context, with their corresponding parameters and their point estimators. The use of a given metric is in some cases required. Therefore, we proposed the original signed distance and the improved versions of it. Using these distances, we estimated different statistical parameters, such as the sample mean, the classical sample moments, the variance, etc. To illustrate these definitions, we provided a simulation study where uncertain random data sets were generated in different setups. The objectives were to understand the influence of increasing the size of the chosen sample on the variance, the skewness, and the kurtosis measures from one side and to investigate the differences of use of the proposed distances in terms of these statistical measures from another one. By this analysis, we found that the size of the sample did not affect the concerned measures. Accordingly, one could settle on a relatively medium sized data base, in terms of the threshold of 1000 observations, for example. For the second test, we saw that the variance and the kurtosis calculated using all the distances are close. No clear interpretations on the difference of use between all the distances can be depicted. However, for the skewness measure, the situation is different. By the non-directional distances, we get very close skewness measures, while the ones by the directional distances are very close as well. In addition, the directional distances present a skewness close to the value 0, in contrary with the non-directional ones. We could then see the relevance of the use of such directional distances compared to the ones already existing.

All the abovementioned notions lead us to introduce fuzzy hypotheses testing procedures using the defended distances. The fuzziness of the data and/or the hypotheses can be treated by our methods. We showed a methodology of testing hypotheses based on fuzzy confidence intervals. These latter are usually estimated using traditional expressions defined for specific parameters only and where the distributions have to be a priori defined. We discussed a practical procedure of construction of such intervals by the likelihood ratio method, for which the advantage is that it is in some sense general and it can be used with any type of parameters without the obligation of pre-defining a particular distribution. Using these intervals, we propose test statistics inducing decisions given by fuzzy numbers. One has then to decide whether to reject or not the null hypothesis according to the obtained fuzzy decisions. Since this task could sometimes lead to a difficult interpretation, we propose to defuzzify these decisions by corresponding tools, such as the original signed distance or an improved version of it. In the conventional theory, it is well-known that a test decision can also be made by calculating a p-value and comparing it to the significance level. It is the same for the fuzzy theory. Thus, we presented a methodology of calculation of the fuzzy p-value where the data and/or the hypotheses are assumed to be vague. The decision rule was also given. The fuzzy p-value can accordingly be defuzzified. Both testing approaches were illustrated by applications on real data sets. Moreover, for the p-value calculations, we proposed different additional tests in the purpose of investigating the influence of the shape and spread of the null and the alternative hypotheses (considered as fuzzy) on the obtained decisions. We found that the alternative hypothesis has no influence on the obtained decision. However, it is clear that variating the shape and the spread of the fuzzy null hypothesis could influence heavily the decision. By comparing these decisions to the ones obtained from the classical approach, we concluded that these latter seem to be a particular case of the fuzzy one. In addition, one could deduce that the more the data set contains fuzziness, the more the decision made tends to be farther from the one obtained by the conventional procedure.

For the application part of this thesis, we showed two main cases where the generalized signed distance and some other distances were used. We first proposed a method of assessment of linguistic questionnaires made on two levels: a global and an individual ones. A great asset of our method is that it can take into account the sampling weights and the missing values occurring in a given data base. We also proposed the so-called indicators of information rate related to the global and the individual levels of evaluation. These indicators aim to measure the quantity of missingness occurring, respectively, in the complete data set or in the records of a particular unit. Some empirical studies were after given. A first finding is that the obtained individual evaluations tend to be normally distributed. We could not reject the normality hypothesis of the obtained distribution. In addition, we performed a test that intends to investigate the influence of the presence of missing values on the defended procedure. We saw that for the same data set, a clear linear relation exists between the evaluations with missing values from one side and the evaluations without missing values from another one. This result seems to be promising since the presence of the missing values did not penalize the results of the assessment. Last but not least, we performed a simulation study where we computed our evaluation method using the different distances. We found that different statistical measures based on the individual evaluations were not affected by the variation of the sample size. A threshold of 1000 observations could eventually be postulated. In terms of the skewness measure, it appears that an order for the skewness of the distributions calculated using the different distances exists. These distances could be grouped by L_p spaces. Thus, by the L_3 metrics, we got the smallest values of the skewness measure, followed by the L_2 metrics, and finally by the L_1 ones. In addition, the skewness measures of the evaluations by the improved versions of the signed distance are close to the value 0, which signalizes an eventual relation of the defuzzified data with the standard normal distribution. This fact is known to be important in terms of the hypothesis on the distribution of the data encountered in various statistical analysis. We finally compared our evaluations with the data set obtained by a fuzzy rule-based systems where we considered symmetrical and non-symmetrical modelling fuzzy numbers. Our model seems to be well adapted to both cases in terms of symmetry of fuzzy numbers. Note that the procedure by the fuzzy rule-based systems was drastically affected by the symmetry of the modelling fuzzy numbers.

For the second application, we proposed to consider the multi-way analysis of variance in the fuzzy environment. According to the tested hypotheses, two exploratory methods of decision making were presented: a heuristic decision rule based on preserving the fuzzy nature of the calculated sums of squares and another one based on the Aumann-type expectation using a chosen distance. A particular case by the signed distance was given. Two empirical applications were also shown. From these examples, it appears that the more the fuzziness is higher in the modelling fuzzy numbers, the more we tend to be farther from the results of the classical approach. The opposite is true as well.

We finally presented the implementation from scratch of the content of all the previously mentioned approaches by R functions. These novel functions form a coherent self-sustaining R package with a complete documentation of the use of each function and its connection with the corresponding theoretical procedure. The involved functions present a wide range of calculation methods. They are also validated on several real and synthetic data sets.

It is important to highlight that our main objectives described in the introduction of this thesis are fulfilled. As such, new sophisticated directional and non-directional distances based on the signed distance have been proposed, where the shape and its possible irregularities are taken into account. The application of these distances in some theoretical, empirical, and programming contexts was afterward shown.

A general conclusion of all our previously described methods is that the fuzzy approach seems to be a reproduction of the classical one, but with more flexibility in the treatment of uncertainty contained in different components of the concerned models. In addition, if some other complexities arise in data sets such as irregularities in the shapes of fuzzy numbers, etc., the use of our models is suggested. Yet, despite all the nice mentioned properties of our procedures, several difficulties still persist. Even though fuzzy set theory is mainly proposed for analyzing natural human-based language, mastering the concepts of this theory is mandatory. As such, the defended models cannot easily be within the reach of any analyst without having a clear detailed overview on the different concepts of this theory. In addition, the fuzzy output results are in many situations difficult to interpret. From another side, if a given expert would like to model the opinion of a human written as a linguistic term by choosing himself/herself the modelling fuzzy numbers, this choice is seen as subjective and in many cases can be biased. Note that the choice of the shape and the spread of the fuzzy numbers are often arbitrary. No criteria are fixed on these choices, except the ones related to the theoretical definition of fuzzy sets and fuzzy numbers. Thus, one would absolutely recommend if possible, that each unit of the sample answers by his own fuzzy numbers, to avoid such modelling bias.

From the development of the abovementioned procedures, many challenging ideas for future works have arisen. Despite the fact that our studies are very promising, they are still in some sense preliminary. Further validations of our models have to be expected. In the foreseeable future, since almost all the concepts were considered in a one-dimensional setting, we would be interested to generalize them to the multidimensional case. From another side, the directionality of the defended distances seemed to be an asset in the proposed analyses. We would like thus to explore its eventual "negative" influence in the purpose of clarifying the guidelines of use of such distances. We would like also to deepen our studies on the influence of the choice of a particular distance on a wider range of statistical approaches. Furthermore, it is clear to see that when fuzziness intervenes, it is not evident to highlight the influence of the method of collection of records on the statistical procedures. This point should be from our point of view deeper investigated, specifically because of the lack of discussion of this problem in the literature. In addition, it would be interesting to investigate the considered unknown distributions of the test statistics in the fuzzy context. In this purpose, non-parametric methods could eventually be a pathway. From a technical point of view, we would enhance the code not only in terms of optimization but also in terms of expanding the proposed calculation methods, in order to be able to include empirical distributions. Indeed, we intend to improve our R package FuzzySTs in order to be better within the reach of other researchers.

Acronyms

ANOVA   analysis of variance. 209, 210, 212, 213, 219

FCI   fuzzy confidence interval. 115, 120–122, 124, 125, 130, 132, 133, 136–141, 143, 144, 147, 162–164, 324

FLS   fuzzy logic system. 33

FRV   fuzzy random variable. 85, 86, 88–96, 99, 106, 107, 125, 126

GSGD   generalized signed distance. 3, 57, 60, 61, 63, 115, 141, 149, 151, 152, 158–160, 163, 166, 211, 238, 242, 243

L-R   Left-Right. 17, 26, 218, 223

LR   Likelihood Ratio. xxii, xxv, 126, 127, 133–135, 137, 138, 140, 162–164

MF   membership function. 14, 15, 17–22, 24, 25, 28, 32, 86, 87, 93, 96, 97, 125, 128, 142, 153–155, 334

MLE   maximum likelihood estimator. 134–137, 162, 260, 291, 292

Mult-ANOVA   multi-ways analysis of variance. xxvi, 209, 211, 212, 218, 219, 221, 235, 239–243, 250, 251

Mult-FANOVA   multi-ways fuzzy analysis of variance. xxii, xxvi, xxvii, 3, 5, 209–211, 219–222, 232, 235, 237, 239, 242–246, 248–251, 315, 316, 318, 322, 323

© The Author(s), under exclusive license to Springer Nature Switzerland AG 2021   345
R. Berkachy, *The Signed Distance Measure in Fuzzy Statistical Analysis*,
Fuzzy Management Methods, https://doi.org/10.1007/978-3-030-76916-1

Printed in the United States
by Baker & Taylor Publisher Services